ASTRAL
ASTRAL

CROP MODIFICATION
Biotech and Non-Biotech Approaches

CROP MODIFICATION
Biotech and
Non-Biotech Approaches

CROP MODIFICATION
Biotech and Non-Biotech Approaches

by
H.D. Kumar
Formerly Professor of Botany and Coordinator,
Biotechnology Programme
Banaras Hindu University

2004
DAYA PUBLISHING HOUSE
Delhi - 110 035

ISBN 81-7035-341-6

Published by : **Daya Publishing House**
1123/74, Deva Ram Park
Tri Nagar, Delhi - 110 035
Phone : 27383999
Fax : (011) 27382329
e-mail : dayabooks@vsnl.com
website : www.dayabooks.com

Showroom : 4762-63/23, Ansari Road, Darya Ganj,
New Delhi - 110 002
Phone: 23245578, 23244987

Laser Typesetting : **Classic Computer Services**
Delhi - 110 035

Printed at : **Chawla Offset Printers**
Delhi - 110 052

PRINTED IN INDIA

PREFACE

The narrowing divide and blurring distinction between biological knowledge and mechanical systems in the past few decades have made biology more technical and the non-living human creations more life-like. Life itself, especially agricultural crops, have been engineered, modified or enhanced genetically by gainfully exploiting advances in technology. The apparent barrier between life and technology is collapsing.

No technology is risk free. While genetic manipulation of crops increases yields and contributes to averting famines, it has aroused concerns about a variety of potential risks. We know that recurring claims by some "greenscare" campaigners that genetically modified (GM) crops (foods) can pose serious risks have no known scientific validity. However, knowing that scientific tests have failed to reveal any such risks, does not warrant ignoring such as possibility in view of the well known fact that there can be known unknowns-or unknown knowns! The best course in the face of such a confusing mix of knowns and unknowns may be to avoid both the extreme positions (viz., adopting transgenics without reservation, or rejecting them equally vehemently). They should instead be accepted as a necessary evil. I have attempted to review the work done on genetically modified crops and to point out the real or perceived risks associated with them. Both pros and cons have been reviewed. The affluent and well-off people can afford to buy non-GM foods such as organic or others grown through non-biotech routes-I have explained some of these, notably apomixis. But considering that the global population is now over 6.5 billion and rising, we just cannot do without the potentially risky biotech option if we have to feed the rising global population, especially the poor in the developing world.

For writing this book, I have depended, *inter alia*, on periodical provision of current literature by the Indian National Science Academy, New Delhi and the Institute for Scientific Cooperation, Tübingen (Germany). I thank all those who sent me newsletters, monographs and other gray literature from time to time. Sincere thanks are also due to the United Nations Environment Programme, Nairobi who, by sponsoring my visits to several countries as a Member of the UNEP Environmental Effects Panel, enabled me to consult several large and well-managed libraries and to exchange views with learned experts.

Several drafts of this monograph were ably and patiently composed by Mangla Prasad Dubey.

H.D. KUMAR

Abbreviations and Acronyms

AGRECOL	International Network on Apomixis Research
ASSINSEL	Association Internationale des Séléctionneurs pour la Protection des Obtentions Végétales (International Association of Plant Breeders for the Protection of Plant Varieties)
BAT	Best Available Technology
BMPs	Best Management Practices
BNF	Biological nitrogen fixation
Bt	*Bacillus thuringiensis*
CBD	Convention on Biological Diversity
CBDC	Community Biodiversity and Development Conservation Program
CGBs	Community gene banks
CGIAR	Consultative Group on International Agricultural Research
CGRFA	Commission on Genetic Resources for Food and Agriculture
CIAT	Centro Internacional de Agricultura Tropical
CIMMYT	Centro Internacional de la Mejoramiento de Maiz y Trigo (International Maize and Wheat Improvement Centre)

DUS	Distinctness, uniformity and stability (requirements for seed certification)
EEA	European Environmental Agency
EMBRAPA	Empresa Brasileira in de Pesquisa Agropecuaria
EPA	Environmental Protection Agency
ET	Embryo Transfer
FAO	Food and Agricultural Organisation
FAO	Food and Agriculture Organization of the United Nations
GATT	General Agreement on Tariffs and Trade
GATT	General Agreement on Tariffs and Trade
GE	Genetically Engineered
GMO	Genetically Modified Organism
GPP	Gross Primary Productivity
GRFA	Genetic Resources for Food and Agriculture
Gt	Giga tonnes
GURT	Genetic Use Restriction Technology
HPR	Host Plant Resistance
HR	Hypersensitive Reaction
IARCs	International Agricultural Research Centres
ICARDA	International Centre for Agricultural Research in the Dry Areas
ICP	Insecticidal Crystal Protein
IDRC	International Development Research Centre
IFOAM	International Federation of Organic Agriculture Movements
IITA	International Institute of Tropical Agriculture
ILELA	Information Centre for Low-external-input & Sustainable Agriculture
INM	Integrated Nutrient Management
IP	Intellectual Property

IPCC	Inter-governmental Panel on Climate Change
IPGRI	International Plant Genetic Resources Institute (former IBPGR)
IPM	Integrated Pest Management
IPNS	Integrated Plant Nutrient Systems
IPRs	Intellectual Property Rights
IRRI	International Rice Research Institute, Manila
ISAAA	International Service for the Acquisition of Agri-biotech Applications
ISNAR	International Service for National Agricultural Research
IUCN	The World Conservation Union
LDCs	Least Developed Countries
LDPE	Low Density Polyethylene
LEIA	Low-External-Input agriculture
LEISA	Low External Input Sustainable Agriculture
LMO	Living Modified Organism
NAFTA	North American Free Trade Agreement
NBPGR	National Bureau of Plant Genetic Resources, New Delhi
NGOs	Non-governmental Organizations
NPP	Net Primary Productivity
OECD	Organisation for Economic Co-operation and Development
OPVs	Open Pollinated Varieties
PAR	Participatory Action Research
PBRs	Plant Breeders' Rights
PCR	Polymerase Chain Reaction
PF	Precision Farming
Pg	Peta gram
PGRFA	Plant Genetic Resources for Food and Agriculture
PNAS	Proceedings of National Academy of Sciences

PPB	Participatory Plant Breeding
PVS	Participatory Variety Selection
QTL	Quantitative Trait Loci
RAFI	Rural Advancement Foundation International
SAR	Systemic Acquired Resistance
SAT	Semi-arid Tropics
SINGER	CGIAR's System-wide Information Network for Genetic Resources
TIBTECH	Trends in Biotechnology
TNCs	Transnational Corporations
TREE	Trends in Ecology and Evolution
TRIPS	Trade-Related Intellectual Property Rights
UA	Urban Agriculture
UNCED	United Nations Conference on Environment and Development
UNDP	United Nations Development Programme
UNESCO	United Nations Educational, Scientific and Cultural Organization
UPOV	International Convention for the Protection of New Plant Varieties
URA	Uruguay Round (Trade) Agreement
VAD	Vitamin A Deficiency
VCU	Value for Cultivation and Use (seed certification requirement)
WIPO	World Intellectual Property Organization
WTO	World Trade Organization

CONTENTS

Chapter 1
EXPLOITABLE BIOTECHNOLOGY

Introduction

Traditional and Modern Biotechnologies

The United Nations Convention on Biological Diversity (see Reid *et al.*, 1993) defined biotechnology as 'any technological application that uses biological systems, living organisms, or derivatives thereof, to make or modify products or processes for specific use.' The keywords in this definition are living, technological, and products. The products should further be qualified as commercial or those used on a large scale, by many users. In this sense, even traditional agriculture or growing crops also comes within the meaning of biotechnology. Some typical examples of traditional biotechnologies are the brewing of beer, wine making, the fermentation of milk into various dairy products (see Kumar, 1998), the combination of cereals and pulses to improve nitrogen availability, and the use of ethnoveterinary vaccines to protect cattle.

Recent advances in biological science (biochemistry, molecular biology, genetics) have greatly complemented the traditional forms of biotechnology with modern applications. Modern biotechnologies are based on new scientific knowledge and depend on the availability of capital and skilled human resources. Modern biotechnology has become a specialized affair of great interest to private research funded by multinational corporations. It has generated concerns about possible risks and spurred debates centred on two issues: (a) who determines which applications to develop and where to apply? (b) who benefits from those applications?

Modern biotechnologies fall into four major classes, *viz.*, *in-vitro* technologies, detection technologies, genomics and genetic modification.

In-vitro Technologies

These technologies separate parts of living organisms in closed containers so as to manipulate and maintain them.

Plant tissue culture came into vogue in the 1970s and involves the maintenance of complete plants or their specific organs or cells under sterile conditions in the presence of nutrients. It allows the rapid multiplication of crop plants at a small scale to produce large of quantities of starter material for wide distribution particularly of vegetatively propagated crops. Plant tissue culture also makes possible the cleaning of virus-infected starter material. Another application is to conserve useful crop genetic resources. *In-vitro* plant tissue culture is used to transfer genes across sexual barriers that are not usually crossed in normal field conditions.

These technologies also find application in animal husbandry. Artificial insemination of cattle is an old application, whose modern extensions include *in-vitro* fertilization, in which a sperm and an egg cell are made to fuse, thus speeding up the generation of new breeds; embryo transfer, which allows the use of carrier animals for the new offspring to develop; and "cryopreservation", the storage of valuable starter material at very low temperatures. These applications are used both in animal breeding and for the conservation of animal diversity.

Detection Technologies

These help in detecting the presence or absence of specific traits in individual organisms. A good example is an array of DNA markers which make use of patterns of specific DNA sequences that reveal the genetic difference between two individual organisms. This technology has greatly speeded up plant and animal breeding. Linking of DNA marker sequences to specific traits allow their use to search for the presence of these traits in the offspring of a cross breeding, long before the trait is actually expressed. These techniques do not actually change the DNA (the genes). Rather, they just allow a quick appraisal of what can be found in the offspring of a breeding programme.

Monoclonal antibodies exmplify another good detection technology. Specific antibody-producing cells from the immune system are grown and multiplied *in-vitro* for the production of large amounts of antibodies to be used to detect specific material. Main applications are in plants and animals, and allow accurate pest and disease management.

Genomics

The field of genomics has developed rapidly during the 1990s. It involves large scale sequencing of DNA, including entire genomes (all the DNA of a single individual), and the comparative analysis of the resulting sequences of various species. The genomes of man, and of many micro-organisms of some plants (*e.g.* rice) and of animals have been sequenced. Genomics provides enormous datasets and a complete new science, bioinformatics, that can handle these databases to allow retrieval and analysis of the information they contain (Visser, 2001).

Genetic Modification

Genetic modification or manipulation involves the transfer of genetic information-in the form of DNA sequences-across sexual barriers between species, which normally would not exchange DNA (Visser, 2001). The resulting organisms are termed genetically modified organisms (GMOs) or transgenics. Genetic modification is today being used to introduce a single new trait, probably based on the activity of a single gene, or a small number of genes. The number of genes with known functions that has been isolated is still so small as to allow for more complicated traits or combinations of traits to be introduced using genetic modification. In agriculture, genetic modification has been applied in several important crops. The majority of these applications introduce resistance traits, particularly to herbicides and insects, but some applications involve the quality of the resulting product, *e.g.* the shelf life of tomatoes. Transgenic animals have been produced in laboratories but have not yet been released for industrial application. Although over 20 transgenic crop varieties have already been approved for commercial cultivation, international agricultural biotechnology companies have focussed their business on mostly cotton, rapeseed, maize, soybean, and wheat.

During the last century, agriculture was highly industrialized in developed countries. Crops were made increasingly uniform to allow for mechanized cultivation, harvesting and processing. Also the size of an average ecoomically sustainable farm increased greatly. Agricultural biotechnology strengthens this trend by increasing the dependence of the entire production chain on a small number of crops and improved varieties. Modern agriculture is high-input dependent and may become even more uniform in the future. It is possible that the negative impact of modern agriculture on agrobiodiversity could increase by the large scale introduction of GM crops and farm animals.

In sharp contrast to industrialized agriculture in the developed world, tropical agriculture continues to be dominated by small farms and marginal farmers whose access to agrochemicals is low. GM crops will not be suitable to the diverse agroecological conditions met with in farmers' fields and hence will do no good there. We do not know how the transfer of GMO traits in non-GMO crops can be avoided as a result of uncontrolled crossing, particularly in cross-fertilizing species. The issue is of particular relevance to organic farming sector for which the maintenance of a GMO-free chain is crucial.

Currently, GMOs usually have no direct relevance to small-scale tropical agriculture. This does not follow from the biological nature of biotechnological applications but from the socio-economic context in which these applications are developed (Visser, 2001). This raises the question of whether and under what conditions some other suitable biotechnologies for small-scale farmers are feasible.

Appropriate Biotechnologies

Biotechnology can potentially serve all farmers, big or small, everywhere. Modern biotechnologies that require low investments and that can be applied in-country or even in the community, have the best chance of being appropriate. This may apply to plant tissue culture for raising healthy, high value varieties in large quantities. The use of monoclonals may also enable the farmers to monitor for specific pests and diseases. Artificial insemination may be extended to well-adapted indigenous breeds. Cryopreservation and tissue culture of valuable plant and animal varieties could also potentially serve the small-scale farming sector. However, the various potential possibilities and promises apart, it seems doubtful whether genetic modification can offer anything to small farmers other than their growing dependence on the seed industry. There exists a wide gap in technology between a fairly small number of commercial crops and crops of regional or local importance.

According to Visser (2001), "more than the risks of monster organisms, food safety problems and environmental pollution, the real threat of GMOs might be the socio-economic dependence it creates for its users from the companies selling such GMOs."

Biotechnology and Agriculture

Growing world population and rising income is expected to almost double the demand for food and other agricultural commodities in the coming few decades. Advances in crop productivity during the 20th century occurred largely through the application of Mendelian genetics. In the next half a century if farmers are to respond effectively to the demands of the day, research in molecular biology and biotechnology will need to be focussed on removing the physiological constraints that currently limit crop yields.

Just as the computer and the semi-conductor assumed the leading role across both the manufacturing and service industries in the third quarter of the 20th century, biotechnology seems poised to become the most important new general-purpose technology of the first half of the 21st century (Ruttan, 2001a).

A general feature of several general-purpose technologies is a long lag period between their initial emergence and their measurable impact. This is why the advances in the new biotechnology achieved thus far have not yet increased yield ceilings beyond the levels achieved using the older methods. Nor are they likely to do so in the forseeable future.

The Mendelian Revolution

Prior to the beginning of the 20th century virtually all increases in crop production were achieved by expanding the area cultivated. Selection by farmers facilitated the development of landraces suited to particular agroclimatic environments. But grain yields, even in favourable environments, rarely averaged above 2.0 metric tons per hectare (30 bushels per acre). Efforts to improve yields through farmers' seed selection and improved cultivation practices had only a small impact on yield before the application of Mendelian genetics to crop improvement. It was only after the introduction of hybrids that the corn yield ceiling was broken (Duvick, 1996).

Similar yield increases could be achieved in other crops in the Western world. Since the early 1970s, dramatic yield increases, heralded as the Green Revolution, materialized in many developing countries in Asia and Latin America. By the 1990s some African countries also started experiencing considerable gains in maize and rice yields.

Yield Constraints

By the early 1990s, it was emerging that yields of several important cereal crops, *e.g.*, maize and rice, might again be approaching limits.

The issue of whether crop yields are approaching a yield plateau has aroused much controversy. In a review and assessment of yield trends for eleven crops in the United States, Reilly and Fuglie (1998) reported that an arithmetically linear trend model provided the best fit for five crops while an exponential model provided the best fit for another five; however, none of the differences between the two models were statistically significant.

It appears that advances in the yields of the major food and feed grains are now approaching physiological limits that are not much above the yields obtained by the better farmers in favourable areas, or at experiment station maximum yield trials (see Ruttan, 2001b). For breaking present yield ceilings, improvements in photosynthetic efficiency, particularly the capture of solar radiation and reduction of water loss through transpiration will be required, but researchers in plant physiology are not optimistic about the rate of progress that will be realized in enhancing crop metabolism (Mann, 1999).

The Biotechnology Revolution

The development of *in-vitro* plant tissue and cell culture techniques can allow the regeneration of whole plants from a single cell or a small piece of tissue. Advances may also occur by plant protection through introduction or manipulation of genes that confer resistance to pests and pathogens. It was expected that advances in the development of the new biotechnologies would lead to measurable increases in crop yields by the early 1990s (Sundquist *et al.*, 1982). These early estimates were overly enthusiastic, but some applications were beginning to occur by the mid-1990s. The first commercially successful virus-resistant crop (a virus-resistant tobacco) was introduced in China in the early 1990s. The Calgene Flavr Savr™ tomato, the first genetically altered whole food product to be commercially marketed, was introduced, albeit unsuccessfully, in 1994. The focus of research was mostly on transgenic approaches to the development of herbicide resistance, insect resistance, and pest and pathogen resistance in some crops. DNA marker technology was employed to locate important chromosomal regions affecting a given trait in order to speedily track and manipulate desirable gene linkages. By the 1998 crop year, almost 44 million hectares had been planted worldwide to transgenic crops, primarily herbicide or virus-resistant soybeans, maize, tobacco, and cotton (James, 2000).

The crucial point, however, is that the biotechnology products currently on the market are mostly designed to enable producers to achieve yields that are closer to present yield ceilings rather than to lift yield ceilings. According to Ruttan (2000a), control of insect pests of cotton, primarily tobacco budworm, cotton bollworm and pink bollworm, represents one of the most dramatic, and clearly positive, results of the introduction of a transgenic crop. The introduction of the *Bacillus* bacterium into cotton has resulted in a dramatic reduction in the use of insecticides while substantially enhancing cotton yields. The effect was, however, not to enhance the genetic potential of the cotton plant but rather to enable the plant to come closer to realizing its genetic potential in the field.

Also, the interest was shifting away from mere yield emphasis on quality traits. The objective of the second generation technologies currently being explored at the laboratory level is to create value downstream from production. A high-oil maize, recently introduced by DuPont, though not strictly a biotechnology product, is often cited as an example. Efforts are underway to tailor-make cereals fortified with the critical essential amino acids such as lysine, methionine, threonine, and tryptophan for use in animal feed rations and in consumer products. Oilseeds are likely to be modified to enhance protein quality and their content of fat that is free of trans fatty acids (Kalaitzandonakes, 1998).

The third generation of biotechnologies is concerned with the development of plants as nutrient factories to supply food, feed, and fibre. High-carotene fruits, vegetables, and oils designed to reduce vitamin A deficiency is a good example.

Conceivably, advances in basic knowledge in areas such as functional genomics, for instance, might provide a scientific foundation for a new round of rapid yield increases. This would, in turn, enhance the profitability of private-sector allocation of research resources to yield improvement. But it is not wise to predict that these advances will leave any measurable impact on production in the coming few decades (Duvick, 1996).

One major concern today is that in most developing countries, yields are still so far below existing biological ceilings that substantial gains can be realized from a strategy emphsizing traditional crop breeding combined with higher levels of technical inputs, better soil and crop management, and first-generation biotechnology crop-protection technology (Ruttan, 2001a). As the fastest rates of growth in demand, arising out of population and income growth, are likely to occur in the poorest countries, it is crucial that these countries acquire the capacity to sustain substantial agricultural research efforts (Ruttan, 2001).

Agrobiotechnology and Sustainable Agriculture

Two decades ago, biotechnology used to be a big issue of the future, but this is not so today–it is now embedded in specific social processes The industrial transformation of agriculture and biotechnological developments influence each other. Economic considerations profoundly affect the seeds sector. The relationships both between farmers and seed companies on one hand, and between farmers and food processing companies on the other, are changing. Biotechnology is not only interwoven with these social processes, but also with locally-specific initiatives.

Modern agrobiotechnology has brought a new dimension into agricultural science. The green revolution of the 1970s and 80s is being pushed to a more ecologically just and politically sound paradigm of sustainable agriculture. Current trends in agrobiotechnology are driving a transition from conventional agriculture to sustainable agriculture. Agricultural innovation systems such as agrobiotechnology are now being based on strong inputs and feedbacks of public opinion, regulatory and political safeguards, new partnerships and organizational formats, social concerns and scientific judgements. Agricultural biotechnology is now being impacted by partnerships demanding capabilities in the social sciences to analyse relationships between science, technology and society (Fransman, 2001; Raina, 2003).

Modern biotechnology is concerned with the identification and expression of a gene or gene sequence so as to increase yield or quality of produce or tackle resistance to some pest, disease or weed. Some good examples include research on bollworm resistant *Bt* cotton, sheath blight resistant rice, and vitamin A enriched golden rice. This kind of approach fails to take into account an evolutionary systems perspective and is unable to tackle issues in biodiversity or sustainable development. Biotechnology can, of course, study the genetic expression of organisms and interactions in complex ecological systems, help restore degraded ecosystems, deal with obligate pathogens, and improve rural livelihoods; it can help gain a systems understanding of the diversity of gene pools and ecosystems. It is this long-term systems perspective that serves as the niche through which agrobiotechnology should facilitate the transition from the green revolution research paradigm to one of sustainable agriculture.

Innovations in agrobiotechnology warrant and demand appropriate institutional and organizational changes for participatory research and technology development, deciphering the locational and cultural specificities characteristic of seed markets in developing countries, examining the historical and political basis of regulatory policies and of criteria employed for risk assessment (Damodaran, 1999, Haribabu, 2000; Clark *et al.*, 2002).

Critical evaluations of the environment-agriculture-biotechnology interactions have tended to question some of the assumptions of neoclassical economics commonly used for impact assessment in conventional agricultural research (for instance, assumptions about private sector roles, market equilibrium, monetary valuations, or externalities) (see Fransman, 2001; Raina, 2003).

The development of biotechnology is strongly linked with two long-term historical processes of the industrial transformation of agriculture namely, appropriation and substitution. Biotechnology is not only infuenced by the specific context in which it is developed, but also influences or modifies these processes. Neither a social nor technological deterministic approach should be followed but rather we should adopt an approach in which the interaction between context and technology is given due weightage (Ruivenkamp, 2003). In the context of the biotechnology mediated industrial appropriation of farming (*e.g.* as soil fertility and the reproduction of plants through the supply of chemical fertilizers and hybrids), it has emerged that biotechnology is influenced by this process and it also reinforces and changes it. Through the development of tissue culture, cell fusion and r-DNA techniques the seed developers and seed merchants, by intervening strongly in the genetic structure of plants, are determining where, when and how a crop should be sown, harvested and processed. Instead of making biological activities of the farmers more-industrialized, the global use of new seeds by world's farmers is becoming the core element of a new social organization of food production. Indeed, the historical process of an industrial appropriation of farming activities is changing into the remote control of the farming activities by the seeds companies. Through the supply of these new seeds with specific properties the companies can remote-control the manner crops are cultivated-with or without herbicides, for instance. Just as the technical properties of the high-yielding modern varieties of the Green Revolution either included or excluded certain groups of farmers in certain continents, in the some way transgenic plants could also include or exclude specific classes of farmers and so exert a socially differentiating effect. Indeed the social relations between farmers and input-supplying industries are already changing in such a manner that farmers will increasingly become labourers for an industrializing agriculture and that the seed companies will increasingly become the actors that program where and how these transgenic crops can be grown.

From Substitution to Interchangeability

Several modern agrobiotechnological developments have led to the substitution of agricultural raw materials by industrial raw materials and synthetic products. The development of new enzymes and microbiological procsses has increased the possibilities for substituting one agricultural raw material by another raw material. New biocatalysts have enabled food manufacturers to extract and produce food components from a spectrum of different agricultural and synthetic sources, which have become interchangeable sources of raw materials for the production of food components. Interchangeability of products rather than the substitution of one product by another synthetic product is the current major trend of the global organization of food production. These biotechnological developments have brought it within reach of the food processing companies to liberate themselves from some specific supply of an agricultural or synthetic product. Not only the borders between different crops and agricultural sectors are fading but even the differences both between crops and between their producers are weakening. When enzymes are used industrially, *e.g.*, it no longer matters whether a sweetener, such as glucose, is made from sugarcane, corn, or sorghum. All these plant materials can also be used for the production of beverages. Separation of food products from their traditional agricultural sources makes food production more flexible in regard to its agricultural sources. Evaporated milk may then be made from cows' milk or soya milk. Margarine is already being made from various interchangeable plant oils and microbiologically produced fatty acids. Even more importantly, however, the production of food products from interchangeable sources implies strong political-economic changes, with the price for a 'flexible and liberated' food production system having to be paid by some social groups. The traditional bonds of farmers and workers around specific crops (*e.g.* sugarcane) weaken their political perspective. The concerned agencies/societies lose their power to negotiate prices when biotechnological developments make it possible to substitute, for instance, sugarcane by high fructose corn syrup made from corn or by aspartame produced by microorganisms (Ruivenkamp, 2003).

A debate on the positive and negative impacts of biotechnological developments has been going on for some years. But several organizations have criticized the potential development of genetically modified crops.

An industrial transformation of agriculture has prevailed and resulted in a substantial growth in the area of herbicide-tolerant and insect-resistant crops in the world-from 1.7 million hectares in 1996 to 52.6 million hectares in 2001 (James, 2001). A new food production system is emerging in which nutrients such as fats, proteins and carbohydrates are assembled from a broad range of interchangeable agricultural and biochemical sources to which flavouring and colouring agents are added to assemble a packet of ingredients in the guise of a traditional foodstuff to be sold to consumers. Finally, the patenting of the politicizing biotech products, such as enzymes and plant breeding techniques has continued unchecked despite worldwide opposition to the patenting of life (Ruivenkamp, 2003).

As against the pro- or anti-lobbies in the debate, there also exists a middle route that involves locally-specific initiatives that do not fit in with the dualistic approach of either accepting biotechnological developments or rejecting them completely. Instead of this pro-anti position, some initiatives in certain regions adopt a more constructive approach and aim to develop a balanced approach of integrating and indigenizing some biotechnological innovations, attuned to local and resources endogenous development.

Plant Biotechnology

Plant biotechnology is best considered under three broad areas *viz.*, plant genetic engineering, molecular breeding, and genomics. The first deals with basic molecular biology, isolation of genes and their introduction into plant species to develop novel genotypes. Molecular breeding involves the use of molecular tools in plant breeding programmes such as construction of linkage maps, character tagging, marker-assisted selection and map-based cloning of genes. Genomics deals with the structural and functional dimensions of plant/crop genetic material.

By using various methods to transfer foreign genes to plants, a variety of foreign genes conferring traits of economic, agronomic, pharmaceutical and industrial importance have been introduced into plant species. Three major gene-transfer techniques employed to transform plant species have been: (i) *Agrobacterium*-mediated approach, (ii) protoplast-based approach, and (iii) biolistic approach. These and some other methods developed for gene transfer to plants are listed in Table 1.1.

Table 1.1: Approaches for Gene Transfer to Plants

Major	Minor
1. *Agrobacterium tumefaciens*	*Agrobacterium rhizogenes*
2. Gene transfer to protoplasts	Agroinfection Agrolistics
(a) PEG-mediated	Macroinjection
(b) Electroporation	Pollen tube pathway
(c) Liposome fusion	Seed incubation
(d) Microinjection	Microlasers
3. Biolistic method	Electrophoresis
(a) Microprojectiles	Liposome injection
(b) Microtargeting	

Resistance to Biotic Stresses

Crop productivity is constantly threatened by pests and diseases. Some 30-40 per cent of agricultural productivity is lost to insect pests annually. Chemical management of pests and diseases is effective but not eco-friendly. Extensive and indiscriminate usage of pesticides has adversely affected human health and degraded the environment. Hence, there is a strong need to reduce the consumption of pesticides by making the crops resistant to pests and diseases. Considerable progress has been made recently to develop transgenic crops resistant to insect pests, viruses and fungi (Tables 1.2, 1.3).

Table 1.2: Selected Examples of Genetic Engineering of Plants for Resistance to Insect Pests

Target Insect	*Gene/Protein*	*Source*	*Transgenic Plant*
Cotton bollworms	*cry 1Ab; cry 1Ac*	*Bacillus thuringiensis*	Cotton
Colorado potato beetle	*cry 3Aa*	*B. thuringiensis*	Potato
European corn borer	*cry 1Ab*	*B. thuringiensis*	Corn
Potato tuber moth	*cry 9Aa2*	*B. thuringiensis*	Tobacco
Yellow stemborer	*cry 1Ac*	*B. thuringiensis*	Rice
Sweet potato weevil	Snowdrop lectin	*Galanthus nivalis*	Sweet potato
Boll weevil	Cholesterol oxidase	*Streptomyces spp.*	Tobacco
Cotton bollworm	Stunt virus genome	*Helicoverpa stunt virus*	Cotton
Deroceras reticulatum	Oryzacystatin	Rice	*Arabidopsis*
Tomato moth	Chitinase	Bean	Potato
Brown planthopper	Lectin	*Galanthus nivalis*	Rice
Bruchids	α-amylase inhibitor	Bean	Pea

Table 1.3: Some Transgenic Crops Raised for Fungal Resistance (after Pattanayak and Ananda Kumar, 2000)

Plant	*Gene/Protein*	*Target Pathogen*
Pathogenesis-Related Proteins		
Tobacco	Chitinase from Serratia marcescens	*Alternaria longipes*
	Chitinase from *Nicotiana tabacum*	*Rhizoctonia solani*
Rice	Chitinase from rice	*Rhizoctonia solani*
Brassica napus	Bean chitinase	*Rhizoctonia solani*
	Chimeric endochitinase by fusion of tomato and tobacco chitinase gene	*Cylindrosporium concentricum*, *Sclerotinia sclerotiorum*, *Phoma lingam*
Potato	β-1,3-endoglucanase from soybean	*Phytophthora infestans*
Antifungal Proteins		
Potato	Osmotin from tobacco	*Phytophthora infestans*
Antifungal Compounds		
Tobacco	Stilbene synthase from grapevine	*Botrytis cinerarea*
Genes/Proteins that Induce Cell Death		
Potato	Barnase and barstar	*Phytophthora infestans*
	Glucose oxidase from *Aspergillus niger*	*Phytophthora infestans*
Tobacco	Nettle agglutinin	*Botrytis cinerarea*
Tobacco	Cryptogein from *Phytophthora cryptogea*	*Erysiphe cichoracearum*

Bacterial Resistance

Understanding of plant-pathogen interactions has enabled the use of genetic engineering for producing bacterial disease-resistant plants. The strategies usually employed for developing transgenic plants resistant to bacteria are: production of antibacterial proteins of non-plant origin, inhibition of bacterial pathogenicity or virulence factors, enhancement of natural plant defenses and artificially induced programmed cell death at the site of infection (Pattanayak and Ananda Kumar, 2000).

Antibacterial proteins of non-plant origin include lytic peptides, lysozymes and iron sequestering glycoproteins (Mourgues *et al.*, 1998; Nakajima *et al.*, 1997) (see Tables 1.3, 1.4).

Table 1.4: Selected Examples of Transgenic Plants with Bacterial Resistance (after Strittmatter *et al.*, 1998)

Plant	*Gene/Protein*	*Origin*	*Target*
Tobacco	Cecropin	Giant silk moth	*Ralstonia solanacearum; Pseudomonas syringae* pv. tabaci
Tobacco	Lactoferrin	Human	*Rhizoctonia*
Tobacco	Tabtoxin resistant protein (Tabtoxin acetyl transferase)	*P. syringae* pv. *tabaci*	*P. syringae*
Tobacco	Thionin	Barley	*P. syringae* pv. *tabaci*
Apple	Attacin	Giant silk moth	*Erwinia amylovora*
Potato	Lysozyme	T4 bacteriophage	*E. carotovora*
Potato	Glucose oxidase	*Aspergillus niger*	*E. carotovora*
Rice	Resistant protein Xa21	Resistant rice cultivars	*Xanthomonas oryzae*
Nicotiana benthamiana	Tomato *pto* gene (serine/threonine protein kinase)	Tomato	*P. syringae*

Nematode Resistance

Nematodes, especially plant parasitic nematodes *Meloidogyne* (root knot nematode), *Heterodera* and *Globodera* spp. (cyst nematodes) are important causal agents of crop diseases. Chemical control of nematodes is not only ineffective but also leads to soil contamination by pesticides. Nematode resistance has been reported by expressing proteinase inhibitors. Cowpea trypsin (CpTI) inhibitor expressed in potato reduced fecundity of *Meloidogyne* spp. and switched the population of *Globodera pallida* from most females to most males, which are less harmful (Atkinson *et al.*, 1995). A gene *Mi-1* has been isolated from wild tomato by map-based cloning strategy and expressed in transgenic plants with considerable protection against root knot nematode (Vos *et al.*, 1998).

Herbicide Tolerance

Gene transfer technology helped in raising crops that are resistant to safe and effective herbicides. Incorporation of herbicide resistance traits is based on genes encoding enzymes that catabolize the herbicides, overproduction of the targeted enzyme or by manipulating the target sites. Several crop species have been transformed to become resistant to herbicides (Table 1.5).

Table 1.5: Some Transgenic Plants Tolerant to Herbicides (after Pattanayak and Ananda Kumar, 2000)

Herbicide	*Gene/Strategy Used*	*Source*	*Transgenic Developed*
Glyphosate	EPSPS (overproduction)	*Petunia*	Petunia, Tobacco
	aroA (mutated EPSPS)	*Salmonella typhimurium*	Tobacco, tomato, Poplar
	CP4 (tolerant EPSPS)	*Agrobacterium* spp.	Soybean, Rapessed
	Glyphosate oxidoreductase (detoxification)	*Achromobacter* spp.	Sugarbeet, maize, soybean, rapeseed
Phosphino-thricin	*bar* (detoxification)	*Streptomyces hygroscopicus*	Tobacco, potato, alfalfa, rice, maize
	pat	*S. viridochro-mogenes*	Tobacco
Sulfonylurea	*csr 1-1*	*Arabidopsis thaliana*	Rape, rice
	Sur B	Tobacco	Tomato
Atrazine	*psb A* (mutant)	*Solanum nigrum*	Soybean
Bromoxynil	*bxn*	*Klebsiella ozaenae*	Tobacco, tomato, cotton
Dinitroaniline	α-tubulin & β-tubulin	*Eleusine indica*	Tobacco

Resistance to Abiotic Stresses

Environmental stresses, such as drought, salinity, heat, freezing and flooding have for long hampered agriculture, resulting in poor harvests and even famine. One way to enhance plant tolerance to these factors is to transfer genes encoding protective proteins or enzymes from other organisms. Some approaches are engineered alterations in the amounts of osmolytes and osmoprotectants, saturation levels of membrane fatty acids, and rate of scavenging of reactive oxygen intermediates. A variety of genes coding for proteins/enzymes involved in stress relief have been introduced into various plant species with reasonable degrees of tolerance to abiotic stresses (Holmberg *et al.*, 1997; McNeil *et al.*, 1999) (Table 1.6).

Table 1.6: Selected Examples of Plant Genetic Engineering for Tolerance to Abiotic Stresses (after Pattanayak and Ananda Kumar, 2000)

Stress	*Gene/Enzyme*	*Source Organism*	*Transgenic Produced*
Drought, salinity, freezing	Dehydration response element-binding protein (*DREB1A*)	*Arabidopsis*	*Arabidopsis*
Osmotic	D-pyrroline-5-carboxylate synthetase (*p5cs*)	*Vigna aconitifolia*	Tobacco
Drought, salinity	Mannitol-1-phosphate dehydrogenase (*mltd*)	*Escherichia coli*	Tobacco
Salinity	Glyoxalase 1	*Brassica juncea*	Tobacco
	Choline dehydrogen-ase (*Bet A*)	*E. coli*	Tobacco
Cold	O-3-fatty acid desaturase (*Fad7*)	*Arabidopsis*	Tobacco
Drought	Trehalose-6-phosphate synthase (*TPS 1*)	Yeast	Tobacco
	Levan sucrase (*Sac B*)	*Bacillus subtilis*	Tobacco
Oxidative	Fe Superoxide Dismutase	*Arabidopsis*	Maize
Oxidative	Mn-superoxide dismutase (Mn-SOD)	*Nicotiana plumbaginifolia*	Tobacco Alfalfa
Hypoxia and anoxia	Nonsymbiotic hemoglobin (vhb)	*Vitreoscilla stercoraria*	Tobacco
Hypoxia	Nonsymbiotic hemoglobin (hb)	Barley	Maize
Cadmium	Metallothionein-I (MT)	*Nicotiana glutinosa*	Tobacco

The accumulation of low molecular mass osmoprotectants and osmolytes, such as quaternary amines (*e.g.* betaines and allied compounds), amino acids (*e.g.* proline) and sugar alcohols (*e.g.* mannitol), is an important strategy to overcome environmental stress.

Floriculture

Recent years have seen growing demand for cut flowers and potted plants in many countries. Molecular approaches are being used to introduce such desirable traits as colour, shape, plant architecture and vase life to meet consumers' demand for novelty. A transgenic petunia containing *dfr* gene from Gerbera shows intense and stable colouration, because the Gerbera *dfr* gene has a low GC content and only a few methylation sites and so is less prone to transgene inactivation by methylation (Elomaa *et al.*, 1995). Researchers have succeeded in producing a commercially interesting bright-orange petunia variety through breeding of transgenic petunia lines containing maize *dfr* gene with elite breeding lines (Oud *et al.*, 1995). The future prospects of combining molecular and traditional breeding seem bright.

Among popular cut flower species, only *Freesia* and *Iris* have truly blue varieties. Development of blue varieties in commercially important species such as rose, carnation and chrysanthemum has stimulated great industrial interest. The simultaneous presence of three factors, synthesis of 3', 5'-hydroxylated anthocyanins (delphinidins), presence of flavonol co-pigments and a relatively high vacuolar *p*H, are conducive to formation of blue colour flowers (Yoshida *et al.*, 1995). The first breakthrough to engineer blue flowers was made by cloning flavonoid-3', 5'-hydroxylase (F3'5'H) gene from petunia (Holton *et al.*, 1993). Its homolog has been isolated from gentians. Transgenic violet carnations have been produced by the introduction of the petunia F3'5'H gene and its high expression in the petals. A transgenic violet carnation is already being marketed in Australia and Japan (Tanaka *et al.*, 1998).

Modification of floral colour intensity has also been achieved. White and pink flowered varieties have been produced by introducing sense and antisense chalcone synthase (*chs*) transgenes into petunia, chrysanthemum, rose, carnation, gerbera and torenia (Suzuki *et al.*, 1997). Flavonols have been shown to contribute to copigmentation and flower colour becoming less intense.

It has become possible to modify size and shape of ornamental plants by changing the ratio of cytokinin: auxin in transgenic plants by expressing oncogenes from the Ti pasmid from *Agrobacterium tumefaciens* or the Ri plasmid from *A. rhizogenes* (Tanaka *et al.*, 1998).

Many cut flowers deteriorate rapidly after harvest and short vase-life constitutes a major obstacle in their marketing. The petals exhibit a characteristic 'in-rolling' behaviour during senescence and also in response to exogeneously supplied ethylene. Supression of ethylene biosynthesis should possibly prolong the vase-life. Transgenic carnation containing an antisense ACC oxidase (*aco*) gene show low ethylene production and a marked delay in petal senescence (Savan *et al.*, 1996). Increase in the post-harvest life of flowers has been achieved by constitutive expression of ACC synthase (*acs*) in carnation cultivars (Tanaka *et al.*, 1998).

The Promise and Peril of Plant Biotechnology

The large scale commercialization of transgenic foods since 1996 in some countries has intensified the debate on the benefits and risks of biotechnology related to genetically modified crops. The intervention of politicians and policy-makers in this debate is delaying or even preventing the application of plant biotechnology in tackling the pressing problem of food security, particularly in developing countries.

In fact, plant biotechnology has great potential both for meeting the food needs of the ever-growing human population, and for conserving the world's dwindling land and water resources. Whereas population is growing faster than increases in food productivity, food production per capita, based on cereal grains, has been declining for nearly two decades. Food production may need to be tripled to meet the demand for food of the nearly 12 billion expected inhabitants of the Earth in 2050 (see Vasil, 2003; 2003a). The major challenge facing us is that population is growing faster than increases in food productivity, and the fresh water supplies are declining (UNESCO, 2002); there is less land per capita available for food production; and over 40% of crop productivity is lost owing to various biotic/abiotic factors (Oerke *et al.*, 1994). All this means that increasing demand for food will need to be met primarily by increasing productivity on land that is already under cultivation, with less and less water and under worsening environmental conditions (Vasil, 2003a).

The first generation of commercial transgenic crops was modified for resistance to herbicides (soybeans, canola), insects (cotton, maize) and viruses (papaya and squash). By significantly lowering or eliminating the losses caused by weeds, pests and pathogens, these GM crops increase productivity, conserving land, water, energy and other resources that would be needed otherwise to produce the same amount of food with nontransgenic plants. By 2003, the worldwide acreage of these crops had grown to nearly 200 million acres and is likely to increase more and more in China, India and other countries. In the USA, about 80% of soybeans, 70% of cotton and 38% of maize planted in 2003 were transgenic and the total market for transgenic seed now exceeds $3 billion (Vasil, 2003a).

Plant biotechnology has become a powerful agricultural technology that is raising productivity by minimizing losses caused by weeds, pests and pathogens. It is also contributing to the conservation of biodiversity, arable land, water and energy sources.

Various useful genes have been introduced into diverse crops in recent years so as to enhance their quality and/or nutritional value and resistance to a variety of biotic and abiotic stresses. The characteristics of the second generation of transgenic crops (2005-2015) are outlined below (see Vasil, 2003; Horvath *et al.*, 2003):

1. Resistance to herbicides, pests and pathogens
2. Tolerance to drought, salt, heavy metals and low/high temperatures
3. Improved nutritional quality (protein, oils, vitamins, minerals)
4. Enhanced shelf life of fruits and vegetables
5. Elimination of allergens, and
6. Production of vaccines, human therapeutic proteins, pharmaceuticals

The second generation of transgenic plants is likely to be released commercially in the coming few years.

Progress in the sequencing of plant genomes, and in understanding the structure, function and regulation of genes has enhanced our understanding of the molecular basis of the growth and development of plants. The introduction and manipulation of genes involved in tillering, flowering, seed characters, nutrient assimilation and photosynthetic efficiency by genetic transformation can significantly increase yield (Hayama *et al*., 2003; Jenner, 2003). The third generation of transgenic plants will be better adapted to various kinds of stresses, will give higher yields and more nutritious and healthful food, and will produce a variety of pharmaceuticals for the treatment of human and animal diseases; these plants should become available for human use sometime after 2015 (see Vasil, 2003; 2003a).

Our understanding of the probable ecological impact of genetically modified crops is quite incomplete. The gaps in knowledge point to our failure to address the science underlying sustainable agriculture.

The government of United Kingdom had decided a few years ago not to allow any commercial planting of GM crops until the effects on farmland biodiversity of the herbicide applications associated with the crops' cultivation had been evaluated through critical ecological trials. This decision was taken in the light of a very strong public opposition to GM crops. But even if, as appears likely, the various farm-scale trials in Britain give herbicide-tolerant crops a fairly clean bill of health with regard to biodiversity, they may convey little about the benefits and risks of GM agriculture as a whole. So the UK government is now seized of the issue to consider the merits of transgenic farming.

GM agriculture has been accepted in North America, but many countries have not yet accepted the technology. And the developing world, already groping in the dark, will look for guidance to expert advice from advanced nations.

The extensive literature on the applications of transgenic techniques has not yielded any convincing evidence that GM crops pose risks to human health, or that they will lead to some ecological upheaval. The farm-scale trials in Britain may also reassure that herbicide-tolerant GM crops can be grown without negative effects on farmland biodiversity. However, when we consider the issue of gene flow from GM crops to wild relatives, many studies have shown that transgenes can spread beyond the crops from which they were introduced. But nothing much is new in this finding-it has been known for long that crops do hybridize with related weeds. The crucial point is whether the flow of transgenes poses any undesirable ecological or agronomic consequences. This question can be answered by producing hybrids between transgenic crops and wild relatives, and monitoring their effects on farmland ecology and crop yields when released in field experiments (Anonymous, 2003). Such trials are currently being carried out in USA. The necessary experiments may be performed by using male-sterile plants, which shouldn't breed and pose any lasting hazard. But in Europe, GM crops have been so greatly demonized that it is virtually impossible to carry out the research to determine whether fears about invasive superweeds have any basis.

According to Anonymous (2003), ultimately, the answer should involve considering, *inter alia*, the nutritional needs of the world's growing population while protecting the planet's biodiversity. It is not wise in this context to cast aside potentially useful technologies without testing whether they can be effectively and safely exploited. This not only applies to transgenics but also to enhancements to conventional breeding-the latter are being allowed by our growing genomic and molecular-genetic knowledge. Secondly, in the context of creating new varieties that might help feed the developing world's growing population, amid all the fuss about GM crops, we tend to forget that similar questions about biodiversity and gene flow need also to be asked about conventionally bred varieties. For instance developing a salinity-tolerant variety of rice-whether created by transgenic or conventional means-would allow the cultivation of soils that are now considered as agricultural wastelands. But might it also spawn superweeds that would choke estuarine habitats? There is genuine need for research that can help us in answering this question.

One point needs to be understood clearly–agriculture as practised before the advent of the Green Revolution could not possibly have fed even one-third of the current population of the developing world. India saw several recurrent famines right from the 12 century to as recent as 1943 when about 1.5 million people perished in one famine (the Bengal famine) alone. While adopting the traditional, low-input, chemical-free, animal-based or organic farming may be ideal in theory, it just cannot feed the worlds population of over 6 billion, which is growing. This point is particularly relevant to India and other developing countries, which will need to adopt modern biotechnologies, howsoever risky they may be.

Molecular and Traditional Breeding

While plant biotechnology is very important, it is not a magic bullet that can solve all the problems of food security and the environment. New varieties should continue to be developed by traditional breeding and selection as well. Biotechnology may contribute further improvement of these varieties through the introduction of some traits that cannot be transferred with conventional methods. Indeed, molecular and traditional breeding should both complement and supplement each other.

The biotechnology community and biotech industry must not distance themselves from plant breeding and genetics; they should not present plant biotechnology and transgenic plants as something entirely new. Those who claimed that this technology would replace traditional plant breeding were mistaken, and unwittingly provided the necessary ammunition to the anti-biotechnology lobby to describe transgenic plants as unnatural and dangerous; they also alienated plant breeders. The fact is that transgenic technology is no different from breeding except in being inherently more precise and predictable. Also the first set of transgenic plants released for mass cultivation had been engineered for resistance to herbicides and do not offer any direct or immediate benefit to the consumer. This perpetuated the myth that transgenic technology is largely for the benefit of the agro-chemical industry. But now this situation is being redressed and rectified by the development of healthier, more nutritious transgenic crops and products of immediate interest and benefit to the consumer.

The hitherto clean record of transgenic crops and their products attests to their safety and wholesomeness. As the risks to human health and the environment from transgenic crops seem to be no different from those of plants developed by conventional breeding and selection, logic warrants that risk assessment and regulation of transgenic crops should not differ from those used to evaluate plants developed by traditional breeding.

Many of the transgenic crops have fulfilled regulatory requirements and have been cultivated and/or used for several years without any ill effects on humans or the environment. Examples are herbicide-resistant soybean and canola, insect-resistant maize and cotton, and virus-resistant papaya. Vasil (2003a) has suggested that new transgenic crops with similar genes should not be required to meet the regulatory requirements for more than two years unless there exist definite signs of risks. However, crops with genes that have not been previously tested under field conditions must be monitored for two to five years, and then released for unrestricted cultivation unless proven to be harmful. Crops engineered for the production of drugs and vaccines must, of course, be physically isolated from other crops so as to prevent accidental pollination of nontransgenic plants.

Plant Tissue Culture and Seed Production

The genes of an organism express and interact in different ways in nature to produce its phenotypic characters. Some small variations among individuals in a population are important for survival. But the law of natural selection concerning the survival of the fittest becomes compromised when elite plants are multiplied on a large scale to produce elite descendants.

Whereas the natural mutation process is rare and affects only a few plants, every tissue culture derived plantlet is a potential mutant. Somaclonal variation although different from bonafide mutation, nevertheless is considered as the biotechnology equivalent of mutation breeding. Somaclonal variation arises universally in tissue culture, and has been projected as a useful means of inducing genetic variations of utility in plant breeding.

In practice, production of seed is usually cheaper than the mass multiplication of tissue culture plantlets. Given the advantages of production cost and natural selection endowments of seed, it is the preferred choice of farmers and tissue culture may not be a substitute to traditional seed propagation in the near future (see Ghosh, 1995).

Artificial selection of plants from hybrid and backcross lineages exemplifies a kind of simple genetic engineering that works in harmony with natural selection. It ensures fitness and survival of the descendants and possible better performance than their parents in a variety of agricultural contexts such as yield increase and resistance to biotic or abiotic stresses. The superiority of this process stems from the fact that it exploits the genomic homologies of crop plants and their natural relatives (see Ram and Indira, 2001).

In contrast, the modern genetic engineering being practised by large industrial companies is totally artificial (Ho, 2001) (with regard to the choice of sources of genes, and random with reference to the integration of genes in the crop plant genome), its products have little or no chance of being selected positively by nature. Transmission of engineered genes to the next generation appears to be problematic-breakdown of the target trait expression has been often reported in the seed descendants of transgenic plants (see Ram and Indira, 2001). Further, additional genes influencing particular characters cause suppression of the target character, a phenomenon called co-suppression (Jorgensen, 1995). This calls into question the stability of genetically engineered crop plants in nature.

One major concern about consuming genetically engineered products has been their potential toxicity or allergenicity. Soybean carrying a gene of Brazil nut has been found to be toxic (Ho, 2001). Concerns have also been voiced about IGF-1 like growth factor in the milk of cattle administered with recombinant bovine growth hormone, and vitamin-A in golden rice which may possibly render them risky for consumption. Some likely risks of consuming these foods include prostate, colon or breast cancer (see Shiva, 2001).

According to Ram and Indira (2001), despite its high potential, biotechnology currently cannot replace conventional plant breeding that evolved in tune with natural selection.

Biosafety

The biological safety of genetically engineered products may be jeopardized by possible unwanted effects on living organisms and the environment in general. In recent years, some unpredicted outcomes of several risk assessment studies have generated concerns about the effects of practical applications of genetically engineered organisms (GEOs) (see OECD, 1993; van Dommelen, 1999). This necessitates critical and systematic development of appropriate research programmes for food safety assessment. We badly need a good testing procedure and a careful risk assessment. The latter can involve the following steps (see van Dommelen, 1999a):

1. Identification of hazards; recognition and characterization of possible unwanted impacts.
2. Estimation of their magnitude; assessing the scale and strength of a possible unwanted impact.

3. Estimation of their likelihood; assessing the chance of a specific unwanted impact to occur.
4. Quantification of risk; risk is defined as 'hazard times the likelihood of its occurrence'.
5. Evaluation of risk: assessment of the importance of some estimated risk against the background of possible costs and benefits (see van Dommelen, 1999).

Herbicide-tolerant crops (Tables 1.5, 1.7) represent the first generation of transgenic crops to enter the world markets. While in Europe, these crops have become a test case for how farmers, producers, and consumers generally accept the new technology of genetic engineering, regulation and legislation have so far only focused on each individual product in a science-based, case-by-case, step-by-step manner (see Lutman, 1999).

Table 1.7: Four Common Commercialized Transgenic Herbicide Tolerant Crops

	Common Name	*Latin Name*	*Herbicide*
1.	Canola/oilseed rape	*Brassica napus*	Phosphinothricin, Glyphosate
2.	Cotton	*Gossypium hirsutum*	Glyphosate
3.	Maize (corn)	*Zea mays*	Phosphinothricin
4.	Soybean	*Glycine max*	Phosphinothricin, Glyphosate

All over Europe, public concern about the safety and desirability of genetically engineered plants is mounting. In 1999, it prompted the European Union (EU) to halt developments and bar the entry of more genetically modified crops to the EU market for a few years–a decision that focuses on the *method* of genetic modification. It contrasts with the case-specific judgement of properties of the modified product that had been the usual basis for regulation until then.

Agroecological v/s Technological Crop Management

Some of the environmental problems associated with conventional agricultural practices are erosion, habitat loss, and pollution of water by sediments, nutrients, and pesticides. The fairly popular perception that to feed the world, we have to accept agricultural land as an "ecological sacrifice" has recently been challenged by Jackson and Jackson (2002). According to Jackson and Jackson (2002), the many challenges in managing land for both agriculture and conservation can be tackled. Jackson and Jackson have highlighted the various social and economic possibilities for creating a partnership between agriculture and conservation. There are many discerning and progressive farmers who carefully steward and manage and increase the conservation function of their farms. There certainly exist farms that continue to be productive and profitable while decreasing their negative impacts on the environment and providing habitat for the conservation of nature. Managing land for agriculture and conservation is not always a simple process and is not the sole responsibility of the farmer-it requires an effective collaboration among different groups of people, such as farmers, conservationists, ecologists, policy makers, and ultimately the consumers.

In view of the multifarious challenges facing the farmer and the environment, sustainable agriculture can best be achieved with a partnership between farmers, conservationists, and policy makers. But we should certainly recognize farmers as competent conservation partners, and address their limitations and constraints.

Agroecological management requires the adaptation of management practices that are in tune and harmony with the conditions and crops rather than applying a fixed set of practices. The latter (technological) management is based on the assumption that the critical determining factors are chiefly genetic; in contrast, the agroecological approach to crop production emphasizes the management of interactions between genetic potential and environmental conditions. In most crops, especially rice, there exists a substantial genetic potential which can be exploited by proper tuning of the agronomic management practices. In fact, much recent work on rice in Madagascar, China and elsewhere has demonstrated that research in genetic engineering is not as effective in raising yield levels as proper agronomic management (Uphoff, 2001). There exists considerable genetic potential in rice varieties that is amenable to exploitation through agroecologically sound practices, implying that genetic modification efforts may not be necessary in those cases where increasing food production and lowering costs of production are the main objectives.

Furthermore agroecological methods of crop production appear to give better results with genetically-improved varieties. Some varieties have a greater potential for tillering, root growth and grain filling than others in response to wide spacing, aerated soil, and other agronomic practices. Conceivably some varieties probably also have higher pest and disease resistance, or greater drought tolerance, when grown with improved agronomic practices. Of course, we need to differentiate clearly between the issues of whether transgenic research is justified or needed and whether genetic improvement should be carried out. There may be risks with the former that are as yet inadequately assessed and one may justifiably object to those companies which judge success on the basis more of private than on public interests directing the process of genetic modification. But this kind of broad-spectrum objections do not apply to conventional genetic improvement of agronomically adapted varieties.

Plants That Protect Other Plants: Alternative to Pest-resistance in GM Crops

Some plants can be used as biological control agents. There exist hundreds, perhaps thousands, of plant species that are useful for the management of populations of harmful insects. The most important applications include use as insecticide, repellent, nematicide and fungicide, all exemplifying ecological pest management. Most of the research work so far has focused on the rescue and technical validation of many of these plants. Several plants have proved effective in regulating pests and diseases. Extracts from a wild forest plant *Argemone sufusiformis* has effectively regulated *Spodoptera frugiperda* population (army worm) in their larvae stage; and *Nerium oleander* has been found to be a potent fungicide, especially to control *Fusarium solani* and *Rhizoctonia solani*; in Peru, horsetails (*Equisetum* spp.) are used to regulate the incidence or presence of *Phytophthora infestans* in potatoes. The high silica content of the horsetail helps check the fungal multiplication (Osorio, 2001). *Crotalaria* spp. can control nematodes. Furthermore, the oil of some plants kills pests that damage grain in storage.

Besides these, some plants have been very widely known and thoroughly investigated for their biological control properties as an alternative for pest control and have great potential for commercial application. The best known example is, of course, neem (*Azadirachta indica)* (family Meliaceae). This tree is very well known all over the world as an effective biopesticide. It is widely used in India for natural crop protection. Neem seeds yield the insecticide. The active substances of neem (Azadirachtin, Salanin, Nimbin, Nimbidin, and Meliantrol) have repellent effects. Upon ingestion, they disturb or inhibit the metamorphosis of insects, preventing their growth and development. These substances are non toxic to humans, mammals, birds, reptiles or fish. In proper concentrations they do not affect the beneficial flora or fauna in the cultivated crop fields.

Biological control represents a good alternative to the use of GM crops. It reduces the dependence of farmers on external inputs such as pesticides, and is conducive to healthier living and working conditions.

Ethics and Public Acceptance of Biotechnology

With a first generation of genetically modified (GM) crops being already established in some parts of the world, surveys are being made of the perceptions and expectations that have developed in the public domain on agricultural gene technology in recent years. Some argue that the genetic modification of crops is just an extension of conventional plant breeding techniques. Others view it as representing a fundamental, dangerous departure from conventional breeding. Because of such polarized opinions, the application of these techniques for food production cannot be considered as a purely technical matter that can be dealt with by scientists and agronomists alone. There are many different and often conflicting stakeholders' perceptions and expectations of genetic modification techniques in agriculture.

Some factors that have influenced public perceptions and expectations of biotechnology as applied to agriculture in various countries over the two decades include strategic research and development decisions by private firms, international and national public research policies, governmental regulations for safety in biotechnology, campaigning activities by non-governmental organisations and the way the media cover biotechnology. Surveys of public attitudes and perceptions of biotechnology in advanced countries have clearly shown that the public wish to be better informed on the subject. Also the levels of public confidence in GM food have been declining. In North America concerns about genetically modified crops are concentrated within a minority of the population and are long-standing and consistent. They are rooted in philosophical objections to the manipulation of nature, and in opposition to private control and exploitation of the natural world. In pressing their case, however, the opponents of genetic engineering have adopted other arguments, less philosophical or overtly political in nature. They have had only moderate success (Charles, 2001).

The current global system governing biotechnology has largely been shaped by Western hopes and fears and more research is needed into how developing countries view this new technology.

Whilst public perceptions in the North can be assessed through surveys, in the South, information about public perceptions and the way these are formed are often distorted or mediated through stakeholders and opinion leaders with their own agendas (see Aerni, 2001). The corporate and promotional literature routinely circulated to farmers, seed stockists and extension officers by agricultural companies have a much greater impact in such situations. Promises of higher yields, reduced expenditure on external inputs and greater food security for farmers caught in economic or ecological crises (see Dhar, 2001) can be quite appealing and can overshadow long-term environmental and health risks that affect future generations.

Farming is, by its very nature, routine and often located in isolated areas, away from mainstream sources of information. Although developing countries may be able to make better use of information technology (IT) in the future, this is not always the case today.

Bioethics appears to have acquired status in the speed and turbulence that has surrounded the genetic revolution, but it remains to be seen whether bioethics is capable of acquiring a common language which can address the issues raised by this new technology as it moves from the laboratory to the field.

It is being increasingly acknowledged that bioethical issues are central to the societal controversy generated by recent developments in biotechnology. Human dignity, ownership of life, indigenous rights, animal welfare, the intrinsic value of the environment are all issues that divide society, and link biotechnology to ethics. Some biotechnologists think that bioethics is a product of the 1990s but in reality it has a much longer tradition. As with medical ethics, professional bioethics goes back to early ethical codes of conduct such as the Greek Hippocratic oath. This was the first explicit acknowledgement in the 'life sciences' of professional responsibility requiring that physicians should not harm their patients (Jonsen, 1998). It was, however, only in the 1960s that bioethics started moving beyond the doctor patient relationship. Bioethics then developed into a coherent professional discourse and discipline, encompassing a variety of ethical issues raised by progress in medical science and technology. Moreover, it slowly began to encompass other ethical issues, such as our relationship towards animals and the environment. It is in this broad sense that the term bioethics is currently used, although many still reserve it for (human) medical ethics alone (Paula, 2001).

In the second half of the 20th century, a growing awareness of the vulnerability of fauna and flora supported the already growing ethical status of non-human animals and the environment. It helped pave the way for the recognition that non-human animals are also worthy of moral consideration for and by themselves, and even that species and ecosystems are deemed morally relevant for their own sake. Consequently, a broad variety of human practices traditionally regarded as 'normal' suddenly became morally questionable and debatable.

Dimensions of Risk

Beck (1992) convincingly demonstrated the political nature of risks caused by technologies in post-modern industrialized society. He argued that in our times, the production of wealth ('goods') is systematically accompanied by the production of risks ('bads'). Modern risks, like ecological disasters and nuclear fallout, for example, typically inflict irreversible harm and are only indirectly visible, based on (scientific) causal interpretations. Since the consequences of these risks have global effects, they render national jurisdictions inadequate and harm even the rich and powerful. They, therefore, also raise the issue of democratic accountability and control since, as Beck (1992) puts it, "safety issues are illegitimately decided by the corporations of engineers of our high-risk civilisation, under cover of the empowerment formula-state of science and technology."

Once these 'late modernity' risks and their societal consequences were recognised, as for instance in Western societies, what was once considered to be apolitical became political. Hence, the risks engendered by biotechnology have turned into intense political debate. In Europe this has led to biosafety policies in which ethical political concepts like the precautionary principle demand that both scientists and regulators apply a new, explicitly moral, approach to biotechnology risks (Paula, 2001).

Several surveys of public attitudes in Europe have shown that an acceptance of exposure to risks depends on whether an application is considered useful and morally acceptable. Moreover, moral concern acts as a veto regardless of perceptions of risk and use. If risk itself is less significant than moral acceptability in shaping public perceptions, then public concerns are unlikely to be alleviated by technically inspired assurances and other policy initiatives that deal solely with scientific risks. It is this hard lesson that has finally caused industry, regulators and politicians, to face the fact that the ethical dimension of risk, and ethical issues in general, lie at the root of the societal controversy about biotechnology.

Professional Ethics

The persistent societal unease with biotechnology and growing impact of the influence ethics exerts on public perceptions has made the biotechnology comunity realise the importance of communicating a concern for responsible 'bioethical' conduct to society. This became quite critical in the mid-1990s, when for the first time, industry had to convince consumers to support biotechnology by buying its genetically modified harvest. It was at that time that both professional codes for individual biotechnologists, such as the European Federation of Biotechnology's Code of Conduct, as well as codes for bioindustry organisations, such as BIO's Statement of Principles, were developed. Voluntary codes traditionally have both internal and external functions. Internally, they provide the means to build a professional ethos, improving the quality of the services delivered to clients. Externally, they also serve as a means to build trust with society. They are the 'certificate' that shows that the profession as a whole will act responsibly towards both clients and society (Paula, 2001).

Opinion surveys in Europe have repeatedly revealed that biotechnology suffers from a lack of public trust and acceptance of its products. This is why the biotechnology community has recognised the importance of having a professional ethical code. A professional code is also a way for the sector to stay a step ahead of compulsory, more restrictive codes, governmental laws and regulations. This second external function is a major reason why such codes have become popular with both industry and regulators. They provide mechanisms for a self-regulation that is less consumptive of time and resources and which is therefore usually preferred by both sides. By designing such codes the biotechnology community as a whole has acknowledged for the first time that there are more perils to manage in biotechnology than mere technical risks. In doing so, the biotechnology community itself has helped to put bioethics on the public agenda.

Typically these codes do not restrict themselves to merely stating an ethical respect for patients, animals and environment. They often go further by clearly acknowledging the fact that biotechnology does indeed raise social issues that requires industry to respect the regulatory ideals of transparency, accountability and public participation. Thus, these codes themselves focus the spotlight on all sorts of socioeconomic and socio-cultural consequences only indirectly tied to science and technology. It is the types of 'globalization' issue that have evoked the fiercest protests in the streets and have been hardest to put on the agenda of international bodies such as the World Trade Organization (WTO) and the Organization for Economic Cooperation and Development (OECD).

By putting biotechnology in the context of globalization, the societal debate on biotechnology inevitably has shifted further towards discussing the ethical and social impacts of biotechnology (Paula, 2001).

References

Aerni, P. *Public attitudes towards agricultural biotechnology in developing countries: A comparison between Mexico and the Philippines.* Center for International Development at Harvard University, Cambridge, Mass. (2001).

Anonymous. Missing the big picture. *Nature* 421: 675 (2003).

Atkinson, H.J., Urwin, P.E., Hansen, E., and McPherson, M.J. Designs for engineered resistance to root-parasitic nematode. *Trends Biotechnol.* 13: 369-374 (1995).

Beck, U. *Risk Society.* Sage, London (1992).

Charles, D. *Lords of the Harvest.* Perseus Publishing, Cambridge, Mass. (2001).

Clark, N., Yoganand, B., Hall A.J. New science, capacity development and institutional change: the case of the Andhra Pradesh-Netherlands Biotechnology Programme (APNLBP)'. *Internat. J. Technol. Management and Sustainable Develop.* 1 (3): 196-212 (2002).

Damodaran, A. Regulating transgenic plants in India-Biosafety, plant variety protection and beyond. *Economic and Political Weekly*. 34(13): A-34 - A-42 (1999).

Dhar, B. Regulations, negotiations and campaigns: introducing biotechnology into India. *Biotech. and Develop. Monitor* 47: 19-21 (2001).

Duvick, D.N. Plant breeding, an evolutionary concept. *Crop Science* 36: 359-548 (1996).

Elomaa, P., Helariutta, Y., Griesbach, R.J., Kotilainen, M., Seppanen, R., Teeri, T.H. Transgenic inactivation in *Petunia hybrida* is influenced by the properties of the foreign gene. *Mol. Gen. Genet.* 248: 645-649 (1995).

Fransman, M. Designing Dolly: interactions between economics, technology and science and the evolution of hybrid institutions. *Research Policy*, 30 (2): 263-273 (2001).

Ghosh, P.K. Impact of industrial policy and trade related itnellectual property rights on biotech industries in India. *J. Sci. Indus. Res.* 54: 217-230 (1995).

Haribabu, E. Cognitive empathy in interdisciplinary research: the contrasting attitudes of plant breeders and molecular biologists towards Rice. *J. Biosci.* 25(4): 323-330 (2000).

Hayama, R., Yokoi, S., Tamaki, S., Yano, M., Shimamoto, K. *Nature* 422: 719-722 (2003).

Ho, M-W. Perils amid promises of genetically modified food. *Financing Agriculture* 33(3): 5-9 (2001).

Holmberg, N., Lilius, G., Bailey, J.E., Bulow, L. Transgenic tobacco expressing Vitreoscilla hemoglobin exhibits enhanced growth and altered metabolite production. *Nat. Biotechnol.* 15: 244-247 (1997).

Holton, T.A., Brugliera, F., Tanaka, Y. Cloning and expression of flavonol synthase from *Petunia hybrida*. *Plant J.* 4: 1003-1010 (1993).

Horvath, H. *et al. PNAS* (USA) 100: 364-369 (2003).

Jackson, D.L., Jackson, L.L. (eds.) *The Farm as Natural Habitat: Reconnecting Food Systems with Ecosystems.* Island Press, Washington, D.C. (2002).

James, C. *Global Review of Commercialized Transgenic Crops.* ISAAA Briefs No. 21 (2000).

Jenner, H.L. *Trends Biotechnol.* (*TiBTECH*) 21: 190-192 (2003).

Jonsen, A.R. *The Birth of Bioethics.* Oxford University Press, Oxford (1998).

Jorgensen R.A. Cosuppression, flower color patterns and metastable gene expression states. *Science* 268: 686-691 (1995).

Kalaitzandonakes, N. Biotechnology and identity-preserved supply chains: A look at the future of crop production and marketing. *Choices* (Fourth Quarter), 15-18 (1998).

Kumar, H.D. *A Textbook on Biotechnology.* 2ed. Affil. East West Press, New Delhi (1998).

Lutman, P.J.W. (Ed.). *Gene Flow and Agriculture. Relevance for Transgenic Crops.* British Crop Protection Council, Nottingham (1999).

Mann, C.G. Genetic engineers aim to soup up crop photosynthesis. *Science* 283: 314-316 (1999).

McNeil, S., Nuccio, M.L., Hanson, A.D. Betaines and related osmoprotectants. Targets for metabolic engineering of stress resistance. *Plant Physiol.* 120: 945-949 (1999).

Mourgues, F., Brisset, M-N., Chevreau, E. Strategies to improve plant resistance to bacterial diseases through genetic engineering. *Trends Biotechnol.* 16: 203-210 (1998).

Nakajima, H., Muranaka, T., Ishige, F., Akutsu, K., Oeda, K. Fungal and bacterial resistance in transgenic plants expressing human lysozyme. *Plant Cell Rep.* 16: 674-679 (1997).

OECD. *Safety Evaluation of Foods Derived by Modern Biotechnology: Concepts and Principles,* OECD, Paris (1993).

Oerke, E.-C., Dehen, H.-W., Schöbeck, F., Weber, A. *Crop Production and Crop Protection.* Elsevier, Amsterdam (1994).

Osorio, L.G. Plants protecting other plants. *LEISA* Magazine 17(4): 23-24 (2001).

Oud, J.S.N., Schneiders, H., Kool, A.J., van Grinsven, M.Q.J.M. Breeding of transgenic orange *Petunia hybrida* varieties. *Euphytica* 84: 175-181 (1995).

Pattanayak, D., Ananda Kumar, P. Plant biotechnology: Current advances and future perspectives. *Proc. Indian Natn. Sci. Acad. (PINSA)* B66: 265-310 (2000).

Paula, L. Ethics: The key to public acceptance of biotechnology? *Biotech. and Develop. Monitor* 47: 22-23 (2001).

Raina, R.S. Innovation capability for agrobiotechnology: Policy issues. *Biotech. Develop. Monitor* 50: 29-31 (2003).

Ram, A.S., Indira, M. Agricultural biotechnology - the science and fiction. *LEISA* India 3: 1-2 (Dec., 2001).

Reid, W.V., Laird, S.A., Meyer, C.A., Gamez, R., Sittenfeld, A., Janzen, D.H., Gollin, M.A., Juma, C. *Biodiversity Prospecting*. World Resources Inst., Washington, D.C. (1993).

Reilly, J.M. and Fuglie, K.O. Future yield growth in field crops: What evidence exists. *Soil and Tillage Research* 47: 275-290 (1998).

Ruivenkamp, G. Monitoring biotechnological developments: Looking back for finding new perspectives. *Biotech. Develop. Monitor* 50: 3-5 (2003).

Ruttan, V.W. Biotechnology and agriculture: A skeptical perspective. *Transactions* (Wisconsin Academy) 89: 83-92 (2001a).

Ruttan, V.W. *Technology, Growth and Development: An Induced Innovation Perspective.* Oxford University Press, New York (2001b).

Savin, K.W., Baudinette, S.C., Graham, M.W., Michael, M.Z., Nugent, G.D., Lu, C-Y., Chandler, S.F., Cornish, E.C. Antisense ACC oxidase dalays carnation petal senescence. *Hort. Sci.* 30: 970-972 (1996).

Shiva, V. Biotechnology: The failed miracle. *Focus WTO* 1(6): 5-11 (2001).

Strittmatter, G., Goethals, K., Van Montagu, M. Strategies to engineer plants resistant to bacterial and fungal diseases. In: *Subcellular Biochemistry* 29: 191-213, (eds.) Biswas, B.B., Das, H.K. Plenum Press, New York (1998).

Sundquist, W.B., Menz, K.M., Neumeyer, C.F. *A Technology Assessment of Commercial Corn Production in the United States.* Bulletin 546. University of Minnesota Agricultural Experiment Station, St. Paul (1982).

Suzuki, K., Zue, H., Tanaka, Y., Fukui, Y., Mizutani, M., Kusumi T. Molecular breeding of flower color of *Torenia fourieri. Plant Cell Physiol.* 38: 538-540 (1997).

Tanaka, Y., Tsuda, S., Takaaki, K. Application of recombinant DNA to floriculture; in *Applied Plant Biotechnology* pp. 181-235 (eds.) Chopra, V.L., Malik, V.S., Bhat, S.R. Oxford & IBH, New Delhi (1998).

Tanaka, Y., Yonekura, K., Fukuchi-Mizutani, M., Fukui, Y., Fujiwara, H., Ashikari, T., and Kusumi, T. Molecular and biochemical characterisation of three anthocyanin synthetic enzymes from *Gentiana triflora. Plant Cell Physiol.* 37: 711-716 (1996).

UNESCO. *Vital Water Graphics. Water Use and Management.* United Nations Educational Scientific and Cultural Organization, (UNESCO) Paris (2002).

van Dommelen, A. A transgene-centred approach to the transgenic herbicide-tolerant crops. *Biotech. Develop. Monitor* 38: 6-7 (1999a).

van Dommelen, A. *Hazard Identification of Agricultural Biotechnology: Finding Relevant Questions.* International Books, Utrecht (1999b).

Vasil, I.K. (ed.) *Plant Biotechnology 2002 and Beyond.* Kluwer Academic Publishers, Dordrecht (2003a).

Vasil, I.K. The science and politics of plant biotechnology-a personal perspective. *Nature Biotech.* 21: 849-851 (2003b).

Visser, B. Biotechnology. *LEISA* Magazine 17(4): 9-11 (2001).

Vos, P., Simons, G., Jesse, T., Wijbrandi, J., Heinen, L., Hogers, R., Fritzers, A., Groenendijk, J., Diergaarde, P., Reijans, M., Fierens-Onstenk, J., de Both, M., Peleman, J., Liharska, T., Hontelez, J., Zabeau, M. The tomato *Mi-1* gene confers resistance to both root-knot nematodes and potato aphids. *Nature Biotechnol.* 16: 1365-1369 (1998).

Yoshida, K., Kondo, T., Okazaki, Y., Katou K. Cause of blue petal colour. *Nature* 373: 291 (1995).

Chapter 2

TRANSGENIC CROPS IN EUROPE AND USA

Introduction

Traditional methods of plant breeding were instrumental in the twentieth century in increasing the productivity of several crops such as wheat, rice and maize (corn). In the new (21st) century, the genomic DNA sequence information about two flowering plants, *Arabidopsis* (a weed) and rice became known besides the genome sequences of a large number of microorganisms. Genome sequencing was made possible through development of such technologies as recombinant DNA, vectors, polymerase chain reaction (PCR), high throughput instrumentation and computational analysis of voluminous data. The totipotency of somatic cells in plants has been exploited to advantage as this property, along with the discovery of molecular mechanisms of crown gall disease, in which the bacterium *Agrobacterium tumefaciens* pushes a segment of its DNA into the plant somatic cells, has spurred attempts at the introduction of foreign DNA into many crop plants. Genes and genetic information can be transferred from an organism to some other species, so increasing the scope of genetic exchange far beyond that allowed by normal sexual reproduction.

In some countries at least, genetically modified (GM) crops have become widely adopted and popular; knowledge and information, which only a few decades ago used to be public property, are now privately patented and a great deal of public research has been privatized. Despite these changes however there is still need for reliable and objective information that can articulate the relevance, process and impact of biotechnology especially on developing countries.

Monitoring and Networking

Critical monitoring is a must for giving people the knowledge and capability to influence certain processes that could otherwise widen the gap between the industrial transformation of agriculture and the social context in which it occurs. Appropriation and substitution are two important long-term processes of industrialization that are particularly important in biotechnology as they lead to the industrial transformation of agriculture.

Networking is the central pillar which provides affected groups in developing countries the means and knowledge to acquire a definite, determining role in the future development of biotechnology. One such network in place is called the Tailor-made Biotechnologies Network; it is instrumental in building up local capabilities and knowledge.

Online technologies may possibly be exploited for biotechnology and development. The Community Biodiversity Development and Conservation Program is a global network that is planning to use interactive technologies to build up knowledge and capabilities in the South, including the development of an online course on biodiversity and conservation. As interactive and new technologies facilitate development of new channels and forging of new networks and alliances, they cross traditional boundaries of space, time and scope.

Transgenic technologies have, however, attracted much controversy also. Some view these technologies as a boon that can markedly enhance productivity and concomitantly enhance the sustainability of agriculture. But others perceive them as a threat to human health, biodiversity, sustainable development and the livelihood of small farmers.

In recent years although some transgenics have been developed by crossing exotic transgenic lines with breeding lines developed in India, hardly any gene discovery work is being done in India.

Global Area Under Transgenics

The global area under transgenic crops has been steadily increasing but the increase has not been even or uniform; only two countries–USA and Argentina–accounted for 90% of the total area under transgenics in 2001. Transgenics have been chiefly developed for soybean, corn, cotton and rapeseed 'Canola', much less work has been done on other crops. And only two traits–herbicide resistance and insect resistance–have attracted priority. However, it is expected that many new and novel products will be commercialized in the coming few years.

Worldwide transgenic varieties are being developed for such traits as resistance to herbicides, pollination control, insect, virus, or fungal resistance, nutritional improvement, senescence retardation, resistance to various abiotic stresses, and for production of high value pharmaceuticals and secondary metabolites.

Crop biotechnology broadly includes transgenics, structural and functional genomics, and marker-assisted breeding. All these areas can potentially generate major breakthroughs required to sustainably improve both quality and quantity of crops. In spite of the on-going debate on GM crops in countries of the European Union, millions of large and small farmers everywhere are planting GM crops for their substantial benefits. It has been noticed repeatedly that farmers planting herbicide-tolerant and insect-resistant *Bt* crops have been able to manage their weed and insect pests much more effectively than those planting non-*Bt* varieties. Indeed, transgenic crops exemplify promising technologies with a high potential to enhance global food, feed and fibre security.

Male sterile (due to *barnase* gene) and fertility restorer (due to *barstar* gene) plants have been produced in *Brassica napus* and *Brassica juncea,* with a view to producing their hybrid seed without manual emasculation and controlled pollination as practised in maize. This approach has abolished the need for adopting any cytoplasmic male sterility (cms) or fertility restoration system in crops, where hybrids are raised for higher yields.

Until the mid-1980s transgenic plants had been produced mainly in tobacco, petunia and tomato, but since then several other plants (see Table 2.1) belonging to various families have been raised.

Table 2.1: A Short List of Transgenic Plants Produced So Far

Popular Name	*Botanical Name*
Tobacco	*Nicotiana tabacum*
Petunia	*Petunia hybrida*
Tomato	*Lycopersicon esculentum*
Potato	*Solanum tuberosum*
Thale cress	*Arabidopsis thaliana*
Sunflower	*Helianthus annuus*
Oilseed rape (canola)	*Brassica napus*
Cotton	*Gossypium hirsutum*
Sugarbeet	*Beta vulgaris*
Soybean	*Glycine max*
Cucumber	*Cucumis sativus*
Muskmelon	*Cucumis melo*
Carrot	*Daucus carota*
Sweet potato	*Ipomoea batatas*
Papaya	*Carica papaya*
Grape	*Vitis vinifera*
Apple	*Malus sylvestris*
Pear	*Pyrus communis*
Poplar	*Populus* sp.
Neem	*Azadirachta indica*
Walnut	*Juglans regia*
Rye	*Secale cereale*
Rice	*Oryza sativa*
Wheat	*Triticum aestivum*
Corn	*Zea mays*
Oats	*Avena sativa*
Banana	*Musa acuminata*

Table 2.2 lists some secondary metabolites produced by plants that have been implicated in resistance to insect attack. The biosynthesis of these metabolites occurs through a series of steps each of which is controlled by a distinct gene. These genes express

in tissue-specific manner, which makes the production of transgenic plants difficult.

Table 2.2: Selected Examples of Secondary Metabolites from Legume Seeds with Demonstrated Insecticidal Properties

Metabolites	*Legume*	*Insect*
Non-protein Antimetabolites Alkaloids		
Castanospermine	*Castanospermum australe*	*Callosobruchus maculatus*
Non-protein Amino Acids		
p-Aminophenylalanine	*Vigna* sp.	*Zabrotes subfasciatus*
Sapanins	*Phaseolus vlugaris,*	*Callosobruchus chinensis,*
	Glycine max	Callosobruchus chinensis
Polysaccharides	*Phaseolus vulgaris*	*Collosobruchus chinensis,*
	P. vulgaris	Acanthoscelides obtectus,
	Vigna spp.	*Callosobruchus chinensis*
Protein Antimetabolites		
Lectins (phytohaemagglutinins)	*P. vulgaris*	*Callosobruchus maculatus*
a-Amylase inhibitors	*P. vulgaris*	*Zabrotes subfasciatus*
	P. vulgaris	*Callosobruchus chinesis*
Protease-inhibitors	*Vigna unguiculata*	*Collosobruchus maculatus*

The Gene Revolution

There is hardly anyone, from producers to consumers, who is not touched by the latest technology in agricultural development-genetic engineering of crops and animals. This subject is arousing much debate and controversy all across the globe, and at all levels of society, and especially from the vantage point of small farmers and sustainable agriculture. It is still uncertain as to what benefits small farmers will gain from GM crops and what risks are involved. This precludes a full focus or all development efforts on genetic engineering. Rather, some of the many alternatives to GM crops should be explored.

Genetic engineering is often touted as a humane technology and one that feeds more people with better food but for this claim there is no warrant. With a few exceptions, the purpose of genetic engineering seems to be to increase the sales of chemicals and bio-engineered products to dependent farmers.

The genetic engineering revolution has made it possible to break through natural species barriers by systematically moving genes from one species to another-genes that do not normally combine in nature. The transfer of genetic material, for instance, is engineered from bacteria to plants. Practitioners of genetic engineering (GE) believe it will provide new plants and animals that would lead to a more environmentally-sound agricultural production with crops that produce their own pesticide thus reducing the use of chemical pesticides (Anon., 2001). They hope to raise crops that produce medicine, plants tolerant to salt and drought, and enriched food to restore micro-nutrient deficiencies. Many view GE as "the solution to hunger, poverty and several health problems."

It is conceivable that GM plants might not have attracted so much opposition if only the first such plant had been developed to prevent, say, wrinkling or to slow aging. Many Europeans happily consumed paste made from GM tomatoes for years as its quality was better than that of non-GM paste. But as soon as genetic engineering of plants was used to reduce farmers' inputs (and hence perceived to increase profits of Western farmers and their suppliers), self-righteous opposition took hold (Tester, 2002). The new technology was perceived to give little apparent benefit to well-off consumers in the West and it was thought to carry unknown risks; so it was opposed.

While many environmentalists oppose this new technology, some 80% of the world's population (who earn only 11% of the wealth) are being denied the opportunity of even trying GM foods as another tool for agricultural improvement. Providing better information can certainly benefit agriculture. Such a step is essential and urgently needed for all countries (the majority) where agricultural self-sufficiency is a social and economic imperative (Tester, 2002; Lurquin, 2002).

But when one looks beyond the promises, some important questions emerge:

1. Who benefits from genetic engineering and who loses?
2. What are the risks and who will bear them?
3. Are there alternatives to genetic engineering?

Genetically modified crops have far-reaching implications on sustainable agriculture in general, and farmers' livelihoods in the South in particular. Genetic engineering is sometimes presented as just another step in a continuous process of agricultural development but this argument is not justified. Genetic engineering is very different from previous technologies - it allows the movement of genes between different species across natural boundaries, which makes the risks unpredictable.

Ecological Concerns

Despite reassuring statements by many companies and some governments, many concerns about the implications of GM crops remain. Major concerns relate to the consequences for the ecosystems into which they are being introduced. These concerns are often neglected by the GM seed industry, the authorities approving their access to the market and the farming communities making use of the proposed technologies. For instance, the insertion of *Bt* (*Bacillus thuringiensis*) genes was once claimed to be a silver bullet, a permanent solution to insect problems. But the model of "one pest-one solution" does not work forever, as is the case with pesticides; sooner or later resistance builds up. Similarly, building of herbicide resistance in plants is troublesome as it unleashes basic ecological reactions. Excessive use of herbicides as a major or only tool of weed management eventually reduces the sensitivity of weeds to herbicides and creates an even worse weed problem. Yield decline in GM soybean, for instance, has been traced to reduced root development, nodulation and nitrogen fixation.

Another effect may be the unexpected impact of gene transfer and its consequences. In one illustration from USA, genes were transferred from a bacterium *Xanthomonas* to another soil bacterium, *Klebsiella planticola*. The new organism was meant to ferment stubble into alcohol, thus providing farmers with an extra source of income from what would otherwise be wasted by burning. However, it turned out that wheat planted in the soil containing the new organism was killed by it.

The unique aspect of the development of GE in the history of agriculture is that it is almost fully controlled by private companies. Transnational corporations (TNCs), often with their roots in the production of agrochemicals, carry out the laboratory research, field trials, production and sale of GM crops. They spend enormous amounts of money on developing herbicide-resistant crops that are being sold to farmers as a package inclusive of both the herbicide and the seeds. Through patents, these TNCs keep competitors at bay. It appears that GE technologies are not being developed because of their problem-solving capacity, but because of the patent-and thus profit-it can bring to the companies (Anon., 2001).

Companies usually promote high-input, highly industrialised monoculture systems which force farmers to buy packages of inputs from just one company. In this context it is surprising that, in 2001, the US government generously funded biotechnology research and development in agriculture with a budget allocation of US$310 million, whereas support for organic farming was less than US$5 million. Farmers were greatly concerned about these developments as can be concluded from the citizen's juries conducted in many parts of the world. It also appears that GM crops have very little to offer to farmers in risk-prone, diverse and complex agriculture. It is expected that GM crop research will be very slow in responding to the needs of low-input agriculture.

Contamination

According to van Bueren and Osman (2001), the organic movement does not consider GM crops as organic. It accepts conventional breeding and the new technologies available to assist it, but finds manipulation at the cellular and subcellular level unacceptable. The debate on GE has led organic farmers to reconsider their dependence on seed companies that focus on high-input agriculture.

The problem here is how can farmers be sure that they grow GM crops, considering that seeds and pollen are spread by wind, water, birds and insects? Large areas can be contaminated by the introduction of GM crops by a single farmer. In the US, contamination by GM crops is now such a serious problem that organic farmers find it almost impossible to get GM-free seeds. Tests have shown that "organic crops" from the US are often contaminated with engineered genes despite farmers' efforts to stay GM-free.

With the introduction of agricultural genetic engineering, the contamination costs and costs of reduced market shares are being imposed on farmers, consumers and the environment as a whole–

not only in advanced nations, but also in the South! Will GM seed companies bear the risk of releasing these crops in the South? What will happen if things go seriously wrong, *e.g.* a GE crop turns out to have negative health effects or becomes a serious ecological threat? Banning of the GM crop in question does not mean it will stop existing. This situation is not comparable to that of an agricultural chemical which turns out to have unanticipated side effects after a number of years. The effect of such chemicals will eventually disappear from the environment. This is not so with GM crops that are likely to survive in the wild and spread their genes through crossing with other plants. In Mexico wild relatives of maize have already become contaminated with genes from GM crops. Mexico being the centre of maize diversity, such contamination constitutes an irreplaceable loss. The wide variety of genes in wild plants and in traditional agriculture is the chief insurance we have to cope with new demands on crops–whether caused by new pests and diseases, increased salt levels or changing climates. In Southern Brazil, despite the ban on GM crops, 30% of the soybean acreage may already be contaminated, threatening Brazil's GM-free status (see von der Weid and Tardin, 2001).

The risk of an unintended introduction of GM crops is more threatening in those countries where no legal framework exists, *e.g.* in many African countries.

The question is: do we really need GM technology to combat malnutrition, to improve local production and to make agriculture more productive? Has the introduction of GM crops contributed to the reduction of poverty? According to a recent report of the FAO "for the world as a whole there is enough, or more than enough, food production potential to meet the growth of effective demand, *i.e.* the demand for food of those who can afford to pay farmers to produce it." This implies that any residual hunger problems are largely related to poverty, rather than to production, which means that reaching the goal of food security for all should be based on a premise other than genetic engineering. Alternative approaches to agricultural production are, therefore, essential.

There exists a wealth of agroecological, low-external-input alternatives to agricultural production. The potential of LEISA is far from exhausted. The case of natural crop protection from the Andes indicates that there are many plants in nature, which provide us with clues for better pest management. Agricultural research in India is only making use of 3% of the total of 3000 rice varieties that are known. Research done in Thailand indicates the potential in nature for selecting and breeding varieties with desired characteristics such as salt tolerance. Many ecological principles that are still being overlooked, underestimated or sidelined, deserve greater attention as they provide relatively cheap, controllable and low external input solutions to many problems that farmers face. The System of Rice Intensification (SRI) is an example of the many roads to sustainable agriculture that are hardly explored (Uphoff, 2001). Moreover, these approaches are not accompanied by the many economic and ecological risks that GM crops are posing (Anon., 2001).

Organic cotton production in Senegal and zero-tillage no-herbicide soybean cultivation in Brazil are two examples of ecologically-sound alternatives that already exist. They are not a danger to the environment, nor do they make the farmers dependent on agricultural supply companies. Third world farmers will certainly be much better off if research efforts and resources are dedicated to agroecological approaches that have wide-ranging possibilities (Anon., 2001).

Table 2.3 lists the 13 countries that planted transgenic crops in 2001. Of the top four countries that grew 99% of the global transgenic crop area, the USA grew 68%, Argentina 22%, Canada 6% and China 3%, the balance 1% being planted in the remaining nine countries (see Fig. 2.1).

The country portfolios of deployed GM crops have diversified in 2001, with several new crop/trait introductions, including herbicide-tolerant corn in Argentina, herbicide-tolerant cotton as well as the stacked *Bt*/herbicide-tolerant cotton in Australia, herbicide-tolerant soybean, *Bt* white corn and herbicide-tolerant cotton in South Africa and *Bt* cotton in Indonesia (James, 2003).

The distribution of the global transgenic crop area for four major crops is plotted in Fig. 2.2 for the period 1996 to 2001. Soybean (transgenic) dominated, occupying 63% of the global area of transgenic crops in 2001; all of this is herbicide-tolerant. In 2001, the global hectarage of the herbicide-tolerant soybean increased by 7.5 million hectares, registering a 20% increase. Globally, transgenic soybean occupied 33.3 million hectares in 2001; transgenic corn occupied the second place at 9.8 million hectares, transgenic cotton was third, at 6.8 million hectares, and canola at 2.7 million hectares (Table 2.4).

Table 2.3: Global Area of Transgenic Crops in 2000 and 2001: by country (million hectares) (after James, 2003)

Country	*2000*	*%*	*2001*	*%*	*+/-*	*%*
USA	30.3	68	35.7	68	+5.4	+18
Argentina	10.0	23	11.8	22	+1.8	+18
Canada	3.0	7	3.2	6	+0.2	+6
China	0.5	1	1.5	3	+1.0	+200
South Africa	0.2	<1	0.2	<1	<0.1	+33
Australia	0.2	<1	0.2	<1	<0.1	+37
Mexico	<0.1	<1	<0.1	<1	<0.1	-
Bulgaria	<0.1	<1	<0.1	<1	<0.1	-
Uruguay	<0.1	<1	<0.1	<1	<0.1	-
Romania	<0.1	<1	<0.1	<1	<0.1	-
Spain	<0.1	<1	<0.1	<1	<0.1	-
Indonesia	-	-	<0.1	<1	<0.1	-
Germany	<0.1	<1	<0.1	<1	<0.1	-
France	<0.1	<1	-	-	-	-
Total	**44.2**	**100**	**52.6**	**100**	**+8.4**	**+19%**

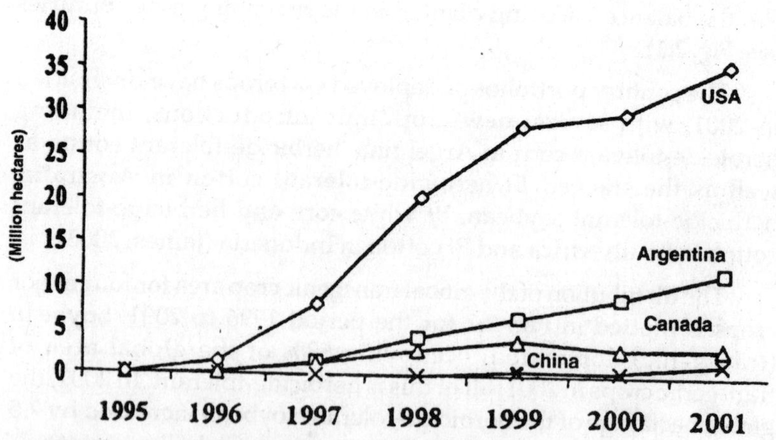

Fig. 2.1: Global Area of Transgenic Crops, 1996 to 2001: by country (after James, 2003)

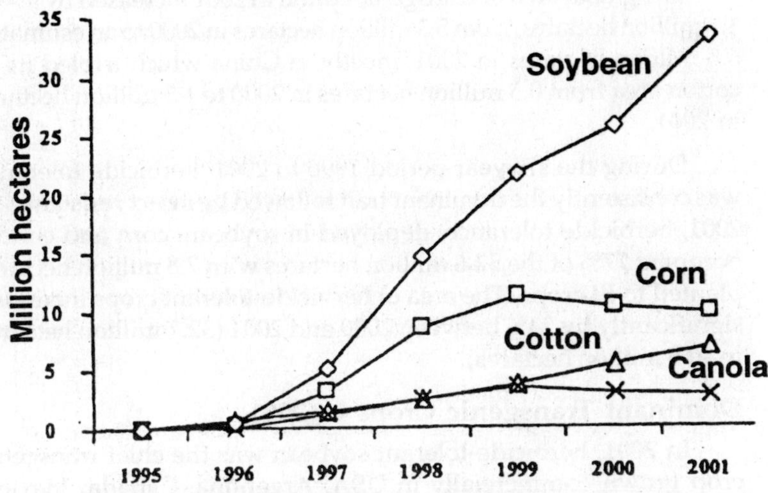

Fig. 2.2: Global Area of Transgenic Crops, 1996 to 2001: by crop (after James, 2003)

In contrast to soybean, the area planted with transgenic corn in 2001 decreased globally by about 500,000 hectares (Table 2.4) with all the reduction taking place in the USA. However, declines in transgenic corn in the USA were offset by significant increases in transgenic corn in Canada, Argentina and South Africa (James, 2003).

Table 2.4: Global Area of Transgenic Crops in 2000 and 2001: by crop (million hectares) (after James, 2003)

Crop	*2000*	*%*	*2001*	*%*	*+/-*	*%*
Soybean	25.8	58	33.3	63	+7.5	+29
Maize	10.3	23	9.8	19	-0.5	-5
Cotton	5.3	12	6.8	13	+1.5	+28
Canola	2.8	7	2.7	5	-0.1	-4
Potato	<0.1	<1	<0.1	<1	<0.1	-
Squash	<0.1	<1	<0.1	<1	(-)	-
Papaya	<0.1	<1	<0.1	<1	(-)	-
Total	44.4	<1	<0.1	<1	(-)	-

The global area of transgenic cotton in 2001 increased by about 1.5 million hectares, from 5.3 million hectares in 2000 to an estimated 6.8 million hectares in 2001, mostly in China which tripled its *Bt* cotton area from 0.5 million hectares in 2000 to 1.5 million hectares in 2001.

During the six-year period 1996 to 2001, herbicide tolerance was consistently the dominant trait followed by insect resistance. In 2001, herbicide tolerance, deployed in soybean, corn and cotton, occupied 77% of the 52.6 million hectares with 7.8 million hectares planted to *Bt* crops. The area of herbicide-tolerant crops increased significantly by 24% between 2000 and 2001 (32.7 million hectares to 40.6 million hectares).

Dominant Transgenic Crops in 2001

In 2001, herbicide-tolerant soybean was the chief transgenic crop grown commercially in USA, Argentina, Canada, Mexico, Romania, Uruguay and South Africa (see Fig. 2.3, Table 2.5). Herbicide-tolerant soybean occupied 33.3 million hectares, representing 63% of the global transgenic crop area of 52.6 million hectares for all crops. The second most dominant crop was *Bt* corn which occupied 5.9 million hectares.

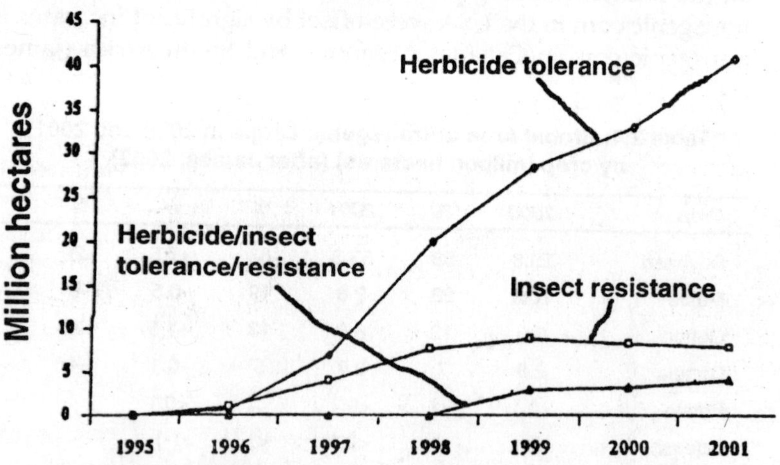

Fig. 2.3: Global Area of Transgenic Crops, 1996 to 2001: by trait (after James, 2003)

Table 2.5: Dominant Transgenic Crops 2001 (after James, 2003)

Crop	*Million Hectares*	*% of Global Transgenic Crop Area*
Herbicide-tolerant soybean	33.3	63
Bt maize	5.9	11
Herbicide-tolerant canola	2.7	5
Herbicide-tolerant cotton	2.5	5
Bt/herbicide-tolerant cotton	2.4	5
Herbicide-tolerant maize	2.1	4
Bt cotton	1.9	4
Bt/herbicide maize	1.8	3
Total	**52.6**	**100**

Table 2.6: Transgenic Crop Area as Per Cent Global Area of Principal Crops, 2001 (million hectares) (after James, 2003)

Crop	*Global Area*	*Transgenic Crop Area*	*Transgenic Area as Per Cent of Global Area*
Soybean	72	33.3	46
Cotton	34	6.8	20
Canola	25	2.7	11
Maize	140	9.8	7
Total	**271**	**52.6**	**19**

The above data clearly show that GM crops have fulfilled the expectations of most farmers planting transgenic crops in both industrial and developing countries. The rapid adoption of transgenic crops reflects the substantial benefits gained by large and small farmers. Improved weed and insect pest control can be achieved with transgenic herbicide-tolerant and insect-resistant *Bt* crops that also require lower input and production costs. More specifically, the use of transgenic crops results in:

1. Greater sustainability and resource-efficiency in crop management practices requiring less energy and fuel.
2. More effective control of insect pests and seeds.

3. A reduction in the quantity of pesticides used in crop production.
4. Less dependence on conventional pesticides that often pose health hazards to producers and consumers.
5. *Bt* maize, having reduced levels of mycotoxins, yields safer and healthier food and feed products.
6. Increased operational flexibility in the timing of herbicide and insecticide applications.
7. Conservation of soil moisture, structure, nutrients and control of soil erosion through no or low-tillage practices; also improved quality of ground and surface water with less pesticide residues (see James, 2003).

According to James, several new and novel products with input and output traits will become available for commercialization in the near future. These products should be planted/used in an integrated manner in which both conventional and biotechnology methods are applied to attain the daunting goal of global food security. Society must continue to gain from important contributions from plant breeding using both conventional and biotechnology tools-improved crop varieties should continue to be the most cost effective, environmentally safe and sustainable way to ensure global food security in the future (see James, 2003).

Transgenic Crops in Europe

Although genetically modified (GM) crops have found considerable acceptance in many countries, in Europe people are highly suspicious about them, owing largely to the various scares from salmonella, *Escherichia coli* and bovine spongiform encephalopathy (BSE) (the mad cow disease). These scares were prominently highlighted by the media and contributed to a general abhorrence of any genetic manipulation. It is also argued that while consumers are at risk by eating GM food, the biotech companies profit greatly from commercializing such products.

The vehemence of opposition to GM crops in Europe seems surprising and ironical in the face of Europe's willingness to gladly adopt biotechnology for medical and other uses. No significant ethical concern has been aroused at the introduction of genetically engineered insulin for treating diabetes, or genetically engineered enzyme chymosin for cheese-making. The case of chymosin is particularly ironical: chymosin is usually extracted from stomachs of calves. Many Hindus and Buddhists are vegetarians and many of the non-vegetarians also do not eat beef or veal. The genetic modification of chymosin extracted from calves has strangely made cheese more acceptable to several vegetarians largely because cheeses are not usually labelled as GM! In reality, the production of these cheeses greatly depends on GM technology. In the various debates on the potential benefits of GM crops, as well as the possible risks, greater attention is usually given to the risks, so that many people are mistakenly believing that GM offers little or no advantage (Williams, 2003). It is only very recently that the potential benefits of applying GM technology to the production of crops are being highlighted. Evidence has been gathered about the possible benefits to wildlife of growing GM sugar beet, and it has been reported that GM cotton in India may well register substantial increases in yield along with cuts in pesticide use. Further work should focus on the impact of GM crops such as cotton on the diversity of plant and animal life, and duly assessed against the problems associated with the intensification of agricultural practices.

GM oilseed rape is now widely grown by Canadian farmers. Around 85% of this crop is being planted to herbicide-resistant GM varieties. Farmers using the new varieties have benefited on average by $14 per hectare, and herbicide use has been reduced by 6,000 tons; 32 million litres of fuel needed for conventional crop spraying have been saved (see Williams, 2003).

The GM debate is now entering a new phase with focus on specific applications of GM technology and a balanced assessment of the potential risks and benefits in specific cases. In some instances the risks may be quite unacceptable but in others the risks associated with GM technology will be found to be so small that they are outweighed by the likely benefits.

Some experts feel that biotechnology can usher in a 'doubly green' revolution for African farmers, who were deprived from enjoying the first green revolution. This could be done by achieving desperately needed productivity gains at much lower environmental costs in the form of pesticide pollution and water and energy use (Taylor, 2003). By introducing genes, biotechnology can place the valuable agronomic performance trait inside the plant rather than having to manipulate the environment outside the plant.

Biotechnology has been quite eagerly accepted in medicine-the direct benefits are enjoyed by the same person who experiences any real or perceived risk. In contrast, food biotechnology does not provide any direct consumer benefits–the benefits accrue largely to the technology provider and farmers, while any real or perceived risks are borne by consumers (Taylor, 2003). In Europe, neither farmers nor consumers gain any direct benefit from American-produced GM crops. But consumers, even in Europe, do benefit indirectly from food biotechnology such as the environmental benefit of reduced pesticide use and the social benefit of improving agriculture and food security in developing countries. Although people do value these benefits, it is not known if they are inclined to subject themselves to any real or perceived personal risk in the hope of benefiting in some remote way.

Proponents of organic agriculture are strongly opposed to GMOs. In France, demand and sales of organic products have increased by some 25 per cent over recent years. In Sweden and Austria also, a sizeable fraction of their agricultural land is used for organic growing and the figure is rising in most countries. All these countries are opposed to GM crops, but the situation varies from crop to crop: we may try to segregate GM crops from organic plantings by sufficient distance to avoid any potential cross-pollination-particularly if labelling of GM products is enforced.

Although in Europe commercial growing of genetically modified crops has now been legally allowed and looks set to go ahead, persuading consumers about their merits is likely to prove difficult. The European Parliament has finally given approval to the commercial growing of genetically modified crops across the continent.

In Britain, the sites for the first set of trials for the coming season involving genetically modified beet, sugar beet and oilseed rape have been announced–such openness backfired on some previous occasions when environmental protesters destroyed the trial crops.

But anti-GM campaigners still harbour serious concerns. They fear that the use of antibiotic-resistance marker genes will only be phased out gradually after commercial crops have been planted and that pollen from GM crops might adversely affect neighboring conventional crops (Williams, 2001).

It also appears that consumers are not yet happy to accept the new foods. There is distrust of the food industry and official regulators, following numerous scares from *Salmonella*, through *Escherichia coli*, to bovine spongiform encephalopathy (BSE or mad cow disease). Opponents argue that, although consumers may be taking risks by eating genetically modified food, the benefits go into the pockets of biotech companies.

Earlier this year, the two UK supermarket grocery chains Tesco and Asda announced they would not sell meat or milk from any animal fed with genetically modified soya or maize. Some other companies have already acted to remove GM ingredients in animal products.

The new European laws also face strong opposition from other European nations. The French are trying to block all new licences for commercial growing of GM crops. According to them, the issue of legal liability of biotech companies for any damage done by the new crops has not been resolved and there is concern about 'traceability'-provisions to ensure that consumers know what they are eating. Greece, Denmark, Austria and Luxembourg are expected to support the French view.

Some regions in Europe are planning to declare themselves GM-crop free zones, before the planting of such crops is granted approval. With uncertainty looming large over who would be legally liable if genetic contamination takes place, it is very unlikely that any European farmer will be eager to plant GM crops, even though it may become legal to do so in the near future.

In Britain, Devon, Dorset, Lancashire, Warwickshire, Cumbria, Somerset and several other regions are GM-free zones. Elsewhere in Europe, some parts of Italy, France, Germany and Austria have decided not to allow GM crops to be grown in their areas. They consider it wiser to market their produce as 'GM free'.

The vehemence of opposition to trials of GM crops seems surprising in the face of Europe's willingness to embrace biotechnology for medical and other uses. There has, for example, been little ethical concern about the introduction of genetically engineered insulin for treating diabetes, or a genetically engineered version of the enzyme chymosin for making cheese.

Many researchers feel optimistic about golden rice. Unlike many GM crops which primarily benefit the owner or grower, golden rice is targeted at the consumer. By introducing three genes into rice to make it rich in beta carotene, a precursor of vitamin A, the new rice could be a valuable source of the vitamin for millions of people whose staple food is rice but whose diets are deficient in the vitamin. Irrespective of whether or not new GM crops may win over people, the organic sector agriculture remains implacably opposed to GMOs. Demand for organic products has risen dramatically in recent years, with BSE being widely seen as a watershed spurring this new demand. People are now realizing that merely ensuring that a culinary end product is safe to eat is not a good enough approach. We must look at the entire process by which food is produced. In France also sales of organic products have increased by 25% over the past two years as BSE cases have been confirmed in that country. In both Sweden and Austria, more than 10% of agricultural land is now being used for organic production and the area is growing across countries in Europe. GMOs face a very tough battle ahead (Williams, 2001).

In the United Kingdom, the debate about genetically modified (GM) food has not involved any reasoned evaluation of the evidence. Rather, the public has been scared by headlines about "Frankenstein foods" and has increasingly distrusted scientists since the BSE debacle, preferring to rely on information from such groups as Greenpeace and Friends of the Earth. Yet the evidence suggests that scares about the safety of GM foods are unfounded (Taverne, 2001).

McHughen (2000), while claiming to chart a middle course between the extreme claims of the proponents and opponets of GM technology, in fact could not name any risks. There could be a risk to health if an allergenic protein were inserted into a novel food, but McHughen was confident that it would be eliminated long before it reached his kitchen. About the only potentially dangerous example he cited is the insertion of a Brazil nut gene into a soybean. Actually, no such bean has ever been marketed.

McHughen focused mainly on current myths: that a transgenic plant, or gene transfer, is 'unnatural'; that conventionally farmed food, or even organic food, is safer; that transgenic crops are more dangerous to biodiversity or are likely to create 'super-weeds'; that labelling, especially of so-called 'GM-free' products, gives the consumer a meaningful choice; and that, in testing food, we should be concerned with the process rather than the product. But one important subject he underplayed is the potential of transgenic crops to fight hunger and disease in the developing world. He only made a passing mention of the benefits to children in the developing world from the insertion of a vitamin A gene into rice.

Manning (2000) dealt with food in the developing world, and what may happen when the green revolution runs out of steam. He realized the growing importance of transgenic crops in the agriculture of the developing world and felt convinced of its potential benefits. According to Manning (2000), "The green revolution at its most fundamental level treated all the world the same. The lessons being learned in agriculture now are local."

The appropriate revolution where yields have stagnated must build on local practice, indigenous knowledge and local communities. According to Taverne (2001), transgenic crops are likely to play a vital part in feeding the extra two billion people to be born in the next quarter-century, but only if it is applied in a way that enhances local communities, in tune with local practices and cultures, and focuses on complexity and diversity. There is the danger that GM technology could lead to a greater degree of monoculture. But it could also mean that the next green revolution avoids the uniform approach of the last one and the damage this caused to many small farmers.

The timing of marketing of the American GM soybeans in Europe happened to be unfortunate in the sense that it coincided with the UK and European food safety crisis involving mad cow disease. The government officers, regulators, and scientists who erroneously proclaimed that mad cow disease could not pass from cows to humans, told consumers that the soybeans were safe. Naturally the public refused to accept their view. In contrast in the USA, the FDA attracts greater respect and public trust and its views about the safety of biotechnology foods are better received.

To enhance public acceptance, what the biotechnology companies should do is to develop products that directly benefit consumers. At the same time, to broaden environmental benefits, the government could create economic and regulatory incentives for reducing pesticide use; carefully evaluate and report the environmental benefits of current biotech crops; and focus more and more on environmentally beneficial applications of biotechnology (Taylor, 2003).

In developing countries, multinational biotechnology corporations have no economic incentive to invest in new commercial seed products for the benefit of poor farmers. We need programs that match local germplasm with the traits required to solve local farming problems.

The developed countries should adopt technology transfer policies that encourage government-funded biotechnology research and tools in the public domain for developing country purposes while also encouraging private sector sharing of patented technologies to help developing country farmers (Taylor, 2003).

In the USA, the regulatory system has ensured that the biotechnology crops and foods on the market today are quite safe for human health and the environment. But there is need for a rigorous post-commercialization monitoring of biotechnology crops for unanticipated environmental impacts.

According to Taylor (2003), "In the end, biotechnology will fulfill its full potential in a society only if it is freely chosen by that society." Whereas in the EU labeling of GM foods and GM-derived materials is the preferred approach, the United States relies more on organic standards and the optional use of labels identifying non-GM foods.

Although several European countries oppose the commercialization of GMOs, Italy went to the extreme by banning all research involving GMOs. The ban evoked strong protest from over 1000 Italian researchers and was supported by the International Agbiotech community (*Nature Biotechnol.* 18: 1229, 2000).

The research situation is particularly bad in Italy because, according to Silvio Garattini, director of the Mario Negri Institute in Milan, it is trapped by two extremes, the green one and the religious one. (Research on cloning and stem cells is at risk through politicians' desire to appease the Vatican). Italy has not produced a new drug in the past 6 years.

As a result of the protest, Italian politicians are now strongly supporting biotechnology and are distancing themselves from the

Green anti-GMO campaigns that, till hitherto have been shaping Italy's agbiotech policies.

However, the happenings in Italy are unlikely to have any effect on the political impasse that is blocking commercial planting of GMOs in Europe where consumer attitudes are more crucial factors. Indeed, in February 2001, the European parliament imposed mandatory monitoring and risk assessment of GM crops and setting guidelines on labeling. However, the environment ministers of France, Italy, Austria, Luxembourg, Greece, and Denmark say they will continue to block the approval of GM products until there is legislation that ensures they can be traced through the entire production chain. With the Commission's refusal to fight the envirocrats and enforce the directive, the legislation is irrelevant and the *de facto* moratorium remains (see Meldolesi, 2001).

Transgenic Crops in USA

As GM crops and their products are generally considered to be safe by American citizens, the USA has been challenging the European *de facto* moratorium on approval of GM food crops in the World Trade Organization, and have campaigned for gaining acceptance in Europe of various GM crops and foods. According to Taylor (2003), the US government and biotechnology industry have been trying very hard to make people more receptive to biotechnology on US terms. While agricultural biotechnology can be supported on its scientific merits, the public attitude to biotechnology usually has very little to do with science. The social dimension is far more important for public acceptance than science alone.

No doubt, biotechnology has helped US farmers to grow corn, cotton and soybeans more efficiently, often with less use of toxic insecticides; but the global concerns surrounding biotechnology threaten its future adoption for food purposes everywhere.

The storm of protests in Europe over GM crops has begun to move into the United States. American surveys show increasing consumer concern about food biotechnology. Major food companies have decided to purchase only non-bioengineered products. Recent finding of StarLink corn in taco shells has generated strong criticisms about the adequacy of American regulations of biotechnology. StarLink, a genetically-altered corn variety produced by Aventis Crop

Science, had been approved for animal but not human consumption because of fears that it might contain an allergen.

Currently, the area planted with GM crops is increasing. More than half of the world's soybean crops and about one third of the corn crops are transgenic. In 2000, four countries-US, Argentina, Canada and China-grew 99% of the global transgenic crop area.

In the USA over 40 GM food crops have completed all the federal regulatory requirements and have made their way into innumerable consumer items. The approval of such products comes under the jurisdiction of three American government agencies: the Department of Agriculture (USDA), the Environmental Protection Agency (EPA), and the Food and Drug Administration (FDA) (Bonetta, 2001).

The FDA, whose regulations have been the primary focus of the 'public debate' over biotechnology, oversee food and feed derived from new plant varieties. The FDA can remove a food from commercialization if it is found to be unsafe but it does not require safety testing prior to commercialization. Pre-market approval is required only for food additives that are significantly different from substances currently found in food.

Till now determination of the regulatory status of a product with the FDA has been voluntary. Under a new proposed rule, the FDA will make this process mandatory.

Proponents of the GM technology point out that there has been no evidence of any negative effects. Some have however cautioned that it is premature to declare a total absence of adverse effects, since no systematic monitoring for such effects has been done in either the USA or Canada. This implies that the doctrine of treating the approval of GM crops as though they were much the same as conventionally grown crops, may not be valid.

In fact both the testing and monitoring of GM food crops present technical and financial challenges and it seems unlikely that any solution will satisfy both sides of the debate (see Bonetta, 2001).

Many biotechnology proponents believe that genetic engineering is the key to ending world hunger. They accuse biotechnology opponents of selfishly denying life-saving benefits to developing countries. But in international negotiations, developing countries have demanded protections against uncontrolled trade in these potentially dangerous products. Some countries have raised issues concerning the influx of genetically engineered (GE) foods and seeds, and have desired cross-border non-governmental organization (NGO) collaboration.

Both corn and its progenitor teosinte originated in Mexico. For generations, individual farms and small farming collectives in Mexico have bred and preserved corn varieties that are adapted to specific local conditions within the country. Because of the skill with which Mexican farmers have tended these varieties, Mexico is also home to great genetic diversity in cultivated corn.

In the United States, concerns about genetically engineered corn include the possible contamination of conventional foods, changes in soil chemistry, and harm to nontarget organisms. But in Mexico, there are even greater risks since GE corn could contaminate corn's progenitor species or the local varieties developed by farmers over millennia. Such contamination could alter the range of genetic variability available to farmers, and threaten the future food security of millions of people around the world who depend on this crop. Genetic diversity in corn is crucial not only for traditional farmers but also for the industrialized agriculture that produces much of the food in the industrialized countries.

Mexico currently produces genetically engineered cotton, soybean and tomatoes, and has allowed field tests of a variety of genetically engineered crops, including corn and potatoes. Commercial (as opposed to experimental) cultivation of GE corn is not allowed in Mexico at this time, but large quantities of corn are imported from the U.S. for use as food and animal feed; some of the imported varieties were genetically engineered. Many small farmers in Mexico purchase corn to use both as food or fodder and for planting. So they may have unknowingly planted GE corn in their fields-potentially contaminating surrounding fields and wild corn relatives.

In the USA, management protocols have been designed to prevent the emergence of insects resistant to *Bacillus thuringiensis* (*Bt*) toxins, but surveys have revealed that almost 30% of farmers are failing to comply with these protocols. However, there is no evidence of *Bt* crop resistance, and second generation *Bt* crops may render the management plan obsolete.

Bt corn has been genetically modified to express *Bt* toxin proteins, killing insects that feed on it. According to the Environmental Protection Agency (EPA), 20% of the total acreage on farms should be planted with non-Bt corn varieties, the idea being that these refuges provide a breeding ground for insects that would mate with any *Bt* -resistant strains emerging in adjacent fields, preventing resistance from spreading. The EPA also stipulates that cotton-growing areas must have 50% non- *Bt* corn refuges because some cotton pests also feed on corn (Dove, 2001).

A panel of scientists convened by the Union of Concerned Scientists (UCS; Cambridge, MA) in 1998 pleaded for refuges significantly larger than the ones required by the EPA. According to UCS member Margaret Mellon, the current strategy is not a conservative refuge program to begin with, and so there is very little margin for error, making the nearly 30% rate of noncompliance among farmers particularly troubling. The UCS believes that 100% of farmers would have to comply with the current plan in order to provide reasonable protection from the development of *Bt*-resistant pests.

Happily, there is still no evidence of insect resistance to *Bt* crops. Although *Bt*-resistant insects have been observed in areas where *Bt* toxins are sprayed on crops-a strategy used extensively by organic farmers-in transgenic crops, the definitive proof has not been established yet.

Moreover, although the UCS wishes the EPA to enforce compliance, the Biotechnology Industry Organization (Washington, DC), says this would raise a heavy new government bureaucracy, to administer a system that the evolution of new and improved products is quickly going to render obsolete. Second generation *Bt* crops will express multiple insecticidal toxins under inducible promoters that farmers will only turn on when an infestation becomes economically significant-giving pests very little opportunity to develop resistance through chronic exposure to the toxins, and rendering complicated refuge strategies infructuous and redundant.

Production of Biopharmaceuticals in Transgenic Plants

Traditionally, cultured mammalian cells or certain micro-organisms have served as production systems for the production of biopharmaceuticals even though the yields with cultured mammalian cell systems are low and the resulting products are very costly. Bacterial and fungal systems are more robust than mammalian cell systems but they are also not suitable for producing mammalian proteins because of differences in metabolic pathways, protein processing and codon usage. These problems, coupled with the high production costs, directed attention to transgenic plants as a better medium for the production of many pharmaceuticals for use by humans and livestock.

This has prompted some biotechnology companies to develop transgenic plants as factories for the mass production of biopharmaceuticals.

As for edible vaccines, the biopharmaceuticals raised from transgenic plants can be stored and distributed as seeds, tubers and fruits, which are either directly used for oral intake or for extraction of the biopharmaceutical. This approach can make the immunization work in developing nations quite cheaper and easier to administer, since the delivery by direct ingestion of modified plant product eliminates the need for product purification; the latter is usually avoided as it adds to the production costs greatly. Also transgenic plants as production systems for the biopharmaceuticals give comparatively higher yields at lower cost and lesser health risk.

Some biopharmaceuticals shown to express in transgenic plants include erythropoietin, insulin, enkephalins, α-interferon human serum albumin, and glucocerebrosidase (Table 2.7). Many other therapeutic proteins are expected to become available soon.

Biopharm Crops

One major aim of many farmers is to provide consumers with safe and cheap value-added products. The same is true of biopharmaceutical products. Tests conducted by the U.S. Department of Agriculture, the Environmental Protection Agency (USA) and the Food and Drug Administration have revealed the best bioproducts and the new technology to produce them to be quite safe.

Plant-grown pharmaceuticals have been allowed to be grown in test plots in Colorado where the new plant biopharmaceutical crop Plant Pharma has been seen to prevent and treat diseases. But there is need to ensure safety and the integrity of agricultural products. Producers of biopharmaceutical crops, and the regulatory

Table 2.7: Selected Examples of Using Transgenic Plants for Producing Biopharmaceuticals

Plant	*Protein*	*Application*
Tobacco	Human serum protease	Protein C pathway
Tobacco, oilseed rape	Human hirudin	Indirect thrombin inhibitors
Tobacco	colony-stimulating factor	Anaemia
Arabidopsis thaliana (oilseed)	Human enkephalins	Antihyperanalgesic (opiate activity)
Tobacco	Human epidermal growth factor	Wound repair/control of cell proliferation
Rice, turnip	Human interferon-a	Hepatitis C and B treatment
Potato	Human serum albumin	Liver cirrhosis
Tobacco	Human haemoglobin	Blood substitute
Tobacco	Human homotrimeric collagen	Collagen
Rice	Human a-1 antitrypsin	Cystic fibrosis, liver disease, and haemorrhage
Maize	Human aprotinin	Trypsin inhibitor for transplantation surgery
Tobacco/tomato	Angiotensin-1-converting enzyme	Hypertension
Tobacco	Glucocerebrosidase	Gaucher's disease
Rice	Daffodil phytoene synthase	Provitamin A deficiency

agencies governing them, must ensure food safety and protect the integrity of the food and grain marketing systems.

Many opponents have considered the contamination of soybeans with pharmaceutical corn in Nebraska as a serious incident. However, the system worked and the problem was tackled before the affected material could enter the distribution system.

Another likely issue is the contamination of soil and water. The corn is usually grown for its protein. All plants produce proteins. If some residues were left in a field, the protein would simply break down naturally and end up as carbon, nitrogen, sulfur and other nutrients required by other plants. Regulatory agencies should use a scientifically sound, risk-based approach to monitor introduced proteins into plant pharmaceutical and industrial crops.

It is advisable to plant the biopharmaceutical corn at least a month later than is generally considered normal for corn, so as to ensure that cross-pollination with non-biocorns does not occur. Additionally, the tassels need to be removed from each plant so that the bio-pharmaceutical plant produces no pollen.

The best safety precaution could be a spatial separation between test plots and adjacent farmland. Tests have shown that essentially no pollination occurs at distances beyond about 1000 feet.

Biopharming (molecular pharming) is an important emerging area that uses transgenic plants for making vaccines, human therapeutic and prophylactic proteins and pharmaceuticals. These include drugs for the treatment of cystic fibrosis, hepatitis B, non-Hodgkin lymphoma, diarrhoea, cholera, diabetes and other diseases. Most of these pharmaceuticals are already being produced, commonly in maize and tobacco. Biopharming lowers the cost and effort for the production of drugs.

Health-enhancing Nutraceuticals

The two early GM plant products were the slow-ripening Flavr Savr tomato and the herbicide-tolerant Roundup Ready soybean. They both reached the American and European markets in the mid-1990s, since when the development and adoption of transgenic technology has progressed rapidly in the United States but slowly in Europe. About 68% of global GM crops acreage is now planted in the United States (James, 2000).

In Europe, one reason for the public rejection of GM crops was the lack of apparent benefit to the consumer-most existing modified traits serve agronomic purposes, which benefit only the farmer. In contrast, GM crops with consumer-oriented traits would be more acceptable to the general public. Manufacturers of GM crop-derived products who wish to market their products in the United States stand to face a regulatory environment very different from their competitors in Europe, especially for those growing GM plants modified to contain health-promoting substances, that is, nutraceuticals and pharmaceuticals (Kleter *et al.*, 2001).

The genetic modification of common food crops can also be beneficial for the consumer in ways other than their basic function as food. Genetic engineering can help to lower the concentrations of life-threatening allergens in a food, or to overexpress health-giving vitamins and nutrients (Table 2.8) (Ye *et al.*, 2000; Muir *et al.*, 2001).

Table 2.8: Examples of Health-enhancing Transgenic Crops (after Kleter *et al.*, 2001)

Compound	*Benefit*	*Crop (transgenic)*
Provitamin A	Anti-oxidant vitamin A supplement	Rice
Vitamin E	Anti-oxidant	Canola
Flavonoids	Anti-oxidant	Tomato
Fructans (indigestible polysaccharide)	Low calorie	Sugarbeet
Iron	Iron supplement	Rice

Functional foods and nutraceuticals (health enhanced foodstuffs) are becoming popular among consumers. Examples include cholesterol-lowering margarine spreads containing plant sterols, and cereals fortified with iron and vitamin B_{12} suitable for vegetarians. The global market for such foods is growing rapidly.

GM crops that express medicines are also being created: in some, the host crop is used as a food, such as the edible vaccines (Kong *et al.*, 2001). In others, the plant is an intermediate host from which the pharmaceutical substance of interest is extracted. Such pharmaceuticals in GM plants are exemplified by recombinant proteins such as antibodies, hormones, and blood-clotting factors, whereas in functional foods and nutraceuticals they are primary or secondary metabolites. Although much more complex, it is possible to engineer an entire metabolic pathway, such as the generation of provitamin A-containing kernels in "golden rice" (Kleter *et al.*, 2001).

According to Kleter *et al.* (2001) the novel food status or food additive status for food items derived from GM crops may provide access to a broad market, but the labeling requirements can deter European manufacturers. Regulators can demand rigorous data, depending on the equivalence of the new food or additive to conventional products. For food supplements, no specific requirements for GM-derived products exist as yet, but this option is only open to a limited number of (purified) substances, such as vitamins and minerals (see Tables 2.8, 2.9).

Performance in Natural Habitats

Crawley *et al.* (2001) reported the results of a long-term study of the performance of transgenic crops in natural habitats. Four different crops (oilseed rape, potato, maize and sugar beet) were grown in 12 different habitats and monitored for over a decade. They found that the genetically modified plants were no more invasive or more persistent than their conventional counterparts.

A decade ago, three conjectural risks associated with GM crops were: that they would become weeds of agriculture or invasive of natural habitats; that the introduced genes would be transferred by pollen to wild relatives, whose hybrid offspring would then become more weedy or invasive; or that GM plants would pose direct hazard to humans, domestic livestock or beneficial wild organisms, for example by being toxic or allergenic. The study of Crawley *et al.* assessed the first two of these fears. They tested all the four crop species and GM constructs that were available in 1990; oilseed-rape and maize plants expressing tolerance of the herbicide glufosinate, sugar beet tolerant of glyphosate ('roundup'), and two types of GM potato expressing either the insecticidal *Bt* toxin or pea lectin. They found that there were no significant differences in average recruitment between conventional and GM plants for any of the four crops. None of the crops–GM or conventional–increased in abundance at any of the sites.

Population sizes of all crops declined after the first year as a result of increased competition from native perennial plants. In no

Table 2.9: Differences in Regulations for Food and Medicinal Applications of GM Plants Between the US and EU (after US FDA, 2001; EC, 2001; Kleter *et al.*, 2001)

Item (GM)	*European Union*	*United States*
Foods & ingredients	"Novel foods" that require safety assessment (Regulation 258/97)	Not considered different from other technologies for production (FFDCA), focus on differences in product; should be "generally recognized as safe," except for food additives and plant pesticides.
Food additives	Requirements do not differ from those for other additives except for labeling (Regulation 50/2000)	Introduced foreign gene products may be considered "food additives" (FFDCA; FDA, 1992).
Food supplements	No harmonized EU legislation	GM food supplements are not distinguished from other supplements (DSHEA)
Medicines, chemical substances	Applications for GM medicines only through centralized procedure (Regulation 2309/93)	No specific regulation for GM medicines (FFDCA)
Medicines, biological substances	Applications for GM medicines only through centralized procedure (Regulation 2309/93)	"Genetic" GM biological medicines still under discussion (FFDCA)

case did the GM lines persist any longer than their conventional counterparts. For oilseed rape, seedling establishment was significantly lower for GM plants as compared with conventional lines in six out of 12 cases, and were not significantly greater in any case. For maize, seedling establishment was significantly lower for the GM line in two out of 12 cases, and significantly greater in no case.

For potato, survival of planted tubers to the end of the first growing season was significantly lower for GM lines in one case out of eight, and significantly higher in one instance. The few cases of increased GM-plant survival that were significant in the short term did not translate into long-term differences in persistence. Survival of perennating potatoes was significantly lesser for GM lines in one case out of eight, and never significantly higher.

For sugar beet, the genetic background (inbred or outbred) turned out to be more important than whether the plants were genetically modified; outbred lines outperformed inbred lines in three cases out of four.

These experiments involved GM traits (resistance to hericides or insects) that were not expected to increase plant fitness in natural habitats. The results of Crawley *et al.* indicate that arable crops are unlikely to survive for long outside cultivation.

Transgenic Plants and Pest Management

Genetically engineered crop plants showing resistance to insect pests offer an environmental friendly method of crop protection. Positive results have been obtained with the expression of *Bacillus thuringiensis (Bt)* and other toxin genes in several crops. However, both exotic and plant-derived genes have some performance limitations, and some failures have occurred in insect control through transgenic crops. The production and deployment of transgenic crops for pest control should carefully consider the issues related to impact of the transgenic crops on the insect pests, ecological cost of resistance development, effects on the nontarget organisms, availability and distribution of the alternate host plants, and the potential for introgression of genes into the wild relatives of crops.

In the near future, most of the increase in food production will have to come from increased yields of major crops grown on existing arable lands. One way of increasing crop production is to minimize the pest-associated losses, estimated at 14% of the total agricultural production (Oerke *et al.*, 1994). Besides the direct losses, insects also cause indirect losses through their role as vectors of various plant pathogens. Additional costs accrue from pesticides applied for pest control, running into billions of dollars annually. Massive application of pesticides not only leaves harmful residues in the food, but also causes adverse effects on non-target organisms and the environment.

Modern genetic transformation techniques make it possible to insert exotic genes into the plant genome that confer resistance to insects. Amongst these, the bacteria such as *Bacillus thuringiensis (Bt)* and *B. sphaericus* have been used successfully for pest control through transgenic crops on a commercial scale. Insecticidal genes such as *Bt*, trypsin inhibitors, lectins, ribosome inactivating proteins, secondary plant metabolites, vegetative insecticidal proteins and small RNA viruses have been used either alone or in combination with *Bt* genes in transgenic plants for pest control (Sharma and Ortiz, 2000). Several transgenic crops with resistance to the target pests have been developed and these have proved promising in reducing insect damage, both under laboratory and field conditions, and thus reducing the need to use pesticides for pest management. Genes conferring resistance to insects have been inserted into maize, rice, cotton, potato, tobacco and soybean.

Modern techniques of biotechnology provide access to novel molecules, ability to change the level and pattern of gene expression. Development and use of transgenic plants with insecticidal genes can lead to: reduced exposure of farmers, farm labour and non-target organisms to the pesticides; increased activity of natural enemies because of reduction in pesticide sprays; reduced amounts of pesticide residues in the food and food products; and a safer and cleaner environment.

Host plant resistance (HPR) reduces the need to apply pesticides as it is compatible with biological control and other methods of pest management in an integrated pest management (IPM) programme. Pesticides are highly toxic to the natural enemies, pollinators and other non-target organisms. Conventional HPR slows down the rate of increase of pest populations and exposes the pests for prolonged periods to the natural enemies. The introduction of transgenic plants reinforces HPR which can potentially influence the tritrophic interactions (Sharma and Ortiz, 2000). Synergism has been detected between *Bt* toxins and the HPR for *Trichoplusia ni* and it needs to be exploited for important crops and their pests in the semi-arid tropics (SAT) to achieve satisfactory control of the target pests, thereby avoiding the need to use pesticides. There appear to be no adverse effects of transgenic plants on the performance of the natural enemies.

Whereas sprays based on *Bt* formulations are unlikely to displace chemical insecticides because of their limited spectrum of activity and lower efficacy compared to synthetic chemicals, transgenic plants are usually sufficiently effective to either displace chemicals or to be used along with chemical insecticides or other methods of pest control. Simulation models using data from the diamondback moth (*Plutella xylostella*) and the Indian meal moth (*Plodia interpunctella*) have shown that transgenic plants bearing only one *Bt* gene may be more effective than the sprays for delaying the development of resistance to *Bt*. A Colorado potato beetle strain that survives *Bt* sprays cannot develop successfully on transgenic plants, not even on plants showing very low levels of *Bt* gene expression (Sharma and Ortiz, 2000). Simulation models have also suggested that transgenic plants may be much more durable than sprays of similar efficacy when more than two genes are deployed.

The above findings should, however, not mislead us to believe that transgenics are a panacea for solving all the pest problems. With the deployment of transgenic crops: the secondary pests will no longer be controlled in the absence of sprays for the major pests; the need to control the secondary pests through chemical sprays will kill the natural enemies and thus offset one of the major advantages of transgenics; the cost of producing and using transgenics can be quite high; proximity of transgenic crops to sprayed fields and insect migration can reduce the effectiveness of transgenics; and development of resistance in insect populations may limit the usefulness of transgenic crops for pest management and can often bring down the number of pesticide applications significantly even by as much as two-thirds to one half.

Environmental Impact

The number of pesticide applications on a crop such as cotton varies from 10 to 40, most of the sprays being directed against the key pests such as *Heliothis* and *Helicoverpa*. Reduction in pesticide application would stimulate the natural enemies, while some of the minor pests may tend to attain higher pest densities in the absence of sprays applied for the control of major pests. The introduction of transgenic crops can have a major impact on the abundance of some insects - the effects being negative for some and positive for others.

Efficacy of transgenic crops for controlling non-target pests should be determined in each region. Some pests maintain high densities on alternate hosts, *e.g*. cereal stem borers on the wild relatives of sorghum. Potential impact of transgenic crops on the beneficial insects can occur through reduction in the number of eggs and larvae of the natural hosts, which may also affect the activity of natural enemies.

A *Bt* toxin, which is highly effective against one insect species, may be weakly so or ineffective against some other insects. We do not know how effective the engineered crops would be in providing protection against the target pests, and there is no information on the key pests of the economically important crops. The technologies used in the developed countries may not prove suitable for the developing countries as the latter have complex cropping systems and a more diversified farming system. Some issues that need to be considered while introducing transgenic crops for pest control include: (i) effects on population dynamics of target and non-target insects, (ii) evolution of new insect biotypes, (iii) insect sensitivity, (iv) performance limitations, (v) gene escape into the environment, (vi) secondary pest problems, (vii) environmental influence on gene expression and failure of insect control, (viii) effects on non-target organisms, and (ix) impact on natural enemies (Sharma and Ortiz, 2000).

In general, no negative effects of *Bt* proteins have been recorded in mammals. There are no specific receptors for CryIA(b) protein present in the gastrointestinal tract of mammals, including man. Oral exposure to transgenic *Bt* tomatoes poses no additional risk to human and animal health. No significant differences in survival or body weight have been seen in broilers reared on meshed or pelleted diets prepared with *Bt* transgenic maize and similar diets prepared using control maize (see Hedley *et al.*, 1996). Broilers raised on diets prepared from transgenic maize showed better-feed conversion ratios and improved yield of the breast muscle.

The quality of produce from the transgenic plants is broadly similar to that from the nontransgenic plants of the same cultivar. The levels of the antinutrients gossypol, cyclopropenoid fatty acids, and aflatoxin in the seed from the transgenic cotton are similar to or lower than the levels present in the parental variety and other commercial varieties. The seeds from the *Bt* transformed cotton lines are as nutritious as those from the parental and other commercial cotton varieties.

Some proteins involved in the defence mechanisms of food plants are allergens which have potential use in molecular approaches to increase resistance to insect pests. These include α-amylase and trypsin inhibitors, lectins and pathogenesis-related proteins. Several self-defence substances made by plants are highly toxic to mammals, including humans. In these cases, the source of the transgene has no relevance in assessing the toxicological aspects of food from transgenic plants. This may result in a trade-off situation between nature's pesticides produced by transgenic plants, synthetic pesticides, mycotoxins or other poisonous products of pests.

Trypsin inhibitors and plant lectins contribute to a plant's defence mechanism in nature and can be potentially used in developing insect resistant transgenic crops. But these compounds can have some adverse effects. No crop plants expressing these genes have been deployed for commercial cultivation. Rats fed on purified cowpea trypsin inhibitor in a semi-synthetic diet based on lactoalbumin showed a moderate reduction in weight gain in comparison with controls, despite an identical food intake.

Resistance Management

Insect pest populations usually develop resistance to chemical pesticides. Most of the transgenic *Bt* crops express only one toxin gene and lack the complexity of the commercial *Bt* formulations. Also the plants continuously produce the toxins, and the insects are exposed to the *Bt* toxins throughout their feeding cycle; this imposes on the insect population a heavy selection pressure. With the development of resistance to *Bt* toxins, the value of microbial insecticides based on *Bt* proteins declines due to weakening sensitivity of the target pest to the *Bt* formulations. One consequence of this is that the farmers have to return to broad spectrum insecticides, which cause environmental hazards. Most of the transgenic plants produced so far have *Bt* genes under the control of cauliflower mosaic virus (CaMV35S) constitutive promoter-a system that develops resistance in the target insects as the toxins are expressed in the plant. Toxin production may also decrease over the crop growing season. Decreasing levels of toxin production can lead to resistance development not only to the toxin but also to other related *Bt* toxins to which the insect populations may initially be quite sensitive. Low doses of the toxins eliminate the most sensitive individuals of a population, leaving a population in which resistance can develop rapidly.

The ability of insects to overcome host plant resistance poses a strong risk. There are several reports on the development of resistance to *Bt* in many insect species; *e.g.* diamondback moth *Spodoptera exigua, Trichoplusia ni, Ephestia kuehniella, Plodia interpunctella, Christoneura fumiferana, Chrysomella scripta, Leptinotarsa decemlineata, Aedes aegypti, Culex quinquefasciatus, Drosophila melanogaster* and *Musca domestica.* This raises the real challenge of developing a strategy for deployment of transgenic plants for sustained protection of crops from insect pests. Different insect species react to *Bt* toxins differently. The resistance can develop quickly in *Diatraea saccharalis,* as some larvae can survive up to a week on transgenic maize (see Green *et al.*, 1990).

Reduced binding of *Bt* toxins to midgut epithelium is one mechanism of resistance in *Plutella xylostella.* There is also a broad-based mechanism of resistance to *Bt*. Complete degradation of *Bt* toxins by the proteolytic enzymes is the principal mechanism of resistance in some insects. Development of resistance may be due to changes in insecticidal crystal protein (ICP) receptors, and alterations in ICP receptors are a general mechanism by which insects can adapt to *Bt*. The absence of cross-resistance to ICPs other than those present in the selecting agent, and the finding that these ICPs bind to distinct receptors indicate that the use of ICP mixtures or multiple ICPs expressed in transgenic plants may be a valuable resistance management strategy.

Induction of proteinase activity probably is one way by which insects that feed on plants overcome plant proteinase inhibitors (PIs). Herbivorous insects can overcome the activity of PIs by secreting inhibitor-resistant enzymes. The insect's midgut contains a number of different proteins with trypsin-like activity, some of which are susceptible to inhibition by PI, while other trypsin(s) are not susceptible to inhibition. When inhibitor-resistant insects ingest PI, the level of activity of inhibitor-resistant trypsin(s) is enhanced in the midgut, thus allowing the insect to digest dietary protein in the presence of PI. Conceivably a group of PIs are required to inhibit the majority of proteolytic activity in the midgut of the target organism, and thus reduce insect growth and development. Once the PIs have been identified, their genes can be transgenically inserted into plants to enhance phytochemical resistance against herbivorous insects. An adaptive mechanism in *Helicoverpa* elevates the levels of other classes of proteinases to compensate for the trypsin activity inhibited by dietary PI.

Use of transgenic plants should be based on the overall philosophy of IPM, and consider not only gene construct, but alternate mortality factors, and reduction of selection pressure. Populations should be monitored for resistance development to design more effective management strategies. One good strategy to minimize the rate of development of resistance in insect populations to the target genes is through: (i) use of resistance management strategies from the beginning, (ii) gene pyramiding, (iii) gene deployment, (iv) regulation of gene expression, (v) development of synthetics, (vi) refugia, (vii) destruction of carryover population, (viii) control of alternate hosts, (ix) use of planting window and (x) use of economic thresholds and IPM (see Sharma and Ortiz, 2000).

Gene Pyramiding

Many of the candidate genes so far used in genetic transformation of crops, are either too specific or are only weakly effective against the target insect pests. Therefore, to make transgenics effective in pest control (*e.g.* by delaying the evolution of insect populations resistant to the target genes) it is desirable to deploy genes with different modes of action in the same plant. Several genes for trypsin inhibitors, secondary plant metabolites, vegetative insecticidal proteins, plant lectins, and enzymes that are selectively toxic to insects can be deployed along with the *Bt* genes to prolong the durability of resistance. Several advances have been made in introducing and expressing multiple transgenes in crops. For instance, Cry1A(c) and Cry1F can be expressed together in transgenic plants for effective control of some pests to increase the durabiltiy of resistance. Activity of *Bt* in transgenic plants can be enhanced by serine protease inhibitors, or in combination with tannic acid. Transgenic poplars expressing proteinase inhibitor and *CryIIIA* genes show reduced larval growth, altered development and increased mortality compared to the control.

Gene Deployment

Appropriate strategies need to be developed for gene deployment in different crops or regions depending on the pest spectrum, their sensitivity to the insecticidal genes, and interaction with the environment. The deployment of different genes and their level of expression should be based on insect sensitivity and level of resistance development. High levels of Cry1C production can protect transgenic broccoli not only from susceptible or Cry1A(R) diamondback moth larvae, but also from those selected for moderate levels of resistance of CrylC. The CrylC-transgenic broccoli is also resistant to two other lepidopteran pests of crucifers (cabbage looper and imported cabbage worm).

Development of Synthetics

One targetted deployment strategy is the development of synthetics. By incorporating various constitutively expressed Cry toxins into lines adapted for specific environments, synthetics can be formed quickly and prove effective against the pest complex and are compatible with the natural enemies. Once released to the farmers, the synthetics can be maintained as narrow-based populations at the farm level by removing the plants showing insect damage.

Refugia

One of the main strategies to manage the deployment of resistance to *Bt* toxins is high dose and production of refugia, in which a certain percentage of the crop consists of non-*Bt* plants (4-20% in maize, and 20-40% in cotton). The non-*Bt* plants produce the susceptible insects, which have a probability of mating with those emerging from the *Bt* crops nearby, and thus dilute the frequency of the resistant individuals. The growers have a contractual obligation to grow the non-*Bt* crops. The refuges can be sprayed or unsprayed. In the latter case, the area under non-*Bt* crop has to be much larger than that under unsprayed conditions. Refugia improve the durability of transgenic plants. The optimal spatial and temporal scale of refugia is likely to be unique for each insect-plant interaction. Refugia should be located closer to the transgenic plants so that the moths produced in the transgenic plants can mate with the insects produced on the transgenic plants. The refugia also enhance the capacity of biological control agents. For some polyphagous pests which feed on several field crops and alternate hosts in the wild, there may not be any need to maintain the refugia in the SAT.

Separate refuges are usually superior to seed mixtures for delaying resistance.

Use of Planting Window

Recourse to a planting window, that allows the crop to escape pest damage or avoid peak periods of insect abundance, can be useful in maximizing the benefits from transgenic crops or to prolong the life of these crops. Observing a close season and planting the crop with first monsoon rains have been effective in controlling the damage by sorghum shoot fly (*Atherigona soccata*) and sorghum midge (*Stenodiplosis sorghicola*). Similar strategies may be feasible for prolonging the effectiveness of transgenic crops.

Use of Economic Thresholds and IPM

Crop growth and pest incidence should be monitored carefully so that appropriate control measures can be initiated in time. Care should be taken to use control options such as natural enemies, nuclear polyhedrosis virus (NPV), neem or entomopathogenic nematodes and fungi, which do not disturb the natural control agents. Use of pesticide formulations such as soil application of granular systemic insecticides and spraying soft insecticides such as endosulfan may be considered to suppress populations in the beginning of the season. Broad-spectrum and most toxic insecticides should be used only during the peak periods of the pest. Pesticides with different modes of action should be rotated and repetition of insecticides belonging to the same group or the insecticides that fail to give effective control of the pests, should be avoided.

References

Anon. Genetic engineering: not the only option. *LEISA Magazine* 17(4): 4-5 (Dec., 2001).

Bonetta, L. GM crops under new US scrutiny. *Current Biol.* 11: 201 (2001).

Crawley, M.J., Brown, S.L., Hails, R.S., Kohn, D.D., Rees, M. Transgenic crops in natural habitats. *Nature* 409: 682-683 (2001).

Dove, A. Survey raises concerns about *Bt* resistance management. *Nature Biotech.* 19: 293-294 (2001).

European Commission (EC). *Proposals COM 2001-182 final and COM 2001-425 final.* (EC, Brussels, Belgium (25 July, 2001).

Green, M.B., LeBaron, H.M., Moberg, W.K. (eds.) *Managing Resistance to Agrochemicals.* pp. 3-16 ACS, Symposium Series, Washington (1990).

Hedley, C., Richards, R.L., Khokhar, S. (eds.) *Agri Food Quality: An Interdisciplinary Approach.* pp. 23-26. Special Publication No. 179, Royal Society of Chemistry, Cambridge (1996).

James, C. *Global status of commercialized transgenic crops: 2000.* (International Service for the Acquisition of Agri-biotech Applications (ISAAA), Ithaca, NY (2000).

James, C. Global review of commercialized transgenic crops. *Current Science* 84: 303-309 (2003).

Kleter, G.A., van der Krieken, W.M., Kok, E.J., Bosch, D., Jordi, W., Gilissen, L.J.W.J. Regulation and exploitation of genetically modified crops. *Nature Biotechnol.* 19: 1105-1110 (2001).

Kong, Q.X. *et al. PNAS* (USA) 98: 11539-11544 (2001).

Lurquin, P.F. *High Tech Harvest: Understanding Genetically Modified Food Plants.* Westview (*Perseus) Press, Boulder, CO (2002).

Manning, R. *Food's Frontier: The Next Green Revolution.* North Point Press, New York (2000).

McHughen, A. *Pandora's Picnic Basket: The Potential and Hazards of Genetically Modified Foods.* Oxford University Press, Oxford (2000).

Meldolesi, A. Italy performs GMO trial about-face. *Nature Biotech.* 19: 293 (2001).

Muir, S.R. *et al. Nature Biotechnol.* 19: 470-474 (2001).

Oerke, E.C., Dehne, H.W., Schonbeck, F., Weber, A. *Crop Production and Crop Protection: Estimated Losses in Major Food and Cash Crops.* Elsevier, Amesterdam (1994).

Sharma, H.C., Ortiz, R. Transgenics, pest management, and the environment. *Current Science* 79: 421-437 (2000).

Sharma, M., Charak, K.S., Ramanaiah, T.V. Agricultural biotechnology research in India: Status and policies. *Current Science* 84: 297-302 (2003).

Taverne, D. Food for thought and action. *Nature* 409: 559-560 (2001).

Taylor, M.R. Rethinking US leadership in food biotechnology. *Nature Biotechnol.* 21: 852-854 (2003).

Tester, M. Some GM facts. *Science* 298: 1341 (2002).

Uphoff, N. The system of rice intensification: agro-ecological opportunities for small farmers? *LEISA Magazine* 17(4): 15-16 (Dec., 2001).

US Food and Drug Administration (FDA). *FDA announces proposal and draft guidance for food developed through biotechnology.* (FDA, Center for Food Safety and Applied Nutrition, Washington, DC (2001).

van Bueren, E.L., Osman, A. Stimulating GMO-free breeding for organic agriculture: a view from Europe. *LEISA* Magazine 17(4): 12-14 (Dec., 2001).

von der Weid, J.M., Tardin, J.M. Genetically modified soybeans. *LEISA* Magazine 17(4): 19-20 (Dec., 2001).

Williams, N. Europe opens the door to GM crops. *Current Biol.* 11: 199-200 (2001).

Williams, N. Seeking balance in the GM crop debate. *Current Biology* 13: R163-R164 (2003).

Ye, X. *et al. Science* 287: 303-305 (2000).

Chapter 3

Transgenic Crops in Developing Countries

Introduction

The issue of whether or not modern agricultural biotechnology is suitable for developing countries has aroused much controversy (Paarlberg, 2001; Pinstrup-Andersen and Schioler, 2001; Altieri, 2001). Can GM crops, *e.g.*, those developed in the industrialized world, solve the pressing agricultural problems of developing countries? So far, 99% of the global GM crop area is planted to insect-resistance and herbicide-tolerance traits (James, 2002). These technologies chiefly substitute for pesticides (see Wossink *et al.*, 1998; Qaim *et al.*, 2000) but yield effects are usually fairly small. In the United States and China, for instance, yield increases of insect-resistant cotton usually do not exceed 10% on average (Huang *et al.*, 2002). For insect-resistant maize and herbicide-tolerant soybeans in the United States, average yield effects are negligible or even slightly negative (Qaim and Zilberman, 2003). Although the economic and environmental gains of pesticide savings and reduced effort for pest control have been reported (see Qaim and Zilberman, 2003), the potential of GM crops in developing countries seems limited without a substantial yield effect, especially in countries registering high population growth.

The case of *Bacillus thuringiensis* (*Bt*) cotton in India suggests that currently existing GM crops can have significant yield effects in the developing world, especially in the tropics and subtropics.

Bt cotton harbours the gene for Cry1Ac, which provides a fairly high degree of resistance to the American bollworm (*Helicoverpa armigera*), the spotted bollworm (*Earias vittella*), and the pink bollworm (*Pectinophora gossypiella*)-the three major insect pests in India. The American company Monsanto developed and introduced this technology into several Indian hybrids in collaboration with the Maharashtra Hybrid Seed Company (Mahyco). The first contained field trials with *Bt* hybrids in India were made in 1997. In subsequent years, field tests were extended to collect agronomic data and information for bio- and food-safety evaluation. Commercial approval to this technology was granted in 2002 when farmers started adopting the new hybrids.

Field trials initiated by Mahyco were carried out on 395 farms in seven states of India and supervised by regulatory authorities. Three adjacent 646-m^2 plots were planted: the first with a *Bt* cotton hybrid, the second with the same hybrid but without the *Bt* gene (non-*Bt* counterpart), and the third with a different hybrid traditionally planted in the particular location.

Besides the regular trial records, more comprehensive information gathered for 157 farms was used for data analysis. They covered 25 districts in three major cotton-producing states: Maharashtra, Madhya Pradesh and Tamil Nadu. Plot-level input and output data were extrapolated to 1 ha to validate comparisons.

On average, *Bt* hybrids were sprayed against bollworms three times less often than were the two controls. Individual bollworm control applications were still carried out, in view of the fact that especially for *H. armigera*, the Cry1Ac protein does not produce 100% mortality. There was no significant difference in the number of sprays against sucking pests such as aphids, jassids, and whitefly because *Bt* does not confer resistance to these pests. Insecticide amounts, especially of highly hazardous chemicals, such as organophosphates, carbamates, and synthetic pyrethroids, belonging to international toxicity classes I and II on *Bt* plots were reduced by almost 70%. In financial terms, the pesticide savings worth about U.S. $30 per ha were achieved.

The yield gains were much higher than what has been reported for other countries where genetically modified crops were used mostly to replace and enhance chemical pest control. Average yields of *Bt* hybrids exceeded those of the two controls by 80% to 87%. The yield gains were largely due to the *Bt* gene itself. Over the 4-year period from 1998 to 2001, *Bt* hybrids showed an average advantage of 60% (see Qaim and Zilberman, 2003).

Yield impacts of new pest-control technologies suggest that they depend on local pest pressure and damage, availability of alternatives for pest control, and farmers' adoption of these alternatives (NRC, 2000).

It appears that Indian farmers may have to greatly increase their current pesticide use in conventional cotton in order to achieve a level of damage control similar to that provided by *Bt* technology.

Many developing countries are assessing the costs and benefits of importing GM crop technologies from abroad for adaptation and use in their agricultural sectors. Pest pressure and related crop damage vary from region to region and even across individual locations but in general pest pressure in developed countries and other temperate zones is moderate, whereas in tropical and subtropical regions it is high. In the noncommercial and semicommercial crop sectors, pest-related crop losses can often exceed 50% (Oerke *et al.*, 1994).

On the basis of pest pressure and current crop protection, the biggest yield gains have been predicted to accrue in South and Southeast Asia and Subsaharan Africa. It so happens that these are also the regions with highest population growth, so increases in agricultural output per unit area are crucial for poverty alleviation and food security. *Bt* cotton, *Bt* maize, and *Bt* potatoes, already commercialized in some countries, have direct relevance to the developing world, and *Bt* rice and some food crops with other pest-resistance mechanisms will further broaden the portfolio in the coming years (Qaim *et al.*, 2000).

Some doubts about environmental and health risks, along with broader concerns about intellectual property rights and corporate dominance, have limited the acceptance of GM crops among the public and politicians (Paarlberg, 2001). There exists strong evidence on the benefit side but the technological potentials are not widely acknowledged. A more responsible risk management and balanced science communication should help in removing misconceptions or doubts and therefore ensuring sustainable use of GM crops.

In many developing countries, small-scale farmers suffer big pest-related yield losses because of technical and economic contraints. Pest-resistant genetically modified crops can contribute to increased yields and agricultural growth in those situations, as the case of *Bt* cotton in India demonstrates.

Transgenics in India

For India, a strong emphasis deserves to be placed upon resistance to diseases and pests for stabilization of crop yields. For yield enhancement, heterosis breeding in some crops can be highly rewarding. Nutritional enhancement also deserves high priority.

Transgenic varieties resistant to major insect pests of Indian crops can be developed by introducing *cry* or *vip* genes from *Bacillus thuringiensis*. Another approach for engineering crop varieties resistant to insect pests is to make use of genes of antifeedant proteins of plant origin. *Bt* corn and *Bt* cotton that carry cry genes have been released for commercial cultivation in many countries. One chief problem relates to proper management of resistance-insects tend to readily develop resistance to insecticidal molecules. This issue may be addressed by diversifying the gene sources for creating resistant varieties.

Fungal diseases greatly reduce crop yields. Various gene-based strategies can be adopted for breeding disease-resistant plants. These strategies involve either development of transgenics with genes encoding antifungal proteins or metabolites, or development of transgenics with *R* genes that function by inducing a hypersensitive response. The use of antifungal molecules has not been very successful so this makes the use of *R* genes a better approach. It may be possible to mine alleles of *R* gene for resistance from related but otherwise sexually incompatible plants, and exploit these for creating transgenics.

Indian agriculture should greatly benefit from input of transgenic technologies which can enhance productivity and make agriculture more sustainable. Crop diversification and rotation are essential for meeting nutritional requirements and for achieving agricultural sustainability. Critical consideration needs to be given to the current IPR regimes prevailing in the developed countries which are harming the interests of developing countries. India should do everything possible to protect not only its own national interests but also the agricultural needs of developing countries in general by arguing against overzealous patenting and by crusading for strengthening of the CGIAR institutes which have been serving the cause of global agriculture. In the context of future breeding goals, biotic stresses and heterosis breeding deserve the highest priority.

Over 60 per cent of India's population is engaged in agriculture or farming. Over three decades ago, agricultural production in India increased greatly from the planting of dwarf varieties of wheat and rice. Some productivity increase accrued from the use of hybrid varieties of corn, sorghum and millet, but the area under cultivation of these crops has been decreasing in the last decade, and the yields of rice and wheat have also plateaued out (Sharma *et al.*, 2003). The fertile soils of the Indo-gangetic plains are facing (largely due to continuous rice-wheat cultivation) problems of sub-soil water depletion, deficiency of micronutrients and increasing use of pesticides, fungicides and herbicides to control pests, pathogens and persistent weeds. Agricultural production has become more and more dependent on inputs of agrochemicals, thereby increasing costs and damaging both the environment and human health.

While the Green Revolution made India self-sufficient in cereal grains, the yields of dryland crops, mostly grain legumes and oilseeds, remain low; therefore India has to import both grain legumes and edible oils. Around 300 million (roughly 30%) of India's population suffers from malnutrition. Nutritional security in India will warrant more extensive availability of grain legumes, edible oils, fruits and vegetables, dairy, fishery, and poultry products. The daunting challenges of addressing malnutrition, enhacing productivity and crop diversification may be met through better resource management by breeding more productive, more nutritious varieties that require lesser inputs of various resources.

A large number of research projects have been started in India in the last decade. These deal with the development of *in-vitro* regeneration and genetic transformation protocols of important crop species and the development of transgenic varieties using genes of agronomic importance. Some projects aim at development of transgenics for resistance to gemini viruses in cotton, mungbean and tomato, resistance to rice tungro disease, resistance to bollworms in cotton, development of nutritionally improved potato with a balanced amino acid composition using a heterologous protein from *Amaranthus*, and development of molecular methods for heterosis breeding. Other projects are devoted to tagging of quality traits in rice and mustard, resistance to karnal bunt and rust in wheat, and an extensive research network on rice biotechnology including work on genomics, mapping, improvement in nutritional quality and resistance to biotic and abiotic stresses. Tagging of quality traits in wheat by molecular markers has already furnished a few interesting leads and certain markers linked to grain protein content, seed size, bread-making quality, etc. have been identified (Sharma *et al.*, 2003).

Transgenics being developed in the country both by the public and the private sector are listed in Table 3.1.

Increasing rice production forms a major element of the food security. The country is participating in the international rice genome-sequencing programme (IRGSP), and work is underway on the sequencing of a part of chromosome II (region between 57.3 cM and 110.1 cM of the genetic map). This segment has genes for several biotic and abiotic stresses and also those encoding some agronomically important traits. Prospective areas for functional genomics of rice have been identified as follows:

1. salinity and drought stress,
2. biotic stresses, and
3. mapping of loci contributing to high yield.

Some areas for future work are the following:

1. Molecular mapping in rice, wheat and grain legume crops. Molecular maps will be used for mapping of QTLs (quantitative trait loci) for abiotic stresses, plant architecture, and yield and mapping of loci conferring resistance to pests and pathogens.
2. Transgenics of various crops to be tested under appropriate containment conditions and those found useful to be subjected to nutritional and toxicological tests to determine their suitability for cultivation in the farmer's field.
3. Improvement in the nutritional quality of major crops such as rice, wheat and potato. High-yielding rice varieties being planted in India to be either transformed with constructs that would code for high b carotene content or would be

Table 3.1: Major Indian Developments in Transgenic Research and Application in Public and Private Sectors (condensed from Sharma *et al.*, 2003)

Plants/Crops	*Transgene(s) Inserted*	*Aim of the Project to Generate Varieties*
Cotton	*Bt. cry* gene(s)	resistant to lepidopteran pests
Cotton	*Cry1E* and *Cry1C* with terminal altered at C-end	resistant to *Spodopotera litura* and *Heliothis armigera*
Potato	*Bt. cry1A(b)*	resistant to lepidopteran pests
Potato	ACC synthase	for control of fruit ripening
Potato	*Ama-1*	that are nutritionally enriched
Tobacco	*Bt. cry1A(b)* and *cry1C*	resistant to *Helicoverpa armigera* and Spodotera litura
Tobacco	Chitinase, glucanase and *RIP*	resistant to fungal attack
Rice	*bar*	showing herbicide-tolerance
Rice	*Bt cry1A(b) Xa21*	resistant to lepidopteran pests, bacterial blight/disease
Rice	*Pyruvate decarboxylase* and *alcohol dehydrogenase*	tolerant to flooding
Rice	*codA, cor47*	resistant to biotic and abiotic stresses
Rice	*Xa-21, cry1A(b)*	lepidopteran pests and bacterial and fungal diseases
Rice	*Bt. cry1A(b)*	resistant to yellow stem borer
Rice	*Gm2*	resistant to gall midge
Mustard/ rapeseed	*bar, barnase, barstar*	herbicide-tolerant plants, male-sterile and restorer lines for hybrid seed production

Contd...

Table 3.1–Contd...

Plants/Crops	*Transgene(s) Inserted*	*Aim of the Project to Generate Varieties*
Mustard	*Ssu-maize Psy* and *Ssu-tpCrt1*	containing high levels of b-carotene
Brassica/mustard	*Bar, barnase, barstar*	that are superior hybrid cultivars
Tomato	*Cix-B* and *Tcp* antigens of *Vibrio cholerae*	for developing edible vaccine
Tomato	*Bt. cry1A(b)*	resistant to lepidopteran pests
Tomato	ACC synthase	to control fruit ripening
Tomato	*Leaf curl virus* sequence	resistant to leaf curl virus
Tomato	Chitinase and glucanase	resistant to fungal diseases
Wheat	*bar, HVA1, PIN2*	resistant to biotic and abiotic stresses
Cauliflower, cabbage	*Bt. cry1A(b)*	resistant to *Plutella scylostella*
Mustard/rapeseed	*Arabidopsis annexin* gene	having stress-tolerance
Mustard/rapeseed	*Choline dehydrogenase*	having abiotic stress-tolerance
Brassica	Chitinase, glucanase and *RIP*	resistant to fungal attack
Pigeonpea	*Protease inhibitor* and *lectin* genes	resistant to bollworms and aphids
Chickpea	*PGIP*	resistant to fungal pathogens
Muskmelon	*Rabies glycoprotein* gene	to develop edible vaccines
Blackgram	Coat protein and replicase genes of *Vigna mungo* yelow mosaic	viral-resistant virus

Contd...

Table 3.1–Contd...

Plants/Crops	*Transgene(s) Inserted*	*Aim of the Project to Generate Varieties*
Blackgram	*Dianthin* and *barnase* gene or *bar*	that are insect resistant and herbicide-tolerant
Rice	*Cry1A(b), cry9C* and *bar* genes	resistant to lepidopteran pests and herbicide tolerance
Tomato	*Alfalfa glucanase* and *Tomato leaf curl virus* genes	resistant to viral and fungal attacks
Cotton	*CP4 EPSPS*	resistant to herbicide glyphosate
Cauliflower	*Bar, barnase, barstar*	that are superior hybrid cultivars
Cabbage	*Cry1H/cry9C*	resistant to lepidopteran pests
Maize	*Cry1A(b)*	resistant to lepidopteran pests

back-crossed to the so-called 'Golden rice' transgenics previously developed in Switzerland.

4. Transgenic approaches to be applied for senescence retardation in fruits, also to bring down post-harvest losses in fruits and vegetables.

Breeding for Improvement in Indian Crops

The following important field crops are grown extensively in India: rice, wheat, maize, sorghum, pearl-millet, pigeonpea, chickpea, mungbean, cotton, potato, mustard and soybean. Grover and Pental (2003) conducted a survey on these crops by elicting the opinions of a large number of experts. They identified the five top problems for each of these crops and analyzed their relative scores statistically. Except for chickpea, mungbean, uradbean, and mustard, all the top 5 problems were related to diseases caused by various pathogens and other agents.

In chickpea, drought resistance ranked third among the 5 problems, in mung and urad, seed sprouting *in situ* under rain ranked second; and in mustard, varieties with low erucic acid and low glucosilonate ranked as the fourth problem (Grover and Pental, 2003).

Rice

Rice (*Oryza sativa*) is the most extensively grown crop of India. It is grown under diverse agro-ecological conditions as irrigated, upland, lowland and deepwater rice. Top priority is identified as breeding for resistance to various fungal and bacterial diseases and insect pests, particularly for blast (*Pyricularia oryzae*), sheath blight (*Rhizoctonia solani*) and bacterial blight (*Xanthomonas oryzae*). Two insect pests *viz.* stem borer (*Scirpophaga* sp., *Chilo* sp., *Sesamia inferens*) and brown plant hopper (*Nilaparvata lugens*) also cause extensive yield losses in rice and need to be controlled. It is generally felt that germplasm for resistance to blast, bacterial leaf blight, tungro virus, stem borer, brown plant hopper, gall midge and whiteback plant hopper is available. However, resistance sources for stem borer, leaf folder, sheath blight and sheath rot are inadequate. There is an urgent need to develop transgenics for resistance to the biotic stresses caused by the pathogens.

Breeding for higher yield through exploitation of heterosis is also a major challenge. Transgenics for rice need to be developed in respect to:

1. High b-carotene and high iron content.
2. Resistance to diseases and insect pests.
3. Resistance to flooding and drought tolerance. Cold tolerance at maturity in rainy season crop and cold tolerance in general for 'boro rice' also deserve focus.
4. Herbicide resistance is particularly important for upland rice and for allowing direct seeding in place of large-scale transplantation.
5. Quality improvement by incorporation of traits.

Wheat

In wheat (*Triticum aestivum*), breeding for resistance to fungal diseases is the most important overall objective. Leaf rust (caused by *Puccinia recondita*), leaf blight (caused by *Alternaria triticina* and *Helminthosporium sativum*), Karnal bunt (caused by *Neovossia indica*), stripe rust (caused by *Puccinia glumarum*) and loose smut (caused by *Ustilago nuda* cv. *tritici*) pose major problems. Diseases such as stem rust (*Puccinia graminis*), powdery mildew (*Erysiphe graminis*), earcockle (*Anguina tritici*), hill bunt (*Tilletia foetida* and *T. caries*) and flag smut (*Urocystis tritici*) are also important. In the North-West, leaf and stripe rusts, foliar blight and Karnal bunt are the major diseases; leaf rust, loose smut and possibly also stem rust, are predominant in the South and Central peninsular regions; hill bunt and stripe rust are common in hills of North India; powdery mildew is dominant in Northern sub-mountainous regions, and North-Eastern region appear to suffer from leaf rust, loose smut and foliar blight. Karnal bunt is a serious disease and development of wheat varieties resistant to this bunt both by conventional breeding as well as by transgenic technology is very important in view of the possibility of exporting surplus wheat (Grover and Pental, 2003).

Pure line breeding has been fairly successful in wheat. Commendable progress in breeding for disease resistance has been recorded, but since the introduction of dwarf wheat varieties in the 1960s, their yield potential has not increased significantly. In this context, heterosis breeding effort needs to be strengthened. Being a predominantly self-pollinated crop, adequate level of cross-pollination is essential for hybrid seed production. For this a stable male sterility / restorer system is needed.

The overall maintenance of wheat productivity depends on the incorporation of disease resistance in the existing high-yielding varieties. Wheat breeders generally feel that enough variability is available to breed for resistance by conventional pure line breeding methods. Yield can also increase through heterosis breeding. As compared to rice, the need for producing transgenics for meeting major breeding objectives in wheat is less pressing (Grover and Pental, 2003).

Maize

Maize (*Zea mays*) is grown mostly as a rainfed crop in the Indo-Gangetic plains and some parts of Southern India. Resistance to stem borer (*Chilo partellus, Sesamia inferens*) and the fungal diseases Maydis leaf blight (caused by *Bipolaris maydis* and *Cochliobolus heterostophus*) and Turcicum leaf blight (caused by *Exserohilum turcicum*) are the most important breeding objectives. Since resistance to stem borer is only available in the wild relatives of *Zea mays*, it would be difficult to transfer through conventional breeding. Development of transgenics for stem borer is therefore important.

Maize breeding is accomplished by means of composites or hybrids. Composites are mostly developed in the public domain. Companies utilizing inbred lines from other countries have developed most of the hybrids currently available in the market. Being quite rich in starch, such hybrid maize is not good for human consumption, so there is need to develop high yielding single cross hybrids for human use. Unlike wheat, male sterility / restorer system is not required for hybrid seed production as male and female flowers are separate and male florets can be readily removed.

Maize is deficient in two essential amino acids, lysine and tryptophan. Quality protein maize (QPM) that is high in lysine and tryptophan has been developed by conventional breeding methodologies. Herbicide tolerance is particularly useful because, being rainfed, maize tends to be heavily infested with weeds. Waterlogging is the most serious abiotic stress for maize. This is why this crop is not grown extensively in North-Western parts of

India as an alternative to rice. Germplasm with adequate resistance to waterlogging is not known; this warrants transgenics resistant to waterlogging. Drought is also a limiting factor in maize cultivation, and cold tolerance needs to be introduced in winter maize the needed germplasm for cold tolerance could possibly be imported from the temperate regions of the world.

Sorghum

Sorghum (*Sorghum vulgare;* jowar) is a grain-cum-forage crop. It is grown in the rainy and the post-rainy season, and is well adapted to grow in rainfed dryland areas. Hybrids are widely cultivated but, in general, the land area under sorghum cultivation is shrinking. The post-rainy season crop and the crop grown in 'Kharif' are good sources of fodder. Major problems facing cultivation include resistance to shootfly (caused by *Antherigona soccata*), grain mould (caused by a number of fungi), stem borer (*Chilo zonellus*), midge (*Contarinia sorghicola*) and charcoal rot (caused by *Macrophomina phaseolina*). Some germplasm is available for resistance to stem borer, shootfly, charcoal rot resistance, nutritional quality and drought resistance, but these traits have not been transferred to good combiners. Weak resistance to grain mould and shootfly in early to medium maturing, high-yielding cultivars during 'Kharif' and 'Rabi' seasons is responsible for low productivity.

Qualitatively superior disease-resistant forage sorghum hybrids are lacking. Attention is needed on nutritional (lysine) enhancement and improvement in dough quality and in protein content. Incorporation of herbicide resistance into sorghum is not needed. Major objectives for transgenic research in this crop are to develop varieties resistant to grain mould, stem borer and shootfly (Grover and Pental, 2003).

Pearl Millet

Pennisetum typhoides is a crop of dryland areas. The chief yield constraint is susceptibility to downy mildew (*Sclerospora graminicola*) particularly in single-cross hybrids. Ergot (*Claviceps microcephala*), smut (*Tolyposporium penicillariae*), rust (*Puccinia penniseti*), blast (*Pyricularia setarie*) and mycotoxins in the grain due to fungal infections are major problems for this crop. Increased drought tolerance is needed.

Millet hybrids are highly susceptible to diseases; resistance should be incorporated into divergent combiners by backcross breeding. To achieve high productivity, heterosis breeding is vital. Parental lines of elite hybrids can be improved by backcross breeding. Genetic diversification via population improvement needs priority to avoid problems related with narrowing down of the genetic base of the crop. Hybrids can potentially yield 50 quintals/ha. Incorporating the requisite traits, notably moisture stress tolerance, could bridge this gap. Transgenics should be developed for resistance to downy mildew.

Pigeonpea

Cajanus cajan is grown in the rainfed dryland areas of India in the 'Kharif' season. Its productivity has not improved in the past three decades. The most important yield constraint comes from the lepidopteran pest *Helicoverpa armigera*. Another significant insect pest is pod fly (*Melanagromyza obtusa*). Fusarium wilt, stem blight, and sterility mosaic disease are some other constraints on its yield. Development of extra-early varieties and resistance to drought and waterlogging are important breeding targets.

Heterosis breeding needs to be intensified for increasing the yield. Although combiners are known, no good pollination control mechanism is available. There is scope for using molecular methods for producing male sterile and restorer lines for hybrid seed production in this crop.

Transgenic varieties resistant to insect pests and abiotic stresses should be produced.

Chickpea

Cicer arietinum is a prominent crop of dryland rainfed agriculture in North India and peninsular India. There is urgent need for breeding varieties resistant to *Helicoverpa armigera*. Breeding for resistance to wilt and blight are some other important goals. *Fusarium* is a problem both in northern plains and southern regions of the country while *Aschochyta* (blight) is prevalent in North India.

Since chickpea is a self-pollinating crop with a narrow genetic base, there is very little scope for heterosis breeding. Therefore, pure-line breeding for developing new chickpea varieties may be developed (Grover and Pental, 2003).

Development of transgenics in chickpea for resistance to *H. armigera* should attract the highest priority. Transgenics are also needed for resistance to the two fungal diseases, as also for resistance to drought and frost tolerance.

Mungbean and Uradbean

Vigna radiata (mung) and *V. mungo* (urad) are the major legume crops in India. Mungbean yellow mosaic virus (MYMV) is the most serious yield-limiting problem in *Vigna* species. A physiological problem related to sprouting of seeds *in situ* under rains, powdery mildew (caused by *Erysiphe polygoni*) and leaf spot caused by *Cercospora* spp. are some other problems.

In general, there is need for strengthening breeding for resistance to insect pests and diseases by transgenic technologies. Genetic engineering for resistance to pre-harvest sprouting is also desirable.

Cotton

Gossypium hirsutum is the most important fibre crop of India that sustains the livelihood of many villagers engaged in its cultivation, picking and processing. Cotton textiles are exported to several countries and bring in foreign exchange of around rupees 45,000 crores annually. The most significant constraints on productivity are insect pests. American bollworm (*Helicoverpa armigera*) is its most prevalent and damaging pest, with pink bollworm (*Pectinophora gossypiella*) and spotted bollworms (*Earias insulana* and *E. vitteulla*) also causing extensive damage. Sucking insects such as whitefly (*Bemisia tabaci*) and jassids (*Amrasca biguttula biguttula*) greatly reduce cotton yields. Bacterial blight (caused by *Xanthomonas malvacearum*) and cotton leaf curl virus (CLCV; a geminivirus spread by whitefly), have affected the crop in Rajasthan, Haryana and Punjab areas but can spread to other cotton-growing areas. Germplasm is available for early maturity, bacterial blight resistance and high ginning amongst *G. hirsutum*, extra long staple in *G. barbadense* and for low shedding and early maturity in *G. arboreum* lines. Resistance to insect pests is not available in conventional varieties of cotton (Grover and Pental, 2003).

Hybrid seed of cotton is raised by manual emasculation and hand pollination between good combiners. As no adequate CMS/ restorer system is available it should prove rewarding to develop male sterility/restorer system for producing hybrid seed on a large scale. Transgenic approaches to hybrid seed production are feasible for improving cotton productivity. Although several combiners for hybrid seed production have been identified, the area under hybrids has actually declined.

A reduction in the content of gossypol in cotton *seeds* will enhance the quality of cotton meal-gossypol should not be reduced in other parts of the plant which require it for fighting insect pests. Insecticidal proteins obtained from *Bacillus thuringiensis (Bt)* may be used for developing transgenics for resistance to lepidopteran insect pests, but more than one insecticidal gene is required as development of resistance to *Bt* toxins has already appeared in some insect species. Resistance to sucking insects through transgenic technologies is another important target. There is also a need to reduce the unsaturated fatty acid content of cotton seed oil and to increase the contents of mono unsaturated and polyunsaturated fatty acids. Further, increasing oil content above 25% is also an important goal.

As in the case of rice, improvement of cotton also requires a major thrust in the development of transgenics (Grover and Pental, 2003).

Potato

Solanum tuberosum is an important vegetable in India both as a crop and as raw material for processed food industry. Being vegetatively propagated, heterosis is readily fixed by propagating F1s. No male sterility systems are required. Resistance to late blight disease (caused by *Phytophthora infestans*) is the most important breeding objective, followed by mosaic diseases caused by several viruses, bacterial wilt caused by *Ralstonia solanacearum*, black scurf caused by *Rhizoctonia solani*, and aphids. Aphids serve as vectors for viral transmission.

Many of the wild relatives of *Solanum* contain genes for resistance to the major pests and pathogens. There is considerable scope for breeding for nutritional enhancement, such as expression in tubers of provitamin A and enriching them in sulphur containing amino acids. The *ama*1 gene (that encodes a seed storage protein having a balanced amino acid composition) from *Amaranthus* has been mobilized into several potato varieties. Transgenics should be developed to control fungal diseases. Transgenics resistant to tuber moth would be valuable since there is lack of germplasm for resistance to tuber moth. Transgenics resistant to abiotic stresses are also important.

Yield improvement is feasible by utilizing the genetic diversity available in *Solanum tuberosum* spp. *andigena* (for biotic and abiotic stresses) through introgression breeding programmes. Crossing unrelated genetic material by a two-way hybridization approach followed by vegetative propagation can enhance yield.

Indian Mustard

Brassica juncea is the chief oilseed crop of rabi season in Northern India, largely grown as a rainfed crop. While predominantly self-pollinated; considerable cross-pollination also occurs. Research is underway on experimental hybrids based on the use of transgenic male sterile and restorer lines.

The most important breeding objective is the development of lines having resistance to *Alternaria* blight (caused predominantly by *Alternaria brassicae* and *A. brassicicola*), aphids (*Lipaphys erysimi*) and white rust (*Albugo candida*). Resistance to white rust is available within *B. juncea* germplasm but sources of resistance to *Alternaria* are not available. Transgenics resistant to *Alternaria* would be a major breakthrough. Conceivably, genes of *Arabidopsis thaliana* and/or of other alien species conferring disease resistance might be used in mustard for developing resistance to *Alternaria* blight. Aphids greatly reduce yield in mustard. Suitable strategies are not available to control this pest, except to breed for early maturing cultivars at the cost of yield. Transgenic technologies may have potential for developing aphid-resistant cultivars.

Improvement in oil quality by developing 'zero' erucic acid, and meal quality by developing 'zero' glucosinolate cultivars (less than 30 μ moles of glucosinolate per gram of defatted meal) are attractive breeding goals. Resistance to herbicides is not so important for this crop. Heat tolerance for early planting and at seed filling stage might prove valuable, and heterosis breeding could substantially increase mustard yield. A complete male sterility/ restorer based on transgenic technologies has been developed both in the private and public domains and could be used for developing hybrid seeds. As mustard (canola) and other *Brassica* species lend themselves well to genetic engineering, there exists considerable scope for further improvements.

Soybean

Glycine max has become an extremely popular legume crop in USA, Brazil, China and Argentina. In India, soybean is grown in large areas of Madhya Pradesh and other parts of Central India during the 'Kharif' season as a rainfed crop. It is also widely grown in the Northeastern states *e.g.* Nagaland. It is afflicted by several diseases, the most important being yellow mosaic virus (YMV) (a geminivirus, spread through whiteflies), soybean rust (caused by *Phakopsora pachyrhizi*), root rot (caused by *Macrophomina phaseolina*), pod blight (caused by *Colletotrichum* sp.) and bacterial blight (caused by *Xanthomonas campestris* pr. *glycines*). Availability of YMV resistant lines could allow cultivation of soybean in place of rice in the Northern plains. Transgenic lines resistant to YMV and fungal diseases are badly needed. Lack of certified seed and poor agronomic practices also contribute significantly to low yields. As a 'Kharif' crop, soybean suffers heavy weed infestation.

(**Note**: Besides the above crops, two other important crops growing in India are groundnut and sugarcane. These could not be included in the survey made by Grover and Pental.)

Insect-resistance

There exist many insecticidal proteins and molecules in nature that can effectively control agriculturally important pests and yet are harmless to mammals, beneficial insects and other organisms. Insecticidal proteins found in *Bacillus thuringiensis (Bt)* have been effectively sprayed on crops over the past half a century and have been gainfully expressed in many crop species. Since cultivation of *Bt*-crops has aroused concerns about the possible development of resistant insects, attempts have been made to prevent or delay the development of resistance. Besides *Bt*, proteinase inhibitors found in several plant species provide a good source of resistance to insect pests. A combination of proteinase inhibitors can be a viable alternative to *Bt* for managing insects such as *H. armigera*. Expression of multiple insecticidal proteins differing in their mechanisms of toxicity can retard or block development of resistance in the insects.

Use of integrated pest management (IPM) strategies in the cultivation of transgenic crops ensures lasting insect resistance (Ranjekar *et al.*, 2003).

Insect pest menace is undoubtedly one of the most critical factors that negatively impacts crop productivity. Many insects damage standing crops as well as stored seed. The rich biodiversity of agricultural, horticultural and forest species is perpetually under attack by insect pests in the predominantly tropical and sub-tropical climates of India. Breeding for resistance to insect pests is surely the top priority for many important crops. Table 3.2 lists important pests on major crops of India. High-yielding crop varieties respond well to applied nitrogen and other fertilizers but are highly susceptible to pest onslaught. Although control of insect pests by chemicals has protected crops, extensive and, sometimes, indiscriminate usage of chemical pesticides has caused environmental degradation and exerted adverse effects on human health and other organisms. It has also led to eradication of beneficial insects and development of pest-resistant insects. This has made it imperative to resort to safe and environment-friendly strategies in agriculture. A variety of insecticidal molecules have been found to be effective against insects but innocuous to humans and other organisms. These molecules should be harnessed and deployed in crop plants in a safe and sustainable manner.

Insecticidal Proteins from *Bacillus thuringiensis*

The bacterium *Bacillus thuringiensis (Bt)* synthesizes crystaline proteins during sporulation; these proteins are strongly insecticidal even at low concentrations. *Bt* strains and their insecticidal crystal proteins (ICPs) are now all over the world being extensively used in agriculture, horticulture, forestry, animal health and mosquito control as eco-friendly biopesticides. Tools of molecular biology and genetic engineering make it possible to introduce the ICPs of *Bt* in crop plants.

Bt strains are now available that are toxic to lepidoptera, hemiptera, coleopterans, dipterans, lice, mites and even nematodes. The *Bt* crystal toxins are classified on the basis of their amino acid sequence homology. The ICPs fall under 340 different classes with some toxins showing specificity to several insect orders. *Bt*-ICPs are highly valuable commercially. ICP crystals, upon ingestion by the insect larva, dissolve in the alkaline midgut into individual protoxins. The latter are acted upon by the midgut proteases which break them into two halves, the N-terminal half being the toxin protein. The latter fragment has three domains: the first is involved in pore formation, the second determines receptor binding, and the third potects the toxin from proteases. The toxin protein binds to specific receptors located in the midgut epithelial membranes. Upon receptor binding, the domain I inserts itself into the membrane and creates pores. The disturbances in osmotic equilibrium and cell lysis paralyse or kill the insect (see Ranjekar *et al.*, 2003). Although sprays of *Bt* ICPs or of engineered *Bt* and other bacteria are in use, they have short half-life and suffer from physical removal (wind and rain) and inability to reach burrowing insects. Engineered bacteria tend to multiply slowly so that their low numbers cannot kill some of the target insects. These limitations may be overcome if the ICPs are expressed in the plant cells at levels high enough to kill the larvae. Thus transgenic plants containing *cry* genes were developed long ago. Tobacco plants have been engineered with truncated genes encoding Cry1Aa and Cry1Ab toxins and are found to be resistant to the larvae of tobacco hornworm. But the levels of Cry protein expression in the plant tissues were not high enough. Researchers at Monsanto Company (USA) successfully modified the *cry* genes (*cry1Ab and cry1Ac*) for better expression in plant cells (Perlak *et al.*, 1991). Expression of genetically modified genes in crop plants, *cry1Ac* in cotton and *cry3Aa* in potato has conferred significant protection against lepidopteran and coleopteran pests respectively. Subsequently, rice, maize, peanut, soybean, mustard (canola) tomato and cabbage could be transformed with various modified *cry* genes. One important landmark has been the introduction of a native *cry1Ac* gene into the choloroplast genome of tobacco which expresses the Cry protein to a very high level (3-5% of leaf soluble protein) (McBride *et al.*, 1995). Chloroplast transformation not only allows good expression of foreign protein but also ensures maternal transmission of the foreign gene and so eliminates the spread of transgene through pollen. Extension of this strategy to plants such as cotton and rice should prove useful.

Table 3.2 lists some important insect pests of crops in India.

Table 3.3 shows some plants that have been transformed with *Bt cry* genes (Schuler *et al.*, 1998).

Table 3.2: Some Important Insect Pests of Major Crops of India* (after Ranjekar *et al.*, 2003)

Crop	*Common Name*	*Specific Name*
Rice	Yellow stem borer	*Scirpophaga incertulas*
	Brown plant hopper	*Nilaparvata lugens*
Mustard	Mustard aphid	*Lipaphys erysimi*
Chickpea	Gram pod borer	*Helicoverpa armigera*
Pigeonpea	Gram pod borer	*H. armigera*
Cotton	Cotton boll worm	*H. armigera*
Groundnut	Leaf miner	*Stomopterix nertaria*
Potato	Tuber moth	*Phthorimaea operculela*
Tomato	Fruit borer	*H. armigera*
Brinjal	Shoot and fruit borer	*Leucinodes orbonalis*
Cauliflower, cabbage		Diamondback moth *Plutella xylostella*

*(N.B. All insects except *Nilaparvata* and *Lipaphys* belong to family lepidoptera; the exceptional two belong to Hemiptera).

Table 3.3: Some important *Bt*-transgenic Crops* (after Ranjekar *et al.*, 2003)

Crop	*Gene(s)*	*Target Pests*
Cotton	*cry1Ab/cry1Ac*	Bollworms
Corn	*cry1Ab/cry1H/cry9C*	European corn borer
Rice	*cry1Ab/cry1Ac*	Stem borers and leaf folders
Tomato	*cry1Ac*	Fruit borers
Potato	*cry1Ab*	Tuber moth
Potato	*cry3a*	Colorado potato beetle
Canola	*cry1Ac*	Diamondback moth
Soybean	*cry1Ac*	Soybean looper

*(N.B. Other plants carrying various *cry* genes include peanut, alfalfa, apple, white clover, grapevine, walnut, pear, and sugarcane).

There is a need for critical evaluation of the insecticidal efficacy of *Bt* ICPs to such pests as *Earias insulana, Chilo partellus, Spilosoma obliqua* and *Maruca testulalis* as has been done in the case of *H. armigera* and *L. orbonalis.*

Some novel insecticidal proteins are produced by certain isolates of *B. thuringiensis*. Unlike the crystal proteins these new proteins are produced during vegetative growth of cells and are secreted into the growth medium. These proteins are termed vegetative insecticidal proteins (Vip). Sequences encoding for a Vip have been cloned, sequenced and the protein has been expressed in *E. coli* (Estruch *et al.*, 1996).

Vips seem well suited for use in insect management programmes. Individually *vip* gene has been successfully expressed in monocots and dicot plants (see Ranjekar *et al.*, 2003). Attempts are underway to pyramid *vip* in the *Bt*-transgenic crops in several laboratories.

Insecticidal Proteins from Other Sources

Proteinase Inhibitors

Several plants make defense proteins such as the proteinaceous proteinase inhibitors (PIs) and lectins induced in response to insect attack (Table 3.4). PIs are abundant in seeds and tubers. Their role against herbivory was hypothesized due to their abundance and their lack of activity against endogenous proteases. PIs are induced in potato and tomato due to wounding and insect herbivory. The resistance of a cowpea variety to the bruchid beetle, conferred by elevated levels of trypsin inhibitor (TI) in the seeds was shown by Gatehouse and Boulter as early as 1983. PIs are induced as components of many defense cascades under such stress-prone conditions as insect attack, mechanical wounding, pathogen attack, or UV exposure. PIs inhibit insect growth when fed to several pests; they inhibit the gut proteinases of the insect which hinders protein digestion in the gut, forcing the insect to synthesize alternative proteases to compensate for the inhibited activity. This creates deficiency of essential amino acids and exerts physiological stress on the insect whose growth is retarded. This mechanism minimizes the emergence of resistance in the insects and so reduces crop damage (Ranjekar *et al.*, 2003).

Transgenic tobacco plants expressing cowpea TI turned out to be resistant to the tobacco worm (*Heliothis virescens*). Several serine PIs have been expressed in transgenic plants for resistance against lepidopteran insect pests; similarly cysteine PIs have been expressed against certain coleopteran pests (see Schuler *et al.*, 1998). But in many cases, the transgenically expressed PIs have failed to show resistance against insects because the latter often adapt to the ingested PIs by producing proteinases which are insensitive to, or degrade, the PI (see Giri *et al.*, 1998).

Table 3.4: Some Examples of Transgenic Plants Expressing Genes Encoding Proteinase Inhibitors, α-amylase inhibitors and lectins (after Ranjekar *et al.*, 2003)

Crop	*Gene*	*Target Pest*
Tobacco	*Cowpea serine PI, Potato serine PI*	Tobacco bud worm
Tobacco	*Hornworm PI*	Whitefly
Rice	*Cowpea serine PI*	Stemborer
Rice	*Snowdrop lectin*	Brown plant hopper
Potato	*Cowpea serine PI*	*Lacanobia*
Potato	*Oryzacystatin*	Potato beetle
Potato	*Snowdrop lectin*	Potato aphid

In the course of coevolution of plant-insect interactions, insects have adapted to the PIs of their host plants; this makes the non-host plants the best means for identifying effective PIs.

In view of the ready adaptation of insects to a single PI, it might be better to express a combination of PIs for effective resistance. An appropriate combination of PIs targeted to inhibit the complete spectrum of insect gut proteinases may prove useful.

Plant Lectins

Lectins are proteins showing affinity for specific carbohydrate moieties. They bind to glycoproteins in the peritrophic matrix that lines the insect midgut, and inhibit digestion and nutrient assimilation. A lectin from snowdrop (*Galanthus nivalis*), upon expression in transgenic potato, became toxic to aphids (Hilder *et al.*, 1995). Foissac *et al.* (2000) expressed the snowdrop lectin in transgenic rice, with engineered plants becoming resistant to brown plant hopper (*Nilaparvata lugens*) and green leaf hopper (*Nephotettix virescens*). Wheat germ agglutinin, pea lectin, jacalin and rice lectin have been expressed in tobacco, maize, and potato, chiefly against aphids (Schuler *et al.*, 1998). However, many lectins are toxic/ allergenic to mammals, and so could restrict usage in developing transgenics that are safe for human consumption. It may be better to try and test lectins from edible plants that are consumed raw without harming human health, for their efficacy against major pests (Ranjekar *et al.*, 2003).

Amylase Inhibitors

The bean *Phaseolus vulgaris* synthesizes a class of related seed proteins called phytohemagglutinins-arcelin and α-amylase inhibitor (AI). AI complexes with certain insect amylases and seems to have a role in plant defense against insects. Transgenic Azuki bean carrying *α-AI* gene became resistant to certain bruchids (Gordon *et al.*, 1995).

Chitinases

The insoluble structural polysaccharide chitin is found in the lining of exoskeleton and gut of insects. It seems to protect the insect against water loss and abrasion. Chitin is a good target for insecticidal proteins. Its dissolution by chitinase perforates the peritrophic matrix and exoskeleton, thereby rendering insects vulnerable to attack by various pathogens. Proteins that interfere with chitin metabolism may inhibit the growth and moulting of insects. The chitinase produced by insects themselves has been deployed as an insecticidal protein. Expression of cDNA for chitinase obtained from the tobacco hornworm, *Manduca sexta,* in tobacco plants conferred partial protection against *Heliothis virescens* (Ding *et al.*, 1998). The larvae feeding on the transgenic plants suffered growth abnormalities and died prematurely. Adding chitinase protein to insecticidal proteins of *B. thuringiensis* potentiates the effect of these toxins. This has focussed attention on chitinases of fungi also which are being evaluated as potential insecticides in combination with *Bt*-toxins. Exposure of insect larvae to high doses of chitinases in conjunction with *Bt* toxins might enhance their vulnerability to *Bt*-toxins, leading to better control of insect pests.

Insecticidal Viruses

Many viruses are pathogenic to pest insects and so find use in insect pest management programmes. Genomes of small viruses can be introduced into crop plants, which then produce the viral particles and become entomocidal. *H. armigera* Stunt Virus (HaSV) is a tetravirus specific to lepidopteran insects and is remotely related to viruses of plants and animals. It is quite harmless to beneficial insects and the environment and its incorporation in transgenic plants may not pose risks (Gordon *et al.*, 1995). There is need to look for suitable entomopathogenic viruses whose genomes could be exploited.

Genes from Bacteria other than Bt

Photorhabdus luminescens lives inside the gut of entomophagous nematodes which invade the insect hemocoel and release the bacteria from their gut. The bacteria divide and kill the host within 48 h. The nematodes feed on the bacteria and the dead host. The bacteria synthesize protein complexes that are toxic to insects. Although the insect toxicity of the proteins is manifested at nanogram concentrations (as for *Bt* toxins), the mode of action differs from that of *Bt*. Conceivably, this complex may be a potential alternative to *Bt* or that may also be expressed in transgenic plants along with *Bt*.

Management of Resistance

In India, the major objective in planting genetically engineered insect-resistant crops, besides achieving high yields, is to ensure the stability of resistance to insect pests, other stresses, or pathogens. This raises the need for suitable resistance management strategies which can retard the evolution of pest resistance. One such promising ad practical strategy is the 'high dose/refuge' strategy (Frutos *et al.*, 1999).

In a pest population, resistance may be conferred by either genetic or non-genetic component of variation in the population though it is the additive genetic component of this variation coupled with fitness advantage that drives the selection process. Alleles of genes conferring adaptive advantages may always be present in the population or may appear at low frequencies by mutation. In an ideal population, when selection pressure is lacking, the allele and genotype frequencies tend to remain in equilibrium. In many cases resistance to *Bt* toxins in insects is conferred by a recessive or partially recessive allele that initially occurs at very low frequency so that resistant insects are rare and not easily detectable but heterozygous individuals are much more abundant than the homozygous-resistant insects. With selection pressure increasing the fitness value for the heterozygotes *vis-a-vis* homozygous-susceptible individuals, frequency of alleles for resistance would increase rapidly (see Roush,

1999; Mellon and Rissler, 1998; EPA, 1999). Many simulation models based on principles of population genetics have been developed to understand the factors that affect this build-up of frequencies. In the high dose/refuge strategy, high dose is aimed to kill virtually all the heterozygous insects. High dose can also change a partially recessive resistance trait into practically recessive form (Fig. 3.1). This figure shows dose response lines for three different genotypes. High dose plants, by killing all the susceptible and heterozygous-resistant insects, might build up resistance in one step (see Ranjekar *et al.*, 2003). The refuge harbouring non-*Bt* plants is conducive to survival of susceptible insects with which the surviving heterozygous (and any homozygous-resistant) insects would mate to produce heterozygous or completely susceptible progeny. Second generation heterozygous insects will again be killed by the toxin present in the plant. Thus high dose strategy coupled with refugia would generate a more durable or stable resistance (see Ranjekar *et al.*, 2003). EPA (1998) defined a high dose as one that is 25 times

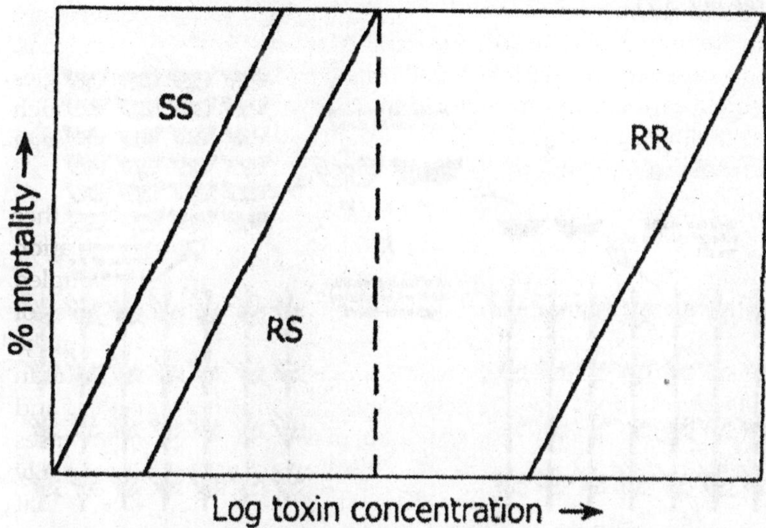

Fig. 3.1: Dose Response Lines Indicating the Mortality of Three Insect Genotypes at Increasing Concentrations of an Insecticidal Toxin. The dotted line indicates the concentration required for a high dose. S, allele conferring susceptibility; R, allele conferring resistance (after Cohen *et al.*, 2000)

higher than that required to kill 99% of homozygous-susceptible insects. If the precise dose mortality response for the purified toxin can be determined for the target insect and the dose required for 99% mortality can be calculated, then high dose could be precisely determined.

Another task is to define a refuge and determine its most suitable pattern and composition. Refuges are non-*Bt* crop plants that help maintain *Bt* susceptible insects in the population. A refuge may be a field of non-*Bt* plants interspersed with *Bt* fields or non-*Bt* plants within fields of *Bt* plants. It is these non-*Bt* plants that harbour susceptible insects and shelter and support them for ready mating with the insects emerging from the *Bt* plants (Fig. 3.2). Suitable conditions conducive to random mating among adults coming from *Bt* and non-*Bt* plants make a refuge effective. In case of a field-to-field refuge the distance between the two fields should not exceed the flight range of the target insects. Mixtures of *Bt* and non-*Bt* plant seeds within the same field may be sown appropriately or by planting some rows of refuge plants within a field of *Bt* plants (see Fig. 3.2).

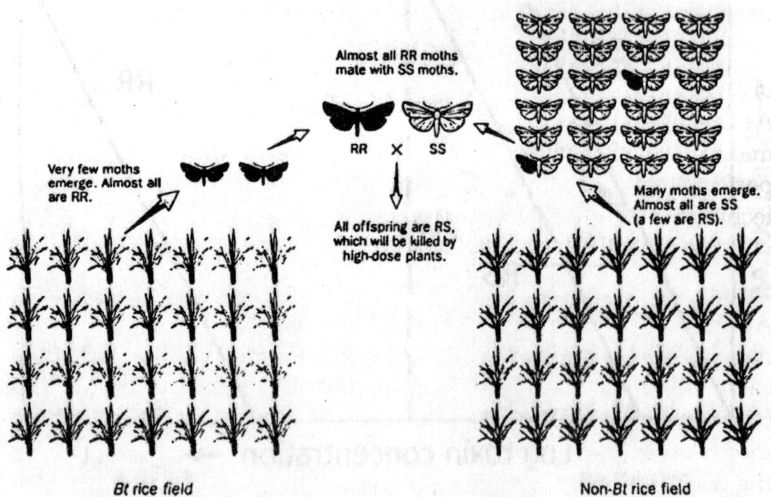

Fig. 3.2: Mechanism of High-dose/Refuge Strategy to Delay the Increase in Highly Resistant (RR) Insects in a Pest Population (after Cohen *et al.*, 2000)

In the USA, farmers growing *Bt* crops are required to plant 4-20% of their land with non-*Bt* cultivars, and these refuge fields must be within approximately 0.8 km of their *Bt* fields (see EPA, 1999). In India, an unstructured refuge could be considered in small land holdings of marginal farmers. Indeed, it is very important to have some refuge fields to fully and sustainably reap the benefits of this useful and environment-friendly technology.

Bt Cotton in India

India uses enormous amounts of pesticides every year to control the dreaded bollworm pests. Any technology that can reduce dependence upon chemicals is obviously welcome. India has the world's largest acreage of 8.9 million hectares under cotton, but is only the third largest global cotton producer, with about 2.8 million tonnes of cotton lint a year. The average productivity of cotton lint is 320 kg per hectare - one of the lowest in the world. The productivity for hybrid varieties ranges from 200 kg to 600 kg per hectare. Many of the land holdings are small-scale and farmers are resource-poor.

Cotton is essentially grown in the rainy season and treated as a perennial crop under rainfed conditions in central and southern India, but under irrigated conditions in north India.

In India, scores of illegal, untested, genetically modified varieties of cotton have been grown by some farmers alongside legal varieties. As the official regulatory controls are weak, such illegal plantings make it extremely difficult to objectively assess the first year performance of GM cotton, and could hinder future progress of the technology in India (Jayaraman, 2002).

Farmers in five Indian states are cultivating Monsanto's *Bt* cotton on over 100,000 acres after India's Genetic Engineering Approval Committee (GEAC) gave approval to three hybrids developed by the company (*Nat. Biotechnol.* 20: 415, 2002). The seeds carry the *Bt* gene and produce a natural pesticide lethal to the bollworm, a scourge of cotton fields. The *Bt* cotton seed is much more costly than the seed of traditional varieties. According to both the Indian government and Mahyco-Monsanto Biotech India Ltd. (Mumbai)-the joint venture between Monsanto and Maharashtra Hybrid Company that sold the seeds-the crop is faring well. But this view contrasts with reports from non-government organizations (NGOs) which suggest the crop is failing. In Andhra Pradesh, *Bt* cotton seems not to be performing well, and in Gujarat, there has been heavy bollworm infestation of *Bt* cotton, which was also found susceptible to leaf-curl virus and root-rot disease. In Madhya Pradesh, *Bt* cotton suffered greater damage due to drought than did traditional varieties.

One of the problems, according to both government sources and NGOs, is that many farmers fail to meet the prescribed technical specifications-such as for refugia management and planting conditions-for *Bt* cotton, a crop that requires high-maintenance. Small farmers, for instance, find it impossible to set aside land for refugia, and only 40% of the total area of cotton is irrigated-which caused serious problems in 2002 because of delayed monsoon.

Thousands of acres in some states have been sown with second and third generations of a Monsanto knockoff known as Navbharat 151. In 2001 these seeds were covertly sold by the Navbharat Seeds Company and planted by Gujarati farmers on over 10,000 acres (*Nat. Biotechnol.* 19: 1090, 2001). The government failed to detect or destroy the illegal *Bt* seeds from previous year's harvest in Gujarat, so these seeds have flooded the market at much cheaper price than legal seeds. First-generation seeds lack the vigour of the originals, with subsequent generations showing even worse quality and yield.

The large quantity of untested and unauthorized *Bt* hybrid seeds can cause losses and make farmers lose faith in *Bt* cotton, which would damage this otherwise useful technology. It seems that this new technology is not appropriate for the poor, dryland small farmers of India.

Pests and Pesticides

Out of some 160 species of insects known to attack cotton at various stages of growth, 15 are important pests. These include jassids (*Amarasca bigutulla*), aphids (*Aphis gossypii*), white fly (*Bemesia tabaci*), spotted bollworm (*Earias vitella*), pink bollworm (*Pectiniphora gossypiella*) and American bollworm (*Helicoverpa armigera*). Important diseases are bacterial blight, *Fusarium* wilt, *Alternaria* leaf spot and grey mildew. Together these pests and diseases result in an estimated loss of some 50% of the potential yield.

Pests appear in successive stages in the growth of the cotton plant. Sucking pests like aphids and jassids appear first, followed by white flies and then bollworms. At the flowering stage, bugs take over. Farmers has to use a combination of expensive chemical pesticides to control pest infestation, and pesticides account for one-third of the total cultivation costs.

Intensive cultivation practices and indiscriminate or careless use of several pesticides have created resistance among some of the key pests. Monocropping and favourable climatic conditions can further accentuate the problem. Dependence on chemicals has occasionally been so strong that farmers apply a mix of several pesticides (so-called pesticide cocktails) and sometime spray more than 30 times per season (Sharma, 2001).

Bt Cotton

In 1995, the seed company, Maharashtra Hybrid Seeds Co. Ltd., (Mahyco) imported 100 grams of cotton plantlets of the American GM variety *Cocker-312*. This variety is modified by the insertion of the *Cry 1 Ac* gene from *Bacillus thuringiensis* (*Bt*). Mahyco crossed the transgenic trait into elite Indian cotton varieties for six generations under controlled greenhouse conditions and then raised stable lines to breed hybrids. At least eight Indian *Bt* cotton hybrids containing the *Cry 1 Ac* gene have so far been developed. In November 1997, their first limited field trials were conducted in Andhra Pradesh, Karnataka, Maharashtra, Tamil Nadu and Haryana. In 1998, experiments were permitted at 40 locations throughout the cotton growing areas of India, followed by some more field trials in 1999. Contained field experiments have been designed to arrest the escape into the open environment of the transgenic plants, plant parts, or seeds borne by the plants. They also create a fairly effective barrier to prevent the escape of the transgenic pollen into the open environment.

Mahyco was permitted to conduct large-scale field trials for seed production and demonstration and to generate environmental safety data on transgenic *Bt* cotton in various agro-climatic regions in India, but not yet allowed to market any seed produced in these trials commercially.

According to Mahyco, the results of the environmental safety studies, including gene flow studies, did not indicate any substantial difference between GM and non-GM plants. Outcrossing was recorded with a ratio varying between 0.42 and 2.1 per cent up to a distance of 2 metres only. These results suggest that the chances of outcrossing with other cotton plants growing in the near vicinity are quite low.

Experimental *Bt* cotton plots required fewer pesticide treatments. In some plots, no sprays were required while in others a maximum of two sprays was enough. In comparison, plots with non-GM cotton required as many as nine to twelve sprays to control the pests effectively.

With regard to its compatibility for cross-pollination with wild relatives, the tetraploid *Bt* cotton pollens, grown in 80% of India, had only travelled short distances at five locations to outcross with the non-transgenic tetrapoloid controls. The possibility of cotton pollens being compatible for outcrossing with any near relative under Indian conditions is therefore thought to be remote.

The seeds, oil and cake of *Bt* cotton have also been evaluated for human food and animal feed safety. Safety for ruminants has been demonstrated. The company claims that the GM seed is safe for mammals, birds and fish.

Based on the studies and field trials, the Department of Biotechnology, Government of India, has become convinced of the economic benefits and environmental safety of introducing transgenic *Bt* cotton and has already progressed to the next phase of large-scale field experiments and seed production. However, the Indian Council for Agricultural Research (ICAR) is not convinced by the results so far presented. It has suggested that further detailed studies should include:

1. Nutritional studies on buffaloes and cows to determine whether transgenic cotton and transgenic cottonseed oil has any effect on animal health, milk production and quality of milk that in turn affects human health.
2. Whether or not there is development of resistance to other plant pests.
3. Toxicity to poultry and fish under Indian conditions.
4. The stability of *Cry 1 Ac* gene.
5. Gene flow and pollen dispersal and an assessment of the impact of such a migration on non-transgenic cotton (see Sharma, 2001). According to Sharma, transgenic cotton is being hastily pushed in India as a viable alternative to the

chemical-intensive cotton farming systems, while its environmental safety implications have not yet been critically evaluated.

Fungus-resistance

Tremendous progress has occurred in recent years in our understanding of the highly complex molecular events involved in plant-pathogen interactions. This knowledge in turn has generated several possibilities and strategies for creating transgenic plants resistant to fungal pathogens. Five of the important steps in plant-pathogen interaction are: (i) the mutual recognition of host and pathogen; (ii) the immediate response of the plant (hypersensitive response); (iii) production of active oxygen species; (iv) a local resistance response in (*i.e.* production of pathogenesis-related proteins and other antifungal proteins); and (v) development of systemic acquired resistance (SAR). Information on all the five steps is being used to produce fungus-resistant transgenic plants in several crop species (Grover and Gowthaman, 2003). Genome sequencing projects on crop plants have also generated some possibilities for raising fungus-resistant transgenic crops.

In India fungal diseases are rated as either the most important or the second most important factor contributing to yield losses in major cereal, pulse and oilseed crops. On the basis of a recent survey (see Grover and Pental, 2003), the impact of fungal diseases in selected Indian crops has been summarized in Table 3.5. Incidence of plant diseases has often been controlled by crop rotation, use of agrochemicals, and by breeding resistant varieties. The use of agrochemicals can be harmful environmentally. The breeding of resistant crops is time consuming, and resistance is often short-lived; new races of pathogens evolve to which crops become susceptible. Even so, in spite of the boom and bust cycles, breeders around the world have often succeeded in protecting some of the major crops (*e.g.* wheat) from fungal diseases.

However, the most promising option for developing resistant varieties is the use of gene isolation and genetic transformation. Improvements in genetic transformation technology have made possible the genetic modification of such important crops as rice, wheat, maize, mustard, pulses and fruits. Many resistance genes encode products involved in recognizing invading pathogens and these products have been identified and cloned (Takken and Joosten, 2000). Some signalling pathways which follow the pathogen infection have also been dissected. Many antifungal compounds synthesized by plants to combat fungal infections have been identified (Does and Cornelissen, 1998). Genome sequencing of *Arabidopsis* has thrown light on some tentative resistance gene clusters (Houlb, 2001). All this knowledge creates the potential to catalyze different strategies for producing fungus-resistant transgenic plants (Grover and Gowthaman, 2003).

Table 3.5: Impact of Fungal Diseases on Selected Crops of India (after Grover and Gowthaman, 2003)

Crop	*Pathogen*	*Disease*	*Yield Loss (%) (Approximate)*
Potato	*Phytophthora infestans*	Late blight	31
Wheat	*Puccinia recondita*	Leaf rust (brown rust)	30
Maize	*Helminthosporium maydis* and *H. turcicum*	Leaf blight	30
Brassica	*Alternaria brassiceae*	Blight	30
Pigeonpea	*Fusarium udum*	Wilt	24
Chickpea	*Fusarium oxysporum*	Wilt	23
Soybean	*Phakospora packyrhizi*	Rust	23
Rice	*Pyricularia oryzae*	Blast	21

The various strategies for producing fungus-resistant transgenics are usually classified into two categories, namely (i) using antifungal molecules like proteins and toxins, and (ii) generation of a hypersensitive response through *R* genes or by manipulating genes of the SAR pathway. Antifungal compounds are exemplified by antifungal proteins from plants and lower organisms, and by metabolites such as phytoalexins. Genes coding for many antifungal proteins which inhibit fungal growth *in-vitro* have been incorporated into plants to make them fungus-resistant. Examples: pathogenesis related proteins, ribosome-inactivating proteins, small cystein-rich proteins, lipid transfer proteins, storage albumins, polygalacturonase inhibitor proteins (PGIPs), antiviral proteins, and non-plant antifungal proteins.

Van Loon and Van Kammen (1970) showed that a set of PR proteins is induced in tobacco plants after their infection with tobacco mosaic virus. Later, PR proteins were found to be induced also by wounding, fungal cell wall elicitors, ethylene, UV light, or heavy metals. PR proteins are induced during HR and also during SAR and therefore probably have some role in natural defense or resistance of plants against pathogens.

Table 3.6 lists selected examples of transgenics developed by constitutively expressing PR-protein genes (see Fritig and Legrand, 1993; Grover and Gowthaman, 2003). Also, transgenic plants expressing more than one PR protein genes in a constitutive manner have been developed (Table 3.7). Such transgenics usually have greater resistance than transgenics carrying a single new gene. The introduction of a desired gene in the host plant under constitutively high expressing promoter often causes silencing of the transgene as well as its endogenous homologue leading to considerable loss of resistance. Therefore, studies on gene silencing have serious implications concerning the use of transgenic plants in combating fungal diseases.

Plant ribosome-inactivating proteins (RIPs) show *N*-glycosidase activity and cut off an adenine residue from 28S rRNA. So the 60S ribosomal subunit cannot bind to elongation factor 2. This blocks protein elongation. Plant RIPs inactivate alien ribosomes of distantly related species and of other eukaryotes including fungi. A purified RIP from barley inhibited growth of several fungi *in-vitro* (Leah *et al.*, 1991). Tobacco plants constitutively expressing a RIP encoding DNA sequence of barley had enhanced resistance to *R. solani.*

A number of small cystein-rich peptides form a distinct class of antifungal polypeptides-some bind to chitin plant defensins, or thionins. Hevein-a non-enzymatic chitin-binding peptide of 43 amino acids-from latex of rubber trees, is rich in cysteins. An agglutinin from *Urtica dioica* is another chitin-binding protein. Both hevein and UDA (*U. dioica* agglutinin) are the only chitin-binding plant lectins shown to inhibit fungal growth *in vitro*.

Thionins are other anti-microbial cystein-rich low molecular weight proteins (about 5-kDa) found in several plant species. They form pores in cell membrane resulting in membrane disruption and cell death. Carmona *et al.* (1993) showed that expression of α-thionin

Table 3.6: Some PR Protein Genes Exploited for Raising Fungus-resistant Transgenic Plants (after Grover and Gowthaman, 2003)

Plant	*PR Protein*	*Donor*	*Fungus Tested**
Alfalfa (*Medicago sativa*)	PR2	*M. sativa*	*Phytophthora megasperma*
Canola (*Brassica napus*)	PR3	*Phaseolus vulgaris*	*Rhizoctonia solani*, Pythium aphanidermatum
Carrot (*Daucus carota*)	PR5	*Nicotiana tabacum*	*Erysiphe heraclei*
Grape (*Vitis vinifera*)	PR3	*Oryza sativa*	*Elisinoe ampelina*
Kiwifruit (*Actinidia chinensis*)	PR2	*Glycine max*	*Botrytis cinerea*
Potato (*Solanum tuberosum*)	PR5	Potato (*S. commersonii*)	*Phytophthora infestans*,
	PR5	Tobacco	*P. infestans*
Rape (*B. napus*)	PR3	Tobacco-tomato (chimaeric)	*Cylindrosporium concentricum*, Phoma lingam, Sclerotinia sclerotiorum
Rice (*Oryza sativa*)	PR3	Rice	*Rhizoctonia solani*, Magnaporthe grisea
	PR5	Rice	*R. solani*
Tobacco (*Nicotiana tabacum*)	PR1a	Tobacco	*Pernospora tabacina*, Phytophthora parasitica, Cercospora nicotianae

Contd...

Table 3.6–Contd...

Plant	*PR Protein*	*Donor*	*Fungus Tested**
	PR2	Soybean	*P. infestans*
	PR2	Alfalfa	*C. nicotianae*
		Barley (*Hordeum vulgare*)	*R. solani*
	PR3	Bean, rice, tobacco	*R. solani*,
			C. nicotianae
	SAR	Tobacco	*Phytophthora parasitica*,
	SAR		*Pythium torulosum*
Tobacco (*N. sylvestris*)	PR3	Tobacco	*Cercospora nicotianae*
Tomato	PR2	Tobacco	*Fusarium oxysporum*
(*Lycopersicon esculentum*)	PR3		f.sp. *lycopersici*
Wheat (*Triticum aestivum*)	PR3	Barley	*Erysiphe graminis*

*(**N.B.** In all cases tested, except one, resistance developed; the exception was tobacco PR1a and PR3 proteins, tested for *Cercospora nicotianae*).

Table 3.7: Two Genes Used in Combination for Making Fungus-resistant Transgenic Plants. Resistance Developed in All Cases (after Grover and Gowthaman, 2003)

Plant	*Gene 1*	*Donor*	*Gene 2*	*Donor*	*Fungus Tested*
Carrot	PR3 (class I chitinase)	Tobacco	PR2 (class I glucanase)	Tobacco	*Alternaria dauci*
					A. radicina
					Cercospora carotae
					Erisyphe heraclei
Tobacco	PR3 (class I chitinase)	Rice	PR2 (class II glucanase)	Alfalfa	C. *nicotianae*
	PR3 (class II chitinase)	Barley	PR2 (class II glucanase)	Barley	*R. solani*
					Alternaria alternata
					B. cinerea
Tomato	PR3 (class I chitinase)	Tobacco	PR2 (class I glucanase)	Tobacco	*Fusarium oxysporum* f.sp. *lycopersici*

gene from barley in tobacco confers enhanced resistance to bacterial pathogens. Plant defensins constitute another class of small cysteinrich proteins; they show structural and functional homology with insect and mammalian proteins having well established roles in host defense (Broekaert *et al.*, 1995). Transgenic tobacco plants producing a defensin showed greater resistance to *Alternaria longipes* (Broekaert *et al.*, 1995).

Lipid transfer proteins mediate the transfer of a variety of lipids through the membrane *in-vitro* and are possibly involved in secretion or deposition of extra-cellular lipophilic materials such as cutin or wax.

Fungal polygalacturonase inhibitor proteins (PGIPS) occur in extracts of pear, tomato and bean (Stotz *et al.*, 1994). They seem to interfere with the host cell wall degradation caused by polygalacturonases. Transgenic tomato fruits constitutively expressing pear PGIPs suffer minimal colonization by *Botrytis cinerea*.

Fungal growth is inhibited *in-vitro* by cell wall degrading chitinases from various fungi. These chitinases interact with PR5 proteins or other membrane affecting compounds and other fungal cell wall hydrolases. An exochitinase gene from bacterium *Serratia marcescens*, when expressed in transgenic tobacco, renders the host plants less susceptible to *R. solani* (see Grover and Gowthaman, 2003).

Various strains of the smut *Ustilago maydis*, a fungal pathogen of *Zea mays*, harbour different double-stranded RNA viruses encoding antifungal proteinaceous killer toxins; *e.g.* three subtypes P1, P4 and P6 of *U. maydis* produce KP1, KP4 and KP6 killer toxins respectively. Strains of *U. maydis* show resistance to the toxin produced within themselves but are sensitive to the toxins of other strains. High level secretion of KP4 or KP6 killer toxin in transgenic tobacco plants makes them resistant to fungal pathogens (Park *et al.*, 1996).

A hen egg white lysozyme (HEWL) gene expressed in transgenic potato and tobacco plants was recovered from transgenic tobacco. It showed antimicrobial activity towards several bacteria and chitin containing fungi (Trudel *et al.*, 1995), but fungi containing chiefly chitosan or cellulose did not suffer growth inhibition by HEWL.

Plants have also been modified genetically to produce antibodies against fungal molecules used by pathogens to infect plants.

Phytoalexins

Phytoalexins are low molecular weight secondary metabolites formed in plants following pathogen attack; being antimicrobial they have a role in plant defense. Transgenics which synthesize new phytoalexins have been developed. Hain *et al.* (1993) introduced the gene encoding stilbene syntheses from *Vitis vinifera* into tobacco plants. Expression of the stilbene synthase (or resveratrol synthase) gene resulted in the production of resveratrol, a stilbene-type phytoalexin. Such transgenics had enhanced resistance to *B. cinerea.* Similar transgenic plants have been raised in rice, tomato, barley and wheat; they show enhanced resistance to *Magnaporthe grisea, P. infestans* and *B. cinerea* respectively (see Grover and Gowthaman, 2003).

Genetic Modification of Plants for Hypersensitive Response

Genes encoding antifungal compounds like PR proteins, phytoalexins, and toxins provide resistance to only a small extent and to only a few fungi. This necessitates strategies that can lead to more durable and broad spectrum resistance in transgenic plants-strategies that depend upon pathogen-induced cell death and general defense responses occurring in plants during incompatible plant-pathogen interactions (Grover and Gowthaman, 2003).

All plants possess certain passive defenses such as cell walls, waxy cuticle and chemical barriers against pathogens. If the pathogen crosses this first line of defense, they can mount a second line of defense via proteins encoded by specific resistance (*R*) genes. This second line is best illustrated genetically by the gene for gene model of Flor (1955). It depends upon a pathogen protein encoded by an avirulence (*Avr*) gene to be recognized by a plant protein coded for by a resistance (*R*) gene. It triggers a series of defense mechanisms, including the HR. In recent years, over 25 genes that determine resistance against a wide variety of pathogens, even nematode and aphids, have been cloned from both monocots and dicots (see Houlb, 2001). Most of these genes show homology to each other and their products are quite similar.

Broad Spectrum Disease Resistance through SAR

One effective strategy for broad spectrum plant disease resistance has involved the SAR route. Several plant mutants constitutively induce SAR (see Grover and Gowthaman, 2003). These lesion-mimic mutations confer resistance to powdery mildew in barley. A daunting task relates to developing transgenic varieties that can express SAR pathway without deleterious side effects. An *RDR* (required for disease resistance) gene termed *Npr1* of *Arabidopsis* confers resistance to the bacterium *P. syringae* and to the fungus *Peronospora parasitica*, when it is constitutively expressed in *Arabidopsis*. Resistance responses in such transgenic plants do not manifest when plants are grown under non-inducing conditions. However, upon infection by a pathogen such as *P. syringae* and *Peronospora parasitica*, the responses are induced at higher levels. There is a need to understand the mechanism by which plants sense pathogens in the absence of the *R* genes. Various defense pathways can be activated by virulent pathogens that are not recognized by *R* genes, suggesting that other pathogen surveillance mechanisms may be operating to attenuate the severity of disease. A good knowledge about these mechanisms should yield a variety of options for creating fungus resistance in crop plants.

Other Approaches

One major event in incompatible plant pathogen interaction is oxidative burst that generates active oxygen species. H_2O_2 triggers production of phytoalexins, PR proteins and other HR-related processes. It is also directly inhibitory to microbial growth. Glucose oxidase (GO) present in bacteria and fungi, oxidises β–d-glucose, yielding gluconic acid and H_2O_2. *GO* has not been reported from animals and plants. Expressing a *GO* gene from *Aspergillus niger* in potato elicited increased level of H_2O_2, and these transgenics showed reduced susceptibility to *E. carotovora* subspecies *carotovora*, *P. infestans* and *Verticillium dahliae* (see Grover and Gowthaman, 2003).

Virus-resistance

In India the resort to various agricultural practices under the Green Revolution has reduced varietal diversity of important crops, leading to the high incidence of viral diseases (see Table 3.8).

Table 3.8: Selected Viral Diseases of Important Crops in India (after Dasgupta *et al.*, 2003)

Crop	*Disease*	*Yield Loss (Approximate) (%)*	*Virus*	*Virus group*
Groundnut	Bud necrosis	>80	Groundnut bud necrosis virus	Tospovirus
Mungbean Blackgram Soybean	Yellow mosaic	21-70	Mungbean yellow mosaic virus	Begomovirus
Potato	Mosaic	85	Potato virus Y	Potyvirus
Rice	Rice tungro	10	Rice tungro badna and rice tungro spherical viruses	Badnavirus and waika virus
Sunflower	Necrosis	12-17	Sunflower necrosis virus	Ilarvirus
Tomato	Leaf curl	40-100	Tomato leaf curl virus	Begomovirus

Control of vector population using insecticides, use of virus-free propagating material, appropriate cultural practices and use of resistant cultivars are some of the common strategies for controlling virus diseases, although the above methods do have certain drawbacks (Dasgupta *et al.*, 2003). This has generated interest in the cloning and sequence analysis of the genomic components of several plant viruses. Many plant viruses have a single-stranded positive-sense RNA genome but some more important viruses in tropical countries like India have single-stranded and double-stranded DNA genomes and RNA genomes of ambisense polarity, *i.e.* genes oriented in both directions (see Hull, 2002).

Two chief approaches are available for developing genetically engineered resistance depending on the source of the genes used. The latter can come either from the pathogenic virus itself or from some other source. The first approach relies on the concept of pathogen-derived resistance (PDR). For PDR, a part, or the complete viral gene is introduced into the plant; this gene or its part interferes with one or more essential steps in the life cycle of the virus. A good illustration is TMV tobacco mosaic virus (TMV) that, when inserted into tobacco, leads to TMV resistance in the transgenic tobacco plants. Non-pathogen-derived resistance, on the other hand, uses host resistance genes and other genes responsible for adaptive host processes, elicited in response to pathogen attack, to raise virus-resistant transgenics (see Varma *et al.*, 2002).

Virus resistant crops have been produced by introducing a sequence of the viral genome in the target crop by genetic transformation. In most cases, resistance is induced due to an inherent plant response, known as post-transcriptional gene silencing (PTGS). Other pathogen-derived approaches exploit the use of satellite RNA and defective-interfering viral genomic components.

Coat Protein (CP)

The use of viral CP as a transgene for creating virus-resistant plants has been successful in many crops which have been transformed to express viral CP and have shown high levels of resistance in comparison to untransformed plants (Tables 3.9 and 3.10).

Table 3.9: Selected Examples of Coat Protein-mediated Transgenic Resistance to Viruses in Crops (after Dasgupta *et al.*, 2003; Sivamani *et al.*, 2002)

Crops	*Viruses**	*Field tested***
Cereals		
Maize	MDMV, MCMV	n.r.
Rice	RSV, RTSV	n.r.
Fruits		
Cantaloupe	ZYMV, WMV2, CMV	Yes
Citrus	CTV	n.r.
Grape	GCMV, GFLV, ToRSV	n.r.
Muskmelon	ZYMV	Yes
Papaya	PRV	Yes
Squash	ZYMV, WMV2	Yes
Vegetables		
Pepper	TSWV	n.r.
Tomato	ToMV, YMV, CMV, TYLCV	Yes
Potato	PVX, PVY, PLRV	Yes

Contd...

Table 3.9—Contd...

Crops	*Viruses**	*Field tested***
Pea	PEMV	n.r.
Cucumber	CMV	Yes
Soybean	BPMV	n.r.

*MCMV, Maize chlorotic mottle virus; MDMV, Maize dwarf mosaic virus; RSV, Rice stripe virus; RTSV, Rice tungro spherical virus; WSMV, Wheat streak mosaic virus; CTV, Citrus tristeza virus; GCMV, Grapevine chrome mosaic virus; GFLV, Grapevine fanleaf virus; ToRSV, Tomato ringspot virus; YMV, Yellow mosaic virus; LMV, Lettuce mosaic virus; PEMV, Pea enation mosaic virus; BNYVV, Bean necrotic yellow vein virus; BPMV, Bean pod mottle virus.
**n.r. indicates not reported.
Source: http://www.bspp.org.uk/mppl/1997/0116fuchs, Sivamani *et al.* (2002).

Table 3.10: Yield Increases Reported for Four Transgenic Virus Resistant Plants (after Moss, 1992; Dasgupta *et al.*, 2003)

Host	*Transgene*	*Yield Increase (%)**
Tomato	TMV CP	40
Potato	PVX + PVY CP	38
Squash	CMV + ZYMV + WMV2 CP	97
Papaya	PRSV CP	90

*Yield increase (approximate) over susceptible non-transgenic plants.

Movement Protein

Movement proteins (MP) are required for cell-to-cell transmission of plant viruses. They modify the gating function of plasmodesmata, thereby allowing the virus particles or their nucleoprotein derivatives to move to adjoining cells. This phenomenon has been used to engineer resistance to TMV in tobacco by producing modified MP partially active as a transgene.

Satellite RNA

The strategy exploits the use of satellite RNA associated with certain viruses. A few strains of CMV encapsidate satellite RNA (sat

RNA) in addition to the tripartite messenger sense, single-stranded RNA genome. CMV sat-RNA requires its helper virus (HV) CMV for replication, movement within the plant, encapsidation and transmission (Dasgupta *et al.*, 2003). The sat-RNA modulates the symptoms induced by the HV and depresses HV accumulation in host species. Transgenic tobacco plants expressing copies of CMV sat-RNA showed attenuated symptoms when challenged with CMV (Baulcombe *et al.*, 1986; Hamilton and Baulcombe, 1999). Further, tobacco plants transformed with anti-sense sat-RNA showed delayed symptom development with the cognate virus (Tousch *et al.*, 1994).

Sat-RNA has been successfully tested as a bio-control agent in field trials in many countries (McGarvey and Kaper, 1993), *e.g.* in tomato.

Transgenics with Non-pathogen Derived Resistance

Non-pathogen-derived strategies utilize genes derived from either the host plant or some other non-pathogenic source. The phenomenon of post-transcriptional gene silencing (PTGS) has been found to be responsible for the inherent ability of many plants to specifically degrade nucleic acids (including those of viruses) in a sequence-specific manner (see Table 3.11). It can prove effective in engineering virus resistance. Some other non-pathogen derived strategies are the utilization of plant disease resistance genes, the ribosome-inactivating proteins, plant proteinase inhibitors, human interferon-like systems, antiviral antibodies expressed in plants, systemic acquired resistance and secondary metabolite engineering (Dasgupta *et al.*, 2003).

Several viruses are able to suppress host PTGS activity. The balance of the pro-PTGS and anti-PTGS activities seems to determine the outcome of virus-plant interaction. Table 3.11 lists some known plant and viral genes inducing or repressing PTGS. None of the genes shown in Table 3.12 has been exploited for plant transformation studies to develop or modulate viral resistance. Once the biochemical steps of PTGS are known, it may be possible to find the appropriate genes and target them to engineer viral resistance (Dasgupta *et al.*, 2003).

Table 3.11: Some Plant and Viral Genes that Induce or Repress PTGS (after Dasgupta *et al.*, 2003)

Genes	*Function*	*Source*	*Possible PTGS-related Role*
		Plant	
inducing PTGS			
Sde1 or *SgS2*	Replication of RNA template	*Arabidopsis*	Synthesis of cRNA, amplification of dsRNA, signalling of methylation, synthesis of systemic signal, viral defense
Ago1	Translation elongation (eIF2C-like)	*Arabidopsis*	Target PTGS to ribosome, signalling of methylation, development
Rgs-CaM	Calmodulin-like protein	*Nicotiana tabacum*	Suppression of PTGS, development
		Viral	
repressing PTGS			
HC-Pro	Replication/proteinase	PVY TEV	Blocks accumulation of 25-mer RNA
P25	Viral movement	PVX	Blocks generation of systemic signals of PTGS
2b	Viral movement	CMV	Blocks initiation of PTGS at the nuclear step
AC2	Virion-sense transcription enhancer	ACMV	PTGS inhibitor

Table 3.12: Some *R* Genes Against Viruses and their Corresponding *avr* Gene Products (after Dasgupta *et al.*, 2003)

Resistance gene	*Source plant*	*Avr product of the virus*	*Pathogen**
HRT, RTM	*Arabidopsis thaliana*	Coat protein	TCV
L2, L3	*Capsicum* sp	Coat protein	PMMV
N	*N. tabacum*	Replicase	TMV
RRT	*A. thaliana*	Coat protein	TCV
Rx, Nx, Nb	*Solanum tuberosum*	Coat protein	PVX
Tm1	*Lycopersicon esculentum*	Replicase	TMV
Tm2	*L. esculentum*	Movement protein	TMV
TuRB01	*Brassica napus*	Cylindrical inclusion protein	TuMV

*PMMV, Pepper mild mosaic virus; BCMV, Bean common mosaic virus; TVMV, Tobacco vein mottling virus; for other viruses, see Table 3.9.

Plant Disease Resistance Genes

Table 3.12 lists some disease resistance genes (R) against viruses of crop plants. They encode products which respond to viral signals (avirulence *(avr)* gene products) leading to emergence of resistance responses in the plant. Many of the corresponding viral *avr* genes have already been identified. Some *R* genes complement the disease susceptibility phenotype in the corresponding cultivars when used as transgenes (see Ellis *et al.*, 2000).

All known *R* genes encode products having two primary functions: to act as sensors for the corresponding *avr* factors/elicitors, and to initiate signalling cascades for the expression of defense-related genes. Some structural features are conserved across several *R* gene products, and are exemplified by leucine-rich repeat (LRR), nucleotide-binding site (NBS), serine-threonine kinase, and leucine zipper. These structural features help in the execution of the above functions.

Many *R* genes tend to be clustered in plant genomes. They can induce resistance to diverse pathogens as exemplified by the *Rx* and the *Gpa2* genes, which are tightly linked, specifying resistance against PVX and nematode (see Dasgupta *et al.*, 2003). Understanding the molecular interactions between the various *R* genes products and their elicitors would allow their effective use in providing resistance against a wide spectrum of pathogens.

A promising approach towards engineering improved resistance to multiple diseases is the development of new *R* genes having multiple specificities (Bent, 1996; Ellis *et al.*, 2000). The *Fen* (for resistance to the insecticide Fenthion) and *Pto* genes occur in the same *R* gene cluster in the tomato genome. They show 86% homology in nucleotide sequence. A functional gene has been created by domain swapping of the two genes (Rommens *et al.*, 1995), thereby increasing the possibility of making a hybrid gene containing multiple specificities (Bent, 1996; Ellis *et al.*, 2000). Another option is a two-component approach to introducing broad-spectrum resistance. This novel strategy involves production of transgenic plants expressing a pathogen *avr* gene under the control of a heterologous infection-inducible promoter. When the plant carries the corresponding *R* gene, it responds with an HR at the site of infection, thereby limiting the pathogen (see Keller *et al.*, 1999).

Ribosomal Inactivating Proteins

Some plants contain antiviral proteins, termed ribosome-inactivating proteins (RIPs) that inhibit translation by catalytically removing a specific adenine base from 28S ribosomal RNA. They are targeted to vacuoles. Because of their specific intracellular localization, RIPs do not affect the endogenous 28S RNA. Presumably RIPs enter cells along with the invading viruses and so damage the host ribosome and possibly viral RNA (see Slusarenko *et al.*, 2000).

Verma (1995) documented antiviral activity of several RIPs. Application of purified RIPs mixed with viruses to plants suppresses virus multiplication and symptom development. Some RIPs inhibit local virus multiplication in RIP-treated leaves and also block viral multiplication systemically. The genes for RIPs have been isolated from several of plant sources. The cDNAs for PAP (Pokeweed), MAP (*Mirabilis jalapa*), Dianthus (*Dianthus caryophyllus*), Momorcharin (*Momordica charantia*), Ricin (*Ricinus communis*), exemplify this. These cDNAs have been used to raise transgenic plants showing broad-range antiviral activities.

Protease Inhibitors from Plants

Many viruses such as poty-, tymo-, nepo-, como-, and closteroviruses need cysteine protease activity to process their own polyproteins for their replication and propagation. This prompts the hope that plants expressing cysteine protease inhibitors might be virus resistant. This idea has been tested by using the cysteine protease inhibitor oryzacystatin of rice to engineer resistance against potyviruses in transgenic tobacco plants (Gutierrez-Campos *et al.*, 1999). Tobacco varieties expressing the rice cysteine protease-inhibitor gene were examined for resistance against tobacco etch virus (TEV) and PVY infection. There was a direct correlation between the level of oryzacystatin message, inhibitor of papain (a cysteine protease) and resistance to TEV and PVY in all tested transgenic lines.

Systemic Acquired Resistance

When plants are infected by viruses, they develop an active resistance which is initially localized only at the site of infection, but later spreads systemically. This systemic acquired resistance (SAR) typically involves the coordinate activation of several genes in uninfected, distal parts of the inoculated plants. It is also characteristically associated with accumulation of salicylic acid (SA), enhanced expression and activation of pathogenesis-related (PR) proteins and activation of phenylpropanoid pathway, leading to the synthesis of higher phenolic compounds, increase of active oxygen species and strengthening of plant cell walls by the deposition of lignin and suberin. Involvement of SA and TMV resistance has been shown by expressing the bacterial salicylate hydroxylase (*NahG*) gene in tobacco plant, which decreases its endogenous salicylic acid, and induces susceptibility to TMV infection (Delaney, 1994).

The fact that SA-binding protein is a catalase whose activity is inhibited by SA prompted the suggestion that SA inhibits the hydrogen peroxide degrading enzyme catalase, so elevating H2O2 levels. Transgenic tobacco plants could be developed that expressed catalase (*Cat*) gene in an antisense orientation and exhibited sharp reduction in catalase activity; they developed chlorosis or necrosis on lower leaves. They also had high levels of SA and PR accumulation as well as greater resistance to TMV (see Dasgupta *et al.*, 2003).

Tobacco has also been transformed with two bacterial genes encoding enzymes that convert chorismate into SA by a two-step process. When the two enzymes were targeted to the chloroplast, the transgenic plants registered upto a 1000-fold increased accumulation of SA and SA-glucoside as compared to control plants. The level of PR-proteins rose and they had high resistance to viral and fungal infection, in a mode akin to SAR in nontransgenic plants.

Secondary Metabolite Pathways

Being involved in viral pathogenesis, some metabolic pathways are key targets for intervention against viral infection. One such step is mediated by *S*-adenosyl homocystein hydrolase (SAHH), a key enzyme in trans-methylation reactions involving *S*-adenosylmethionine as the methyl donor. It probably has some role in the 5' capping of mRNA during replication. The antisense RNA for tobacco SAHH could be expressed in transgenic tobacco plants which turned out to be resistant to infection by plant viruses. These plants accumulated high level of cytokinin which is known to induce acquired resistance; this points to the increased resistance being due to high cytokinin level.

Pre-requisites for Developing Virus-resistant Transgenics

Although viral genes are highly variable, certain geographically distinct isolates of viruses, not showing such variations in symptoms, have been found to contain significant variability in their genes. Under field conditions, most viruses probably exist as collection of variants, or 'quasi-species'.

As with naturally occurring virus-resistance genes, strain specificity and breadth of protection are important issues when considering virus resistance under field conditions. The extent of protection and the relatedness between the challenge virus and virus from which the transgene was derived are often correlated. In transgenic papaya, the level of resistance depends upon the homology between the prevalent viral isolate and the transgene (see Dasgupta *et al.*, 2003). This means that in any viral transgene strategy, sequence of the aggressive prevalent strain of the virus in that region should be used and enough information on the extent of diversity among the biologically indistinguishable viral strains has to be gathered before designing the transgene, particularly in the case of whitefly transmitted geminiviruses, where the evolution of the virus is rapid (Harrison and Robinson, 1999).

The success of any transgenic strategy is determined by the level of resistance to multiple inoculations of the same or related strains, by vector transmission. Frequency distribution of the variants in a given virus population should be assessed to design a transgenic strategy targeting any virus causing an economically important disease.

The use of pathogen-derived genes to induce transgenic resistance has raised several ecological concerns (Hull, 1998), especially related to (i) recombination between viral-derived transgene and non target virus and (ii) transmission/vector host range changes brought about by encapsidation of the genome of non-target virus with the transgenically expressed CP. This warrants sufficient care to avoid any risks due to such heteroencapsidation while designing the constructs.

The success of a transgenic approach varies for any specific host/virus combination. A variety of phenotypes can be recorded amongst the virus-resistant transgenic plants. While CPMR confers broad-spectrum incomplete resistance, Rep-mediated resistance confers immunity against the virus, but to only a few strains. In RNA-mediated resistance, antisense RNA targeting mRNA of DNA viruses has greater potential than against positive-stranded RNA virus. Any antisense RNA/ribozyme strategy should consider the association/dissociation parameters of the molecules. Pyramiding of different transgenes or combination of transgenes with natural resistance targeting different events in viral life cycle is conducive to better management of viral diseases, ensuring stability of resistance at the field level. Durability, broad-spectrum character of the transgene-derived resistance, coupled with higher crop yield of the transgenics, are the essential parameters to be incorporated in any important strategy.

In recent years, ilarviruses that cause severe necrosis and destruction in sunflower and grain legumes, and tospoviruses responsible for severe bud necrosis in groundnut tomato, melons and grain legumes have become widespread as serious pathogens

(Bhat *et al.*, 2001). The host range of these viruses is spreading fast. Viral diseases of horticultural crops such as bunchy top in banana, tristeza virus in citrus, and ringspot disease in papaya have also assumed serious and unmanageable proportions in some parts of India.

For most viral diseases, resistant lines have been raised by conventional breeding. These, along with judicious insecticide sprays to control the vector population, have helped in containing some of the diseases. But in cases where the source of resistance is not available, a biotechnological approach is necessary. For the whitefly-transmitted geminiviruses like ToLCV, CLCuV, ICMV and yellow mosaic virus in legume, encouraging results have been obtained with transgenics containing replication initiation protein. CPMR for ilarviruses, and both CPMR and PTGS for potyviruses, have also shown good results. Characterization of *R* genes associated with the well-established resistant lines, if achieved, may provide a durable solution.

Genetic engineering of crop plants for virus resistance is a potent biotechnological tool that can minimize the losses in crop yield caused by viral diseases. Most of the important viruses have already been identified and the cloning and molecular characterization of their genomic components is underway. However, to successfully develop and test a series of virus-resistant transgenic crops, the major bottleneck of lack of transformation and regeneration systems for the major crops will have to be removed. In India, transformation systems are available for only a few cereals, vegetables, fibre crops and oilseed varieties. A major effort is warranted for pulses and legumes, which suffer heavy losses from viruses. The dominant and virulent strains of each important virus in the country need to be identified for accessing genes for resistance engineering. Emerging techniques of functional genomics need to be exploited for understanding the molecular interactions between the viral pathogen and the resistant and susceptible plants leading to resistance or pathogenesis. This can yield clues to effective disease control.

Weak Regulatory Strategies

Table 3.13 lists some likely effects of biotechnology on developing countries.

Table 3.13: Some Likely Effects of Biotechnology and International Trade of Biotechnology Products and Services on Developing Countries

Positive

1. Increased crop production, more food; Extension of growing area with drought- and salt-tolerant crops.
2. Improved health and production of livestock
3. Increased storage life of foodstuffs, especially useful for the poor who lack refrigeration and storage facilities
4. Decreased dependence on imported fertilisers and chemicals (except herbicides)
5. Improved public health, via cheaper multiple disease vaccines, and genetic diagnosis kits
6. Faster growing trees for biomass production as an energy replacement
7. Preservation of biodiversity as international companies buy exploitation rights in return for conservation
8. Possibility of new products from increased research into genetic resources and from the use of animals or plants as bioreactors.
9. Tissue culture allows continuous production of more reliable high quality products through tissue culture.

Negative

1. Loss of export markets as products are substituted by production of alternatives in industrialised countries
2. Loss of foreign income and employment
3. More cash crops may be grown, to produce high value commodities
4. Strengthening of large agricultural estates and displacement of small-scale landholders and marginal farmers
5. Less manpower and labour needed for cultivation, leading to unemployment
6. Herbicide-tolerant plants may reduce labour market in weeding, and may increase dependence on foreign imports of chemicals
7. Privatisation may favour/promote seeds that require fees for use, so that local farmers may lose control of cooperatives
8. Greater privatisation increases both legal and financial barriers to use of improved varieties
9. Biodiversity may be reduced as more efficient monocrop systems are introduced and larger farms are favoured
10. Loss of natural ecosystems in marginal lands when new crops are introduced to those areas

Jayaraman (2001) reported on the illegal planting of *Bacillus thuringiensis (Bt)* transgenic cotton by Indian farmers. This is a sign of weak regulatory mechanisms found in several developing countries. Such a regulatory weakness can undermine the basic premise of insect-resistance management (IRM) in regard to *Bt* toxins in prolonging the sustainability of *Bt* crops. The basic issue for field deployment of *Bt* crops is whether a *Bt*-resistance management designed for resource-rich farmers will be suitable for resource-poor farmers (Dirie, 2002).

In the United States it is customary to enforce IRM for *Bt* crops by setting aside a 4-50% refuge area (*Bt* toxin-free portion) surrounding *Bt*-planted field. Farmers have to sign contracts to provide such refuges when buying *Bt* seeds from companies; the EPA requires companies to educate farmers on the proper use of the *Bt* crops. But similar IRM strategies cannot be enforced in developing countries where most farmers are smallholders or marginal and operate under very different socio-economic conditions. There major challenges in implementing IRM strategies are:

1. Lack of regulations on IRM for *Bt* crops in developing countries and lack of policy for creating refugia, especially by resource-poor farmers.
2. Limited land resources with small farms being usually less than a hectare in area, unlike farms in developed countries that can cover hundreds of acres. Small-scale farmers are not able to spare 4-50% of their cropland for insects to happily feed on because this would represent a major loss.
3. Lack of trained manpower. There is an acute shortage of entomologists and ecologists with adequate knowledge of the management of *Bt* resistance in developing countries.

The International Rice Research Institute (IRRI), Manila, has done much work on resistance management for *Bt* rice and has proposed practical recommendations for field deployment (see Dirie, 2002). *Bt* crops should only be grown in areas where benefits will be greatest and farming communities as well as extension agents are properly organized and educated. It might be practical to organize farmers into cooperatives in order to implement IRM strategies.

GM Crops in China

In 1988, China's genetically engineered (GE) tobacco became the fist GE crop in the world to be grown commercially. But production of the crop had to be stopped within a few years because of rejection of GE crops in export markets. Undeterred, however, Beijing is moving ahead into the brave new world of biotechnology, testing many genetically engineered crops since 1997.

Transgenic *Bt* cotton is the most widely grown GE crop in China, with estimates ranging from over 700,000 hectares to one million hectares, or about one-third of the total cotton crop. GE proponents in China point to the success of *Bt* cotton in greatly reducing pesticide use, and reducing overall production cost for farmers by 20%. However, there are indications that the cotton bollworm has developed resistance to *Bt* in some parts of China.

Other commercial GE crops include tomato, sweet pepper and petunia. Some other not-yet-commercialized products include two species of *Bt* rice engineered to be resistant to the pyralid moth. China has developed genetically engineered tomatoes and eggplants that can be irrigated with seawater. Additional research is targeting rice, canola and wheat. By transferring genes of salt-tolerant mangrove plants into freshwater crops, Chinese scientists claim that tarnsgenic plants have survived seawater irrigation for four generations.

In 2001, the government issued new regulations for genetically engineered products that require mandatory labeling and safety assessment. So far no GE imports have been banned. Rejection of GE products by European markets and governments prompted the new GE labeling restrictions in China (Price and Hamburger, 2001).

In China, GM plants have already taken over many rice paddies and cotton fields, and benefitted millions of poor farmers. Biotechnologists in China have targeted many crops that have been ignored by large transnational companies in advanced countries. Their move from lab to field is occurring rapidly.

The country has already introduced over 100 genes into some 45 plant species. Unlike Western companies which mostly raise crops that are resistant to herbicides, most Chinese field trials target insect- and disease resistance, reducing the need for expensive and dangerous pesticides. Rice varieties resistant to three major pests have been released and field trials done on GM wheat. Other GM crops on sale include pest- and disease-resistant cotton, tomato and sweet pepper. GM potato, rape, peanut, cabbage, melon, maize, chilli, papaya and tobacco are some other crops in the pipeline (see Pearce, 2002).

The biggest success story is that of *Bt* cotton, which carries a gene for a toxin that kills insects. Over 2 million Chinese cotton farmers now grow *Bt* cotton, in fields covering 7000 square kilometres. Farmers' production costs have fallen by as much as about 30% and the average income has increased substantially. Use of toxic pesticides such as organophosphates has declined by 75% and pesticide poisonings have gone down.

About 5.5 million of the world's farmers, mainly in the US, Argentina, Canada and China-now grow GM crops covering over 50 million hectares. But potential health impacts of GM foods are arousing concerns in UK and elsewhere. The fact that existing GM crops have not harmed anyone so far should not make us complacent. Because the next generation will be more complex, even very small changes in foods could possibly have an impact on people dependent on single food sources-*e.g.*, babies fed formula milk.

As a recent report from the British Royal Society pointed out, inserting genes into plants is not yet an exact science, so unknown side effects on a plant's metabolism are quite possible.

GM Cotton

Today, China is the largest producer of cotton in the world, whereas India is the third largest. Interestingly, China with half the area under cotton production compared to India, produces 1 1/2 times the amount of cotton, has 1 1/2 times the world market share and thrice the yield. The average yield of cotton is China of about 670 kg/ha is about twice the current cotton yield in India (Choudhary and Laroia, 2001). Imperfections in the methods of picking, cleaning, grading, roving and spinning into yarn, inadequate transfer of technology, long-winded procedure for approval of new technologies, and extensive adulteration all have adversely affected India's position in the global cotton market (Table 3.14).

The textile and apparel industries play a pivotal role in the economy of both India and China. This sector accounts for 30% of

India's and 25% of China's total export volume. Unfortunately, sluggish exports, high costs and poor domestic demand are plaguing the Indian textile industry.

Table 3.14: Cotton Production, Yield and World Market Share (1997-98)

Country	*Market share (%)*	*Area (M Ha)*	*Production (MT)*	*Yield (kg/ha)*
China	24.5	4.56	4.30	943
USA	16.5	5.37	4.13	769
India	15.2	8.9	2.86	321
Pakistan	7.5	2.89	1.59	552
Egypt	1.3	0.36	0.32	873
Turkey	4.6	0.71	0.75	1065
Uzbekistan	5.5	-	-	-
World	100	33.82	19.74	584

(Source: Choudhary and Laroia, 2001).

Four cotton species are commonly grown in India, of which two are diploid (*Gossypium arboreum* and *G. herbaceum)* and the other two tetraploid (*G. hirsutum* and *G. barbadense*). Hybrids generated from crossing tetraploid species *G. hirsutum (hirsutum x hirsutum*) are also cultivated in the central and southern cotton-growing zones. The diploid species referred to as the 'Desi' variety accounts for 25-30% of the production. This variety has low productivity and does not respond well to good agronomic practices. Also its fibre is rough and short. Yet, its cheaper seeds make it popular amongst the poorer farmer. The tetraploids account for the remaining 70% of the cotton production in India (varieties 50% and hybrids 50%). These varieties produce fine quality fibre, valued by the textile industry. Although hybrid seeds are costly, their yield can be as high as 800 kg/ha, which however is lower than the 1200 kg/ha yield obtained in the US.

The low yields of cotton in India mainly result from attacks of insects and pests. Cotton is a crop to which 45% of the pesticides and 58% of insecticides used in India are applied (Fig. 5.3). The major culprits are 'bollworm complex' and 'jassids'. Sixty per cent of the insecticide spray is used to control the damage caused by

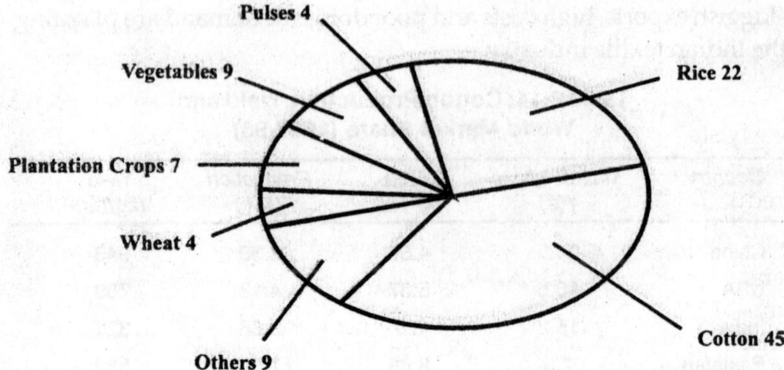

Fig. 3.3: Pesticides Consumption (in %) by Different Crops (after Choudhary and Laroia, 2001)

Lepidopterans such as *Heliothis* sp., *Helicoverpa armigera*, *Spodoptera* sp., *Pectinophora* sp. and *Earias* sp., - collectivley referred to as the 'bollworm complex'.

Although many insecticides and pesticides are available in the market, the damage from pests continues. Some difficulties faced by farmers in effective utilization of these agrochemicals include lack of quality spraying equipments, unawareness of integrated pest and insect management practices, high costs, adulteration of insecticides and pesticides, and over-spraying. Build-up of pest resistance and resurgence of secondary pests have also become serious problems because of indiscriminate use of pesticides.

Broadly, three main factors affect the yield. Genetic make-up (quality of seed) and optimization of gene technology contributes 50-60% to the yield. This includes increasing yield by developing cultivars resistant to pests, herbicides and viruses, as well as those tolerant to salinity and drought. Others include agronomic practices and agricultural technologies (25-30%), and biotic (pest infestation) and abiotic (drought, salinity, etc.) stress-related factors (20-25%).

Bacillus thuringiensis (Bt), a naturally-occurring soil bacterium, produces a protein that is toxic to Lepidopteran insect pests. *Bt* has been used widely by small-scale and organic farmers in conventional sprayings and is an environmentally benign pesticide. The gene for

Bt protein has been introduced into cotton, to produce transgenic *Bt* cotton varieties which are resistant to attack by Lepidopteran pests.

Whereas insecticides control bollworm infestation only in the early stages, the *Bt* transgenic controls bollworm infestation at the later stages. The benefit derived from this variety includes the reduced frequency of spraying of highly toxic synthetic insecticides.

GM cotton has spread across the cotton fields of China, USA, and several other countries. Over the past half a century, some of the most toxic and persistent pesticides have been used to fight cotton insect pests, but cotton still requires the use of more insecticides than any other crop. Insect pest resistance to these chemicals is a serious problem that causes economic hardship to farmers, as well as health and environmental problems. The acreage of GM cotton increased by 40 per cent in 2000 - against as increase of 11 per cent for all GM crops - and now covers 5.3 million hectares or 16 per cent of the 34 million hectares devoted to cotton production globally. This has made cotton possibly the most rapidly expanding transgenic crop. There are three types of GM cotton:

1. Herbicide-tolerant cotton (*Roundup Ready*).
2. An insect resistant, GM variety with a gene from *Bacillus thuringiensis* (*Bt*) producing an effective toxin (*Bollgard* or *Ingard*); and
3. A combination of *Bt*/herbicide tolerant cotton.

Out of these varieties the *Ingard* has shown variable performance, and is not proving attractive to farmers.

In several countries concerns have grown about GM cotton. In China, *Bt* cotton has turned out to be more susceptible to the fungal disease *Fusarium* wilt than conventional cotton.

A report by the World Wide Fund for Nature (WWF) in July 2000 stated that in the USA there had been no significant reduction in the *overall* use of insecticides in cotton despite the fact that almost one fifth of American cotton is produced with *Bt* varieties, possibly because *Bt* controls only some pests. The acreage of *Roundup Ready* cotton has doubled, but herbicide usage has remained almost the same. Individual farmers have reduced their sprays per season but, as the overall usage remains static, the environmental burden has not decreased (Dinham, 2001).

The strongest concerns with GM crops relate to insect and weed resistance and the impacts of widespread use on non-target beneficial insects and plants. Crops genetically engineered to contain *Bt* may speed up resistance, and may also affect beneficial, non-target insects. The genes for herbicide tolerance might be transferred from the GM crops to closely related weeds, making it difficult to control those species that were originally susceptible to the herbicide glyphosphate.

One major worry is that the promotion of GM could deflect attention from alternatives, such as Integrated Pest Management (IPM) and organic cotton (Myers and Stolton, 1999), which offer an approach based on increasing farmers' knowledge and their ability to manage crops ecologically and economically. Training IPM assists farmers to reduce dependence on hazardous pesticides while improving yields. The European Union supports a major new cotton IPM project in several Asian countries *e.g.* India, Pakistan and China. The IPM strategy advocates the managing of pests through biological control, host-plant resistance and appropriate farming practices. This approach minimizes the application of pesticides, assures good yields, reduces costs, is environmentally friendly and contributes to the sustainability of agriculture (Dinham, 2001).

Cotton Biotechnology in China

China is the largest cotton producer in the world, with about 50 million farming households growing cotton. The crop is of primary importance to the Chinese textile industry, which is the largest in the world. Its contribution to China's export volume comprises about 25% of the total.

During the past two decades, China has made great progress in cotton biotechnology and genetic engineering, obtaining first regenerative plants form cotton anther and protoplast culture, and then also obtaining regenerative plants from many domestic elite cotton varieties. After the commercialization, in 1996, of transgenic cotton carrying the insect-resistant (*Bacillus thuringiensis, Bt*) gene, at least ten *Bt*-cotton varieties were planted in China. In 1998, over 100,000 hectares of *Bt*-cotton were planted. Two kinds of insect-resistant transgenic cotton have been obtained. These new insect-resistant cottons carried two insecticidal genes, *Bt* gene and *CpTI* gene, or pea lectin (*P-Lec*) gene and soybean Kunitz trypsin inhibitor

(*SKTI*) gene respectively, and are being commercialized now. Herbicide-resistant varieties for 2,4-D and Bromoxynil are under development and are expected to reach the market very soon. Disease-resistant transgenic cotton is being developed and tested in laboratories and fields, and is also expected to reach the market in 2002. Fibre improvement, stress resistance, and male sterility and fertility for hybrid cotton are the next targets for cotton biotechnology. Several genes for fibre improvement and hybrid cotton are being tested in various laboratories. New genes for insect, herbicide and disease resistance are being sought (Zhang *et al.*, 2000).

In China, work on tissue culture of cotton was initiated in the 1970s, and involved anther culture, and protoplasm culture (Zhang, 1997). The first regenerative plant was obtained in 1986 from cotton somatic cell culture via somatic embryogenesis. Regenerative plants have been obtained via somatic embryogenesis from about 20 cotton cultivars. Somatic cell culture has been used to select somaclonal variants. It is a good system to select the variations against adverse conditions, such as salt stress, low temperature stress, or toxins produced by *Fusarium* or *Verticillium* wilt.

Protoplast culture is required for obtaining somatic hybrid cells. Regenerative plants have been obtained from protoplast culture of *Gossypium hirsutum*. The methods for cotton protoplast culture have recently improved, and the procedure for cotton protoplast culture has been established (see Zhang *et al.*, 2000). Haploid plantlets have been obtained from a wild species, *G. klotzschianum* and regenerative plants obtained from the callus of some varieties of *G. hirsutum* via anther culture.

Compared with somatic cell culture, shoot and meristem culture is an easier method to obtain regenerative plants. Scientists have developed shoot and meristem culture for genetic engineering and conservation of cotton germplasm materials. Transgenic cotton plants have been obtained by transforming shoot meristems by the method of particle bombardment.

Ovule culture has been used to study the development of cotton fibre, to obtain the fibre *in vitro*, and to achieve interspecies hybridization.

Several Chinese institutes have used ovule culture to introgress the fine characteristics of wild species into upland cotton, and obtained hybridization, which had at least one of the following characteristics: pest resistance, disease resistance, resistance to adverse conditions, the glandless seed/glanded plant trait has been raised. Some of these were successfully bred.

Genetic Engineering

The two areas that have advanced most rapidly are (a) insect resistance based on the toxins produced by the bacterium *Bt* and (b) resistance to herbicides.

Insect pests pose a major problem in cotton production. Cotton is attacked by a number of insect pests, *e.g.* the cotton bollworm *Helicoverpa armigera*, cotton pink bollworm (*Pectinophora gosspiella*), and cotton aphid (*Aphis gossypii*) which can cause extensive damage. *H. armigera*, in particular, is a serious threat as it is resistant to currently available chemical pesticides such as synthetic pyrethroids and organophosphates.

Traditional breeding has done much to improve the host plant resistance of cotton to insect pests, and continues to provide new varieties which require less chemical intervention than old varieties. New characters such as glabrousness (absence of hairs on the leaves), frego bract (outward bending, thin bracts around the cotton boll), nectariless (absence of nectar-forming nectary glands on the leaves), and high gland content are being assessed for their capacity to reduce the attractiveness of the cotton plant to insect pests. There are, however, limits to the improvement in natural resistance to insects provided by alterations in plant shape and structure. None of the cotton varieties developed so far has shown any significant resistance. Therefore, control of insect pests in cotton cultivation depends mainly on the use of chemical insecticides that arouse serious public concern for reasons of safety and environment pollution. An attractive alternative is the production of proteins with insecticidal activity by the cotton plant itself. Genetic engineering should enhance the capacity to produce more tolerant plants by accessing a much wider gene pool for novel insect resistance characters not present in any of the *Gossypium* species or their close relatives. The most efficient and cheapest method for protecting cotton from pests may be the utilization of transgenic cotton for insect resistance.

The genes considered to be most useful for cotton are the *Bt* toxin genes which contain a crystalline protein toxin. *Bt* toxins are insecticidal proteins found as parasporal crystalline inclusions in sporulated *Bt* strains. They are highly potent, show specificity towards insect pests, and are fairly safe to non-target insect species and vertebrates, particularly humans.

A large number of *Bt* genes have been identified, cloned and characterized, and a few have been expressed in transgenic tobacco or tomato plants where they have increased tolerance to insect pests. Transgenic *Bt* cotton carrying *cryIA(b)* and *cryIA(c)* genes has been raised from *Bt* spp. *kurstaki* strains HD-1 and HD-73 respectively.

Modified *Bt* genes for several different *Bt* toxins have been transformed into cotton plants by *Agrobacterium tumefaciens*-mediated, pollen-mediated transfer or by the pollen tube pathway (see Zhang *et al.*, 2000). Field performance of transgenic cotton containing *Bt* genes has shown significant caterpillar resistance. Incorporation of *Bt* genes into major commercial cultivars is expected to reduce insecticide applications for lepidopteran pests by more than 60-70%.

Although *Bt* cotton is very effective in controlling lepidopteran larvae *e.g.* of cotton bollworm and armyworm, there is considerable concern among the scientific and environmental groups that a single *Bt* gene may not be sufficient for the long-term protection of cotton. When used like a chemical insecticide, *Bt* toxins rarely involve resistance mechanisms in insect pests because of their short field life, which is mainly a consequence of rapid degradation by ultraviolet light. When expressed in a transgenic cotton plant, however, the insects will be constantly exposed to the toxin and this would favour the selection of resistant individuals. In order to cope with this problem, attempts are being made to transform two different insect-resistance genes into the same cotton plant.

The development of transgenic cotton that expressed CryIA insecticidal proteins from *Bt* spp. *kurstaki* has resulted in new varieties or lines showing improved resistance to key lepidopteran insect pests. Cotton plants expressing modified *cryIA* gene sequences have demonstrated good control of bollworm, tobacco budworm and pink bollworm in greenhouse and field experiments.

Modifying Cotton Fibres

Work is underway on the genes critical to fibre differentiation and development. Some genes relevant to fibre strength and length have been sought. Rabbit hair keratin gene has been cloned and transformed successfully into cotton plants at the Shanghai Institute of Plant Physiology. The length of the transgenic cotton fibre is 60% higher than normal cotton fibres, and fibre elasticity and thermal preservation are better than normal cotton fibre, just like rabbit hair (see Zhang *et al.*, 2000).

References

Altieri, M.A. *Genetic Engineering in Agriculture: The Myths, Environmental Risks and Alternatives.* Special Report No. 1, Food First, Oakland, CA (2001).

Baulcombe, D.C., Saunders, G.R., Bevan, M.W., Mayo, M.A., Harrison, B.D. *Nature* 321: 446-449 (1986).

Bent, A.F. *Plant Cell* 8: 1757-1771 (1996).

Bhat, A.I., Jain, R.K., Varma, A. Naresh Chandra and Lal, S.K. *Indian J. Phytopathol.* 54: 112-116 (2001).

Broekaert, W.F., Terras, F.R.G., Cammune, B.P.A., Osborn, R.W. *Plant Physiol.* 108: 1353-1358 (1995).

Carmona, M.J., Molina, A.,Fernandwz, J.A., Lopez-Fando, J.J., Garcia-Olmedo, F. *Plant J.*, 3: 457-462 (1993).

Choudhary, B., Laroia, G. Technological developments and cotton production in India and China. *Current Science* 80: 925-932 (2001).

Cohen, M.B., Gould, F., Bentur, J.S. *Int. Rice Res. Notes* 25: 4-10 (2000).

Cornelissen, B.J.C., Does, M.P., Melchers, L.S. In *Rhizoctonia Species: Taxonomy, Molecular Biology, Ecology, Pathology and Control* (eds Sneh, B., Jabaji-Hare, S., Neate, S., Deist, G.), pp. 529-536, Kluwer, Dordrecht (1996).

Dasgupta, I., Malathi, V.G., Mukherjee, S.K. Genetic engineering for virus resistance. *Current Science* 84: 341-354 (2003).

Delaney, T.P. *Science* 266: 1247-1250 (1994).

Ding, X. *et al. Transgenic Res.* 7: 77-84 (1998).

Dinham, B. GM cotton-farming by formula? *Biotech. Develop. Monitor* No. 44/45: 7-9 (March, 2001).

Dirie, A.M. Regulatory oversight in developing nations. *Nature Biotechnol.* 20: 547-548 (2002).

Does, M.P., Cornelissen, B.J.C., in *Crop Productivity and Sustainability: Shaping the Future* (eds. Chopra, V.L., Singh, R.B., Verma, A.) pp. 233-244. Oxford and IBH, New Delhi (1998).

Dove, A. *Nature Biotechnol.* 19: 293-294 (2001).

Ellis, J., Dodds, P., Pyror, T. *Trends in Plant Sci.* 5: 373-378 (2000).

EPA (US Environmental Protection Agency), FIFRA Scientific Advisory Panel, Subpanel on *Bacillus thuringiensis (Bt)* Plant-Pesticides and Resistance Management, EPA, Washington, D.C. (1998).

EPA. *Insect Resistance Management in Bt Crops.* EPA, USDA, Washington, D.C. (1999).

Estruch, J.J., Warren, G.W., Mullins, M.A., Nye, G.J., Craig, J.A., Koziel, M.G. *PNAS* (USA) 93: 5389-5394 (1996).

Flor, A.H. *Phytopathology* 45: 680-685 (1955).

Foissac, X., Loc, N.T.,Christou, P., Gatehouse, A.M.R., Gatehouse, J.A. *J. Insect Physiol.* 46: 573-583 (2000).

Fritig, B., Legrand, M. (Ed.) *Mechanisms of Plant Defense Responses.* Kluwer, Dordrecht (1993).

Frutos, R., Rang, C., Royer, M. *Crit. Rev. Biotechnol.* 19: 227-276 (1999).

Gatehouse, A.M.R., Boulter, D. *J. Sci. Food Agric.* 34: 345-350 (1983).

Giri, A.P., Harsulkar, A.M., Deshpande, V.V., Sainani, M.N., Gupta, V.S., Ranjekar, P.K. *Plant Physiol* 116: 393-401 (1998).

Gordon, K.H.J., Johnson, K.N., Hanzlik, T.N. *Virology* 208: 84-98 (1995).

Grover, A., Gowthaman, R. Strategies for development of fungus-resistant transgenic plants. *Current Science* 84: 330-340 (2002).

Grover, A., Pental, D. Breeding objectives and requirements for producing transgenics for major field crops of India. *Current Science* 84: 310-320 (2003).

Gutierrez-Campos, R., Torres-Acosta, J.A., Saucedo-Arias, L.J., Gomez-Lim, M.A. *Nature Biotech.* 17: 1223-1226 (1999).

Hain, R.H.J. *et al., Nature* 361: 153-156 (1993).

Hamilton, A.J., Baulcombe, D.C. *Science* 286: 950-952 (1999).

Harrison, B.D.,Robinson, D.J. *Annu. Rev. Phytopathol.* 37: 369-398 (1999).

Hilder, V.A. *et al. Transgenic Res.* 4: 18-25 (1995).

Houlb, E.B. *Nature Genet.* 2: 516-527 (2001).

Huang, J., Rozelle, S., Pray, C., Qang, Q. *Science* 295: 674 (2002).

Hull, R. *In Methods in Molecular Biology: Plant Virology Protocols.* Foster, G.D., Taylor, S.C. (eds.). pp. 547-555, Humana Press, New Jersey (1998).

Hull, R. *Matthew's Plant Virology.* 4th ed. Academic Press, New York (2002).

Ishimoto, M., Sato, T., Chrispeels, M.J., Kitamura, K. *Entomol. Expt. Appl.* 79: 309-315 (1996).

James, C. *Global Review of Commercialized Transgenic Crops: 2001 (Feature: Bt Cotton).* ISAAA Briefs No. 26 (International Service for the Acquisition of Agri-biotech Applications (ISAAA), Ithaca, NY (2002).

Jayaraman, K.S. *Nature Biotechnol.* 19: 1090 (2001).

Jayaraman, K.S. Poor crop management plagues *Bt* cotton experiment in India. *Nature Biotechnology* 20: 1069 (2002).

Keller, H. *et al., Plant Cell* 11: 223-236 (1999).

Leah, R., Tommerup, H., Svendsen, I., Mundy, J. *J. Biol. Chem.* 266: 1467-1573 (1991).

McBride, K.E., Svab, Z., Schaaf, D.J., Hogan, P.S., Stalker, D.M., Maliga, P. *Bio/Technology* 13: 362-365 (1995).

McGarvey, P.B., Kaper, J.M. In *Transgenic Plants* (ed. Kung, S.D., Wu, R.) pp. 277-296, Academic Press, New York (1993).

Mellon, M., Rissler, J. (eds.). *Now or Never: Serious New Plans to Save a Natural Pest Control.* Union of Concerned Scientists, Cambridge, MA (1998).

Moss, J.P. (ed.). *Biotechnology for Crop Improvement in Asia* (ed.). ICRISAT, Hyderabad (1992).

Myers, D., Stolton, S. (eds.). *Organic cotton: from field to final product.* IT Publications, London (1999).

NRC. (National Research Council). *The Future Role of Pesticides in US Agriculture.* National Academy Press, Washington, D.C. (2000).

Oerke, E.-C., Dehne, H.-W., Schönbeck, F., Weber, A. *Crop Production and Crop Protection: Estimated Losses in Major Food and Cash Crops.* Elsevier, Amsterdam (1994).

Paarlberg, R.L. *The Politics of Precaution: Genetically Modified Crops in Developing Countries.* Johns Hopkins Univ. Press, Baltimore (2001).

Park, C.M., Berry, J.O., Bruenn, J.A. *Plant Mol. Biol.* 30: 359-366 (1996).

Pearce, F. Reaping the rewards. *New Scientist*, p. 12 (Feb. 2, 2002).

Perlak, F.J., Fuchs, R.L., Dean, D.A., McPherson, S., Fischhoff, D.A. *PNAS* (USA) 88: 3324-3328 (1991).

Pinstrup-Andersen, P., Schiøler, E. *Seeds of Contention: World Hunger and the Global Controversy over GM Crops.* Johns Hopkins Univ. Press, Baltimore (2001).

Price, E., Hamburger, J. Genetically engineered crops expand in China. *Global Pesticide Campaigner* 11(3): 7 (Dec. 2001).

Qaim, M., Krattiger, A.F., von Braun J. (Eds.). *Agricultural Biotechnology in Developing Countries: Towards Optimizing the Benefits for the Poor.* Kluwer, Boston, MA (2000).

Qaim, M., Zilberman, D. Yield effects of genetically modified crops in developing countries. *Science* 299: 900-902 (2003).

Ranjekar, P.K., Patankar, A., Gupta, V., Bhatnagar, R., Bentur, J., Kumar, P.A. Genetic engineering of crop plants for insect resistance. *Current Science* 84: 321-329 (2003).

Rautner, M. Designer trees. *Biotech. Develop. Monitor.* 44/45: 2-7 (March, 2001).

Rommens, C.M.T., Salmeron, J.M., Oldroyd, G.E., Staskawicz, B.J. *Plant Cell* 7: 1537-1544 (1995).

Roush, R.T. In *Insecticide Resistance: From Mechanisms to Management* (eds. Denham, I., Pickett, J.A., Devenshire, A.), pp. 101-110 CAB Internat. Publishing, Wallingford (1999).

Schuler, T.H., Poppy, G.M., Denholm. I. *TiBTECH.* 16: 168-175 (1998).

Sharma, D. The introduction of transgenic cotton in India. *Biotech. Develop. Monitor* 44/45: 10-13 (March, 2001).

Sivamani, E., Huet, H., Shen, P., Ong, C.A., De Kochko, A., Fauquet, C., Beachy, R.N. *Mol. Breeding* 5: 177-185 (2002).

Slusarenko *et al.* (eds.). *Mechanism of Resistance to Plant Diseases.* p. 428, Kluwer, Dordrecht (2000).

Stotz, H.U., Contoa, J.J.A., Powell, A.L.T., Bennett, A.B., Labavitch, J.M. *Plant Mol. Biol.* 25: 607-617 (1994).

Takken, F.L.W., Joosten, M.H.A.J. *Eur. J. Plant Pathol.* 106: 699-713 (2000).

Tousch, D., Jacquemond, M., Tepfer, M. *J. Gen. Virol.* 75: 1009-1014 (1994).

Trudel, J., Potvin, C., Asselin, A. *Plant Sci.*, 55-62 (1995).

Van Loon, L.C., Van Kammen, A. *Virology.* 40: 199-211 (1970).

Varma, A., Jain, R.K., Bhat, A.I. *Indian J. Biotechnol.* 1: 73-86 (2002).

Verma, H.N. *Anti-viral Proteins in Higher Plants.* (eds. Chessin *et al.*) pp. 1-37, CRC Press, Boca Raton (1995).

Wossink, G.A.A., van Kooten, G.C., Peters, G.H. (Eds.). *Economics of Agro-Chemicals: An International Overview of Use Patterns, Technical and Institutional Determinants, Policies and Perspectives.* Ashgate, Aldershot, UK (1998).

Zhang, B.H. *Cotton Biotechnology and its Application.* Chinese Agricultural Press, Beijing (1997).

Zhang, B.H., Liu, F., Yao, C.B., Wang, K.B. Recent progress in cotton biotechnology and genetic engineering in China. *Current Science* 79: 37-44 (2000).

Chapter 4

Transgenic Crops: Scientific, Social and Economic Aspects

Introduction

Recent increases in research interest in the potential effects of the release of genetically modified (GM) crops on biodiversity have been prompted by concerns relating to the direct impact of these crops on target organisms and the indirect effects on the general environment. The environmental issue needs to be considered within a biodiversity perspective already affected by the intensification of agriculture. Although, in some cases the introduction of GM crops may not differ from the introduction of some other technology that leads to the further intensification of agriculture, this new technology might offer a particularly rapid increase in intensification.

Watkinson *et al.* (2000) modeled the effects of the introduction of a herbicide-resistant sugar beet (a spring-sown crop grown in Europe and North America) on the population dynamics of a cosmopolitan annual weed, *Chenopodium album*, whose seeds are eaten by farmland birds such as the skylark *Alauda arvensis*. They simulated the effects of the introduction of genetically modified herbicide-tolerant (GMHT) crops on the weed populations and the consequences for seed-eating birds. They predicted that weed populations might be reduced to low levels or practically eradicated, depending on the exact form of management. Consequent effects on the local use of fields by birds might be severe, because such reductions represent a major loss of food resources. The regional impacts of GMHT crops were shown to depend on whether the adoption of GMHT crops by farmers covaries with current weed levels.

It was concluded that the effects of GMHT crops on weed abundance, and hence on bird populations, will depend on the socioeconomic reaction to the new technology as much as on any possible improvement in weed control.

In their model Watkinson *et al.* (2000) grow a conventional (not GM) sugar beet crop in a 5-year rotation: Winter cereals are grown for 4 years and then conventional sugar beet is grown in the fifth year. The seeds of *C. album* are an important food source for farmland birds. These seeds remain dormant in the soil for several years; each year a proportion of this weed seedbank germinates. Watkinson *et al.* assumed that weed control is perfect during the years that cereals are grown so that new weed seeds are produced only during the 1 year in 5 that sugar beet is cultivated. They also assumed that the numbers of skylarks that feed on the seeds is greatest in fields where lots of the weed seeds are available. The authors of the model presume that their model reflects a wider range of farmland birds and the weeds that are their food source. The modelers examined what happens to the population density of *C. album* when GMHT rather than conventional sugar beet is grown in the fifth year of the crop rotation. They calculated how the altered use of herbicides with the GMHT crop can affect the growth of *C. album* and hence its seed production, and how this will impact the skylark that feeds on these seeds (Firbank and Forcella, 2000).

Watkinson *et al.* concluded that consequent effects on the local use of fields by birds can be quite severe if GMHT crops are introduced commercially. But this conclusion seems questionable in light of several experiences with growing GMHT maize, soybean, canola, and sugar beet in the United States (Firbank and Forcella, 2000). Data from the southern United States, where GMHT crops have been grown for the past few years show that there is sometimes less success with weed control when growing GMHT crops than when cultivating conventional varieties. Farmers who grow GMHT soybean do so not because weed management with this crop is more effective, but because it is simpler (fewer types and fewer applications of herbicide are needed; and timing of applications is also not very critical).

According to Firbank and Forcella, the Watkinson model provides a good conceptual framework for assessing the impacts of GMHT crops on farmland birds, but the authors concentrated on assumptions and scenarios that emphasize harmful effects, to the exclusion of benefits to biodiversity. For example, herbicide use in current GMHT crops is usually later than in conventional crops, potentially favouring breeding birds. Also, sowing GMHT crops facilitates minimum tillage, which favours the conservation and biodiversity of soils, tends to keep more weed seeds near the surface, increases weed numbers and, again, changes the weed spectrum. This interaction between GM cropping and tillage practice could imply that any reduction of weed seed production under GM crops can be compensated for in that a higher proportion of the seeds will stay nearer the soil surface to be available to feeding birds. Indeed, both GMHT and non-GMHT systems can be modified to favour biodiversity: Field patches can be left unsprayed to provide food for skylark chicks until they fledge, and winter stubbles and selective spraying may help maintain bird food resources in the cereal phase of the rotation (Firbank and Forcella, 2000).

Hazards of Introducing Genes into Plants

Schubert (2002), who identified three hazards arising from the introduction of genes into plants, concluded that "GM food is not a safe option." According to him an introduced gene may: (i) produce different proteins in different cell types; (ii) produce some proteins that can react with other substances in a cell leading to the biosynthesis of toxic, allergenic, or carcinogenic molecules; or (iii) result in the formation of a new biochemical pathway that generates novel or unexpected products of physiological consequence. But according to Beachy *et al.* (2002), each of these three possibilities can and do occur in nature during the course of the random "natural" gene mutations and rearrangements in the course of evolution. Beachy *et al.* feel that genetic engineering is just like traditional plant breeding, only more specific.

Schubert suggested that all GM plant products destined for human consumption should be tested for long-term toxicity and carcinogenicity before being brought to market. These safety criteria should also be met for many chemicals and all drugs. This suggestion immediately prompts the question why are these concerns not equally applied to new crop varieties that are developed through intensive radiation or chemical mutagenesis? The three unknowns pointed out by Schubert, *viz.* differential post-translational protein processing, unexpected changes in gene expression, and disruption of endogenous enzymatic pathways are as relevant to mutagenesis-derived varieties, or to varieties that are created by widecrosses, as they are to crops developed through recombinant genetic technologies (see Alex Avery's untitled comment in *Nature Biotechnol.* 20: 1196, 2002). Further, current US regulatory rules already account for cases where the genetic modification significantly changes the composition of the foodstuff. In fact, current biotechnology food regulations in the USA at least already contain a strong precautionary component.

Crops produced through biotechnology are subjected to a highly selective screening process that is unlikely to permit unintended or unexpected variation. Any plant variety produced through biotechnology has to go through rigorous safety evaluations and it must grow and perform in the field as well as or better than its conventional counterpart. Safety assessments for biotechnology crops focus on careful analyses of plants.

The various kinds of scientific tests conducted on all biotechnology crops are designed to detect unpredicted effects from all kinds of sources, including alternative splicing of mRNAs and post-translational processing of the target protein, and any other unexpected impacts on plant metabolism that might occur. The gene must express the intended protein in the predicted manner or the plant would not display the desired phenotype or trait and would therefore not attract further consideration. The protein produced in the new host has to be subjected to extensive biochemical characterization so as to confirm that the protein produced is the one and only one intended. Additional tests include acute toxicity and allergenicity evaluations; agronomic, performance and yield analyses, extensive characterization of the expressed protein to assure specificity, compositional analyses of key metabolites and nutrients, and animal nutrition and feed-performance studies (Beachy *et al.*, 2002).

According to Beachy *et al.*, unintended genetic and metabolic events can take place, but that these consequences are more likely to occur in nature than in biotechnology because nature relies on the unintentional consequences of blind random genetic mutation and rearrangement to produce adaptive phenotypic results, whereas GM technology employs precise, specific, and rationally designed genetic modification toward a specific engineering goal.

Schubert repudiates the argument about the analogy of traditional breeding versus genetic modification by pointing out that genes from completely different species (sometimes kingdoms!) are expressed in GM food products-an event that is not likely to happen in nature. No doubt variation and selection are important but only in the context of a normal complement of endogenous genes that, though perhaps different, are allelic. This is quite distinct from GM plants where many copies of a gene are introduced and integrate randomly (see Svitashev and Somers, 2002). Many workers have noted that the response to (trans)genes is unpredictable and not specific. Genetic modification is also distinct from breeding two strains that have been safely consumed for long periods of time.

Shaky Foundation–The Dogma in Distress?

As per the Central Dogma of biology (Crick, 1970), DNA is the master molecule that embodies all genetic information of any living being. DNA alone governs inheritance. It governs the expression of this information in the organism and its transmission to the next generation. It is the DNA that transmits and directs the biological functions of all living beings. The DNA and the Central Dogma are also the basis on which genetic engineering industry is built: if genes make up the universal code of life, they can surely be slotted into plants, animals and humans, to express some desired effect. Techniques to move genes around have resulted in raising pigs with genes from cows producing bovine growth hormone, plants with genes from bacteria producing natural pesticides, and bacteria with human genes to produce insulin.

And yet, it is now emerging that the Dogma does not convey the whole story. Several discordant experimental results contradict this theory. But this new evidence remains virtually unacknowledged, and work with transgenic plants and animals is continuing as before. Recent work on deciphering of the human genome has revealed that the entire human genome consists of 30,000 genes-much less than the number originally estimated by considering the sum total of different proteins and inherited traits that humans have. We have more proteins than genes. So the question arises as to how those proteins which do not have a corresponding gene, are synthesized? The two logical conclusions seem to be that genes code not for a single protein but for many different proteins and traits; and/or some other, unknown regulatory mechanisms are involved in protein production. Recent researches suggest that both these conclusions are true. Proteins themselves help define the functions of other proteins by influencing their three dimensional structure. Also there are many other types of genetic interactions in the cell, including those where proteins feed back information to DNA. More interestingly, it is now known that the parts of DNA that did not seem to code for proteins (and so were called 'junk DNA') do in fact produce molecules that interfere with protein production and therefore constitute an essential part of the cell's regulatory system.

Thus, the 'one gene, one trait' logic is too simplistic and needs to be abandoned. But the genetic engineering industry is holding on tight to the Central Dogma while ignoring the complexities in the functioning of genes as unwelcome distractions. Biotech companies and MNCs need to be able to assure both their clients and regulatory authorities that the transgenic crops and animals they sell will do exactly what they were designed to do-tolerate herbicides, kill off insect pests, or produce specific molecules. They badly need a theoretical basis that explains precisely and predictably how the new gene will behave in the new host; they need the Central Dogma (GRAIN, 2003). But in reality the current race towards transgenic agriculture is really based on an outdated theory of the laws of heredity.

Although many scientists employed by MNCs do believe that genetic engineering needs plant breeding to deliver the seeds to the farmer, and that genetic engineering is just one tool in the toolbox of the plant breeder, the gap between the worlds of genetic engineering and plant breeding is increasing.

The unholy but highly profitable marriage between a simplistic, outdated concept of genetics and a powerful capital-driven conglomerate of industrial giants is pushing scientists and agronomists away from finding solutions in consultation with farming communities and policy-makers to address the food problem.

What we need is to give up our obsession with the gene. It is time to move away from the Mendelian pedigree breeding approach, which focuses on uniform varieties carrying specific genes to the next generation while eliminating others. Instead, we need to refocus on the farmer's field, where desired traits are incorporated into all the plants of a crop, in all its genetic diversity. The entire population from the field needs to be screened to isolate a few elite plants which may be used in the next breeding cycle. This population breeding approach has actually been something that farmers have been adopting for millennia but is usually considered a nightmare by industrial plant breeders who are used to working with uniform pure lines. In fact, this approach successfully delivers a more lasting genetic improvement than single gene approaches, be they genetically engineered or not. An added merit of this approach is that it costs nothing. Farmers can do this breeding on their own farms.

What we really need to address is the whole series of complex issues that peasant farmers face in their food production systems. In many cases their challenges have little to do with agronomy; rather they relate to access to land, markets and credit, or labour issues and gender aspects. Countless farmers in southern and western India who used to be engaged in farming in dryland conditions are now adopting a variety of water and soil management technologies; they have already been able to triple their yields of sorghum and millet to more than 2 tons/hectare. When agronomic questions do become important, it is quite often not the genetic potential of the crops and animals that is the most limiting factor. Instead farmers talk about soil fertility, agroecology, integrated crop management, or water retention and supply (GRAIN, 2003).

The biotechnology industry has repeatedly assured that genetically engineered soybeans have been modified only by the presence of the alien gene, but in fact the plant's own genetic system has been unwittingly altered as well, with potentially dangerous consequence (see Windels *et al.*, 2001; Commoner, 1968; 2003).

Canalization

Biologists still do not understand how genes are connected to the traits seen in whole organisms. All that is known is that the overall appearance and behaviour of an organism are determined by the highly complicated interaction between its genes and the environment. Some sort of developmental buffering ensures that similar traits are produced despite genetic changes (mutations) and environmental perturbation. A better understanding of this buffering effect, known as canalization, may well change how we think about evolution, development, or genetic modification of crops.

Eliminating Hunger

When the genetic engineering corporations found that the scientific basis of their work is shaky, and as they had hardly any practical results to show off, they urgently needed some ideological prop to secure their huge investments in genetic engineering. They found it in the 800 million poor people who lack access to food day. So they started resorting to conquering the markets and the farmers fields in the Third World for transgenic agriculture, arguing that genetic engineering can help in combating hunger. However, the real fact is that solving hunger has never been the business of the transnational corporations engaged in genetic engineering, and never will be.

Box 4.1 brings this out clearly - only a few powerful corporations are developing a few profitable crops in a handful of countries, mostly for animal feed and export markets. This picture cannot possibly solve the world's highly complex food problem (GRAIN, 2003).

Box 4.1

GM CROPS IN 2002*:

- ✔ More than 90% of commercially grown GM crops represented just 4 crops: canola, soybean, cotton and maize - the bulk of which are being grown for export, not for food.
- ✔ More than 90% of commercially grown GM crops in the world are grown in just 4 countries: the US, Canada, China and Argentina - largely serving (with the possible exception of China) the export and cattle feed market.
- ✔ Virtually all commercially grown GM crops come from one corporation (Monsanto) which, together with a few other gene giants (Dupont, Syngenta, Bayer and Dow), command most of all crop transgenic research in the world.
- ✔ Virtually all these crops are engineered for just two traits: resistance to herbicides and the incorporation of the toxic *Bt* gene, supposedly to hive off insects.

(* Source: GRAIN, 2003).

The wonders of genetic modification are claimed to emanate from the premise that the molecular structure of DNA is the exclusive agent of inheritance in all living things-the DNA gene is supreme. The premise of the Central Dogma assumes that an organism's genome should fully account for its characteristic assemblage of inherited traits. The Dogma seeks to reduce inheritance to molecular dimensions. Segments of DNA comprise the genes that, through various molecular processes, give rise to each of our inherited traits. But this premise is false. The Human Genome Project has sounded the death knell for the Dogma. There are far too few human genes to account for the complexity of our inherited traits or for the numerous inherited differences between plants, say, and people. The decoding of the Human Genome, published in Feb., 2001, destroyed the scientific basis of genetic engineering and the validity of the biotechnology industry's widely advertised claim that as its methods of genetically modifying food crops are "specific, precise, and predictable" (see Commoner, 2003), they are quite safe.

Discovery of the DNA double helical structure, followed by the enunciation of the Central Dogma, generated the fallacious proposition that genes have unique, absolute, and universal control over the totality of inheritance in all forms of life. According to Crick (1958, 1970), genetic information originates in the DNA nucleotide sequence and terminates, unchanged, in the amino acid sequence of the protein encoded by the gene. This pronouncement endowed the gene with strong control over the identity of the protein and the inherited trait that the protein creates. In fact, Crick held on to this belief so tightly that he emphatically asserted that "the discovery of just one type of present-day cell in which genetic information passed from protein to nucleic acid or from protein to protein "would shake the whole intellectual basis of molecular biology (Crick, 1970).

This dogmatic view is surprising because it was known even in the 1960s that in living cells proteins come into promiscuous molecular contact with several other proteins and with other molecules of DNA and RNA. Crick's belief finally exploded in early 2002 when the two human genome research teams reported their results. The major result turned out to be unexpected (IHGSC, 2001): instead of the 100,000 or more genes predicted by the estimated number of human proteins, the gene count was only about 30,000. By this measure, people are only about as gene-rich as the mustard-like weed *Arabidopsis thaliana*, (which contains 26,000 genes) and about twice as genetically endowed as a fruit fly or a primitive worm (Venter *et al.*, 2001). These findings were highly surprising and not only contradicted the scientific foundation on which the genome project was undertaken but also dethroned its guiding Central Dogma. Also it fails to explain the vast inherited difference between a weed and a person, pointing to the high probability of there being much more to James Watson's 1990 reference to the Human Genome Project as the, "ultimate description of life" than the genes alone can convey (see Commoner, 2003).

Alternative Splicing of Genes

Venter *et al.* (2001) offered one explanation for the shortfall in the gene count: "nearly 40% of human genes are alternatively spliced". This explanation has shaken the whole intellectual basis of molecular biology. It negates the validity of its application to genetic engineering. Alternative splicing certainly is a startling departure from the orderly design of the Central Dogma, in which a single gene encodes the amino acid sequence of a single protein. In alternative splicing, the gene's original nucleotide sequence is split into fragments that are then recombined in different ways to encode several proteins differing from each other in their amino acid sequence and also from the sequence that the original gene, had it remained intact, would have encoded. Alternative splicing can have an enormous impact on the gene/protein ratio. In the fruit fly, one gene can generate over 38,000 variant protein molecules (Schmucker *et al.*, 2000). The discovery of alternative splicing undermines the supremacy of the gene in the molecular process of inheritance: the gene's effect on inheritance cannot be predicted simply from its nucleotide sequence-the determination of which is one of the main purposes of the Human Genome Project (Commoner, 2003).

One disturbing fact that emerged from this story is that by 1989, when the Human Genome Project was still being debated, its champions should have known that over 200 papers on alternative splicing of human genes had already been published. The shortfall in the human gene count should have been predicted. Surely, at least some of the project's prospective investigators already knew that the discrepancy between the numbers of genes and proteins in the human genome was to be expected, and that the $3 billion project could not be justified by the tall claim that the genome would tell us who we are (Venter *et al.*, 2001; Commoner, 2003).

But alternative splicing is by no means the only discovery that has contradicted basic precepts of the Central Dogma. Some research has eroded the centrality of the DNA double helix itself. In their original description of the structure of the helix, Watson and Crick stated that this structure "immediately suggests a possible copying mechanism for the genetic material." It turns out that this was premature. For it is not the DNA alone that is responsible for biological replication; the precise duplication of DNA is really accomplished by the living cell. In the living cell the gene's nucleotide code can be replicated faithfully only because certain specialized proteins intervene to prevent most of the errors-which DNA by itself is prone to make-and to repair the few remaining ones. Genetic information arises not from DNA alone but through its essential *collaboration* with protein enzymes–a contradiction of the Central Dogma's precept that inheritance is uniquely governed by the self-replication of the DNA double helix!

Another crucial observation is that in order to generate the inherited trait, the newly made polypeptide must be folded up into a precisely organized three-dimensional ball-like structure created by the particular way in which the molecule is folded into that structure. Crick had erroneously assumed that the nascent protein-a linear molecule-always folded itself up correctly once its amino acid sequence had been determined. In fact, some nascent proteins are on their own likely to become misfolded-and therefore remain biochemically inactive-unless a special type of "chaperone" protein properly folds them (Ellis and Hemmingsen, 1989).

The fact that one gene can give rise to multiple proteins also shatters the theoretical basis of the genetic engineering of food crops. In genetic engineering it is assumed, without adequate experimental proof, that a bacterial gene for an insecticidal protein, for instance, when transferred to a maize plant, will produce precisely that protein and nothing else. But in the alien genetic environment of the maize, alternative splicing of the bacterial gene might give rise to multiple variants of the intended protein-or even to proteins bearing little structural relationship to the original one, with unpredictable consequences for ecosystems or human health (Commoner, 2003).

Actually, the delay in dethroning the all-powerful gene may well have led in the 1990s to a huge invasion of genetic engineering into American agriculture.

While DNA has undoubtedly played an important role in the development of life, it does not mean that life can be reduced to a master molecule for the sake of simplistic our emotional understanding. The experimental data points to the irreducibility of the living cell, whose inherent complexity suggests that any artificially altered genetic system will surely give rise sooner or later to unintended, potentially disastrous, consequences (see Commoner, 2003).

Farm Level Impact

The development of agricultural biotechnology offers novel opportunities to increase agricultural productivity (see Betz *et al.*, 2000). Some of the potential gains achievable with this technology in the developed countries include higher yields, enhanced product quality and lower pesticide and labour costs. Such gains are particular important in developing countries, especially in the struggle of alleviating poverty. Ismael *et al.* (2001) reported the findings of an independent survey of smallholder farmers in the Republic of South Africa designed to explore the economic benefits of their adoption of *Bt* cotton. The *Bt* variety generally resulted in a per hectare increase in yields, value of output and reduction of pesticide costs which outweighed the increase in seed costs to give a substantial increase in gross margins. *Bt* cotton was the first commercial release of a GM crop variety in Sub-Saharan Africa. In 1999/2000, 100,000 ha area of *Bt* cotton was grown in South Africa by 1530 commercial farmers and 3000 small-scale farmers mostly under dryland conditions in the Northern province. Small-holder farmers in the Makhathini Flats (KwaZulu-Natal province) have recently been adopting the genetically modified cottonseed variety NuCOTN 37-B with Bollgard. Sixty per cent of these farmers have plots of between 10 and 20ha. Seed companies estimate that in the 2000/2001 growing season, around 95% of the 4000 smallholder farmers in the Makhathini region may have adopted this *Bt* cotton variety.

Rural households in the Makhathini area typically occupy 1-3 ha of land allocated to them by their tribal chiefs. The major crops in the area are beans, maize and cotton. The latter usually occupies most of the farm and is grown as a commercial crop. One advantage for growing cotton is that it needs less intensive management than maize or beans and can better survive fluctuating whether conditions. Two constrains on market-orientated agriculture in the region are the climatic constraint of erratic rainfall, and pest attack (Ismael *et al.*, 2001).

The survey of 100 small farm holders was conducted in November 2000. The questionnaire was designed to obtain information on the physical characteristics of the farm; characteristics that defined the farmer; cropping patterns; input use and costs; cotton output and revenue; and other income sources and assets. The aim of the study was to examine the factors involved in the adoption of *Bt* cotton as well as its impact on yields, gross margins and technical efficiency-notably pesticide use. The analysis compared *Bt* cotton farmers (adopters) with non-*Bt* cotton farmers (non-adopters). The survey covered two growing seasons: 1998/1999 (first year) and 1999/2000 (second year). In the 1998/1999 season, only 10% of the 4,000 farmers in the region had adopted the new *Bt* cotton variety, while by 1999/2000 this had risen to 40%.

The average farm size of respondents was 6 ha, although 62% of the farms included in the survey were less than 5 ha. About 73% of 100 respondents owned livestock such as hens, sheep, goats and 25% had non-farm income sources. The average is 6 ha for the sample, and most farmers own between 0.5 to 3 ha. If a farmers does not use the land allocated to him, someone else can do so. They farm small areas of cotton because of the lack of labour and credit, and cotton is only a cash crop. Some migrants work in nearby towns and most of them move to Johannesburg or Durban in search of work.

Amongst the agronomic problems, 57% of farmers considered pests to be their bigger problem and 62% of these categorized the bollworm as the major pest. Too much rain was a more distant second at 24% followed by weeds, which was ranked highest by only 11%.

Adoption of *Bt* Cotton

The stratified sample consists of 19 adopters and 81 non-adopters in the first year and 60 adopters and 40 non-adopters in the second year. All 19 farmers who grew *Bt* cotton in the first year also grew it in the following year, suggesting that they were satisfied with the variety. Generally, the adopters were more experienced and had larger farms.

When questioned in 2000, after one season of experience, about why they adopted *Bt* cotton, 44% of respondents cited savings on the cost of insecticide as the chief reason, while 24% cited expected increases in yield. Approximately 10% said that the expectation of less time spent spraying *Bt* cotton was critical in their decision to adopt it. Most of the farmers surveyed did not identify any problems with *Bt* cotton. Almost 90% of the non-adopters in 1999/2000 were willing to adopt the technology in the future, but cited high seed cost as the chief reason for not adopting it.

Economic Impact

Table 4.1 summarizes the average yield performance and economic impact of *Bt* cotton. *Bt* cotton gave higher yields per hectare than non-*Bt* varieties, but there was a marked seasonal effect. In 1998/99 the average yield increase of *Bt* over non-*Bt* was nearly 18%, and in the 1999/2000 season it rose to 60%. These increases occurred despite the seeding rate of *Bt* adopters being only 45% of the recommended rate (the equivalent rate for non-adopters was 55% of the recommended rate). Thus, *Bt* cotton gave even higher yields per kg of seed planted than the non-*Bt* crop. The shift in yield differential in the two seasons was probably related to rainfall, which can affect the incidence of bollworm or growth of the crop or even the planting time. In the 1999/2000 season there was unusually heavy rainfall, and average cotton yields fell. However, the *Bt* adopters suffered a fall in yields between the two seasons (of 18%) much less than those who did not adopt (40%). The rain could have affected the incidence of diseases and pests such as bollworm and the length of the growing season. It meant a delay in planting and also affected the germination rate (Ismael *et al.*, 2001).

The use of *Bt* cotton increased seed cost per hectare by over 100% in both seasons. This high seed cost was only partially offset by the fact that *Bt* cotton reduced pesticide costs for *Bt* adopters in both growing seasons. On average, in the first season pesticide costs for *Bt* adopters were reduced by nearly 13%, while in the second season the reduction was 38%, possibly because in the first season some producers might have continued to spray either as a risk reduction strategy in case the *Bt* variety proved not to be as resistant as claimed, or because they were not well informed about the nature of the *Bt* variety and its pesticide attributes.

The average figures in Table 4.1 conceal considerable variation and some interesting differences related to farm size. Table 4.2 shows a breakdown of results according to four categories of cotton area plantings. Not surprisingly, the smaller holdings grew more intensively (with higher seed and pesticide costs per hectare) and achieved higher per hectare yields on average than the larger farms.

Table 4.1: Average Per Hectare Cost and Returns Based on Adoption of *Bt* Cotton and Growing Season (after Ismael, *et al.*, 2001)

Variable	*1998/1999*			*1999/2000*		
	Non-Adopters	*Adopters*	*Differ-ence*	*Non-Adopters (%)*	*Adopters*	*Differ-ence (%)*
Yield (kg/ha)	434 (49)	511 (82)	18	261 (43)	417 (30)	60
Value of output (Rand/ha) [1]	944 (107)	1,111 (178)	18	568 (94)	907 (65)	60
Seed cost (Rand/ha)	100 (8)	202 (14)	102	91 (17)	197 (12)	116
Pesticide cost (Rand/ha)	112 (8)	98 (14)	13	116 (11)	72 (8)	38
Gross Margin (Rand/ha)	732 (91)	811 (150)	11	361 (66)	638 (45)	77
Number of farmers sampled	74	17		32	59	

[1] Price per kg is 2.175 SAR (South African Rand).
( ) Standard deviation.

Despite the variation between the two seasons, the results of this survey provide considerable cause for cautious optimism regarding the impacts of *Bt* cotton at farm level. By the second year the adopters were clearly gaining economically in terms of increased yield, lower insecticide costs and hence a higher gross margin. This perhaps suggests that the advantages of *Bt* cotton become particularly apparent in times of environmental stress, since the second year was very wet-a condition that favours the bollworm. For risk-prone resource-poor farmers this is an important consideration.

Box 4.2

OBSTACLES AND OPPORTUNITIES IN INTER-GOVERNMENTAL REGULATIONS

(Box is reprinted unchanged from Choudhary, 2001).

USPTO

USA law does not consider 'prior art' in a foreign country when granting patents. USPTO can grant patents on compounds made from natural products unless they are in 'prior use' in the USA. Unless India promotes the use of its national 'prior art' in the USA it cannot legally restrict USPTO from awarding patents on the 'prior art' of indigenous and local Indian communities. USA law needs to recognize the use of 'prior' art or knowledge in a foreign country.

TRIPS

Geographical Indications of TRIPS restricts the patenting of a good that originates in a territory of a member country when the quality, reputation or other characteristic of that product is intrinsically associated with the region of origin. It is suggested that the provisions be extended to natural products and indigenous and local community knowledge and practices as well as to genetic resources that have a certificate of origin (*Appellation of Origin*).

The patentability of subject matter should have been reviewed in 1999. Developing countries including India proposed an amendment stating that life forms and living processes should not be patented. During the WTO Seattle Ministerial Conference, the African Group strongly supported by other developing countries submitted a comprehensive proposal in an effort to address, revise and resolve this issue. In particular they recommended the establishment of an international *sui generis* system to protect plant varieties and indigenous and farming community knowledge.

CBD

The major difficulty encountered by the CBD continues to be the legal enforcement of its provisions. Even though it is an international convention, the implementation of its provisions is subject to national legislation. To check biopiracy and illegal bioprospecting, it is necessary to broaden its provisions and create a specific international legal framework. In 1998, the CBD established a working group whose main task is to develop comprehensive guidelines and an appropriate framework to facilitate access to genetic resources and the sharing of the benefits derived from their use.

IT PGRFA

On 3 November 2001 the International Treaty on Plant Genetic Resources for Food and Agriculture (PGRFA) was finally adopted

Contd...

Box 4.1–Contd...

by the UN FAO assembly after seven years of negotiation. Central to the treaty is the establishment of a Multilateral System that facilitates Access and Benefit sharing of Plant Genetic Resources. The treaty has already been criticized at several levels. One of these is that the list of crop genera to which farmers should have free access only contains 35 crops genera and 29 forage species. How the treaty's provisions on farmers' rights and IPRs relate to other agreements is likely to emerge from upcoming negotiations on implementation.

WIPO

The Intergovernmental Committee on Intellectual Property and Genetic Resources, Traditional Knowledge and Folklore is a special WIPO body. WIPO played an advisory role during intergovernmental negotiations on IPRs, and some developed countries suggest it should negotiate the harmonization of CBD and WTO agreements. Other developing countries resist this approach. They see WIPO's mandate as the promotion of IPRs, a more limited field than that of traditional knowledge and *sui generis* options.

UPOV

Instead of using patent legislation, the TRIPS agreement gives countries the option of protecting plant varieties by an effective *sui generis* system. The UPOV convention provides a workable example of a *sui generis* system. The UPOV deals with plant breeders' rights (PBR). It regulates the requirements for obtaining PBR, such as the newness, distinctness, homogeneity and stability of the new variety. It provides rules on the scope of PBR, how to obtain PBR and how to name and protect a variety. Signatories to the convention have to implement UPOV regulations in their national legislation. Countries can choose to sign different versions of the convention: the original 1961 version, the 1978 revision, or the 1991 revision. The latter provides the most protection for breeders, and is most restrictive for farmers. The UPOV convention has been advocated as a *sui generis* system for plant variety protection that developing countries can adopt to fulfil their obligations under TRIPS agreements.

Precautionary Principle and Risk-risk Analysis

The "precautionary approach" or "precautionary principle" warrants that activities threatening harm to human health or the environment require precautionary measures, even if some cause-

Table 4.2: Average per ha Cost and Returns Based on Adoption and Growing Season (of *Bt* cotton) for Different Categories of Cotton Area Planting (after Ismael *et al.*, 2001)

		1998/1999 Season				1999/2000 season			
Cotton Area		*<2.5 ha*	*>2.5-5ha*	*>5-10*	*>10*	*<2.5ha*	*>2.5-5ha*	*>5-10*	*>10*
Output									
Yield kg/ha	Adopters	839	598	354	326	565	480	319	273
	Non Adopters	576	447	36	468	330	327	178	228
Value SAR/ha (2.175/kg)	Adopters	1825	1301	770	709	1229	1044	694	594
	Non Adopters	1253	972	796	1018	718	711	387	496
Cost									
Seed Cost/ha	Non Adopters	232	285	115	152	286	231	139	128
	Adopters	149	116	80	73	149	129	58	95
Insecticide cost/ha	Adopters	108	140	48	50	131	70	52	31
	Non Adopters	189	114	63	105	156	125	77	192
Gross Margin [Output Value - (seed cost + insecticide cost)]									
Gross margin SAR/ha	Adopters	1485	876	607	507	812	743	503	435
	Non Adopters	915	742	653	840	413	457	252	209
Number (N)	Adopters	2	6	5	4	17	22	12	8
	Non Adopters	27	29	15	3	11	15	5	1

and-effect relationships are not clear. This concept has become popular in environmental law as applied to the commercial use of genetically modified organisms. Opponents of the principle argue that its definition and goals are vague, with the result that its adoption by regulatory agencies tends to stifle trade and limit innovation. Auberson-Huang (2002) argued that the precautionary approach fills a gap created by the increasing disenchantment with the limitations of classical risk-assessment methodology for evaluating the wide range of potential impacts in agricultural biotechnology.

Both proponents and opponents of the precautionary principle agree that it substitutes for risk analysis. The precautionary principle calls for reducing, if not eliminating, risks to public health, the environment, or both (Goklany, 2001, 2002). According to the Wingspread Declaration, "When an activity raises threats of harm to human health or the environment, precautionary measures should be taken, even if some cause-and-effect relationships are not established scientifically" (Raffensperger and Tickner, 1999).

The principle itself has been considered as being revolutionary, in offering a sound approach to managing potential risks associated with new technologies (or actions or policies) than the risk-analysis paradigm currently employed by American society and the World Trade Organization. On the other hand, opposition to the precautionary principle has focussed on precisely the point that it seems to reject the risk-analysis approach (Miller and Conko, 2000). According to Goklany (2002), we must, in fact, employ not just risk analysis, but risk-risk analysis.

Assessing risk for the introduction of GMOs into the environment involves four steps: (i) hazard identification; (ii) evaluation of potential consequences; (iii) estimation of damage potential, if the hazard does materialize; and (iv) implementation of risk-mitigation strategies.

The classic approach of assessing hazards individually, based on likelihood of occurrence and extent of damage, has been too limited to cover the complete range of technological, societal, and environmental impacts from the deliberate release of GMOs. The interpretation of the final residual risk by different stakeholders can depend on context and is liable to be easily politicized or misconstrued. Two notable examples are the assessments of the risks of horizontal gene transfer of antibiotic marker genes in transgenic plants, and the risks of *Bt* proteins expressed in transgenic corn pollen and their potential effect, under field conditions, on the mortality of the monarch caterpillar. There are several difficulties in interpreting the meaning of residual risk, stemming partly from the various value systems that are applied in our definition, regulation, and acceptance of plant genetic engineering in crop breeding.

Most policy options reduce some public health and environmental risks while increasing or prolonging others (Goklany, 2001, 2002). Examples are policies that would foreswear the use of either GM crops or dichlorodiphenyl trichloroethane (DDT).

In such ambiguous situations, we need to ensure that a policy is truly precautionary-that is, it reduces net risks or is risk-neutral. For this, the risks of adopting the policy need to be compared against those of not adopting it. This inevitably pushes us into risk-risk analysis. According to Goklany (2001), "all else being equal, the immediacy criterion gives greater weight to threats that are more immediate, the uncertainty criterion to threats that are more certain, the expectation value criterion to those that are larger, and the adaptation criterion to those that are more difficult or costly to cope with."

The two-part public health criterion relies more on ethics than common sense. The first part, the human mortality criterion, holds that the risk of death to a human being outweighs similar risks to members of other species, regardless of the species. The second part, the human morbidity criterion, is less absolute.

Remarkably, with or without this anthropocentric criterion, using this framework to evaluate whether a global ban on GM crops would indeed be precautionary warrants the conclusion that a ban would, in fact, increase net risks to both global public health and the global environment. Thus, any version of the precautionary principle should actually require the use of GM crops, provided due caution is exercised before commercialization of individual GM crops (Goklany, 2001). But the result is at variance with conventional environmental wisdom. The significance of this result may be appreciated by considering that 800 million people worldwide suffer from hunger and undernourishment, and over 2 billion from malnutrition. Hunger and malnutrition annually kill millions of children worldwide. Further poor nutritional habits significantly contribute to diseases of affluence which kill almost 20 million more. To reduce the future toll of hunger, malnutrition, and poor nutritional habits, despite the almost inevitable future increase in human population, requires that the quantity and nutritional quality of food must be enhanced. The quicker this occurs, the fewer casualties there will be. And GM crops can increase the quantity and nutritional quality of food supplies faster than conventional crops (Goklany, 2002).

While a GM crop ban would retard reductions in global hunger, malnutrition, and diseases of affluence, the health effects of ingesting GM crops, if any, are not only much more uncertain, they are not now-and unlikely to be in the future-comparable in magnitude to the global toll from hunger and malnutrition. Therefore, a GM crop ban is likely to increase net harm to public health, pushing large numbers to premature death (Goklany, 2001).

With respect to environmental risks, we know that conventional agriculture, with its high demands for land, water, pesticides, and fertilizers, strongly stresses global biodiversity and it is a significant source of greenhouse gases. We can reduce these environmental pressures more rapidly-and more certainly-with GM crops than with conventional crops because the former are more likely to increase agriculture productivity (in terms of land and water) and to do so faster and with fewer or less toxic chemicals. According to Goklany, therefore, "Public health and environmental policies should attempt to minimize net risks to public health and the environment based on the best available scientific information and their net anticipated costs to society."

According to Stirling *et al.* (1999), the shortcoming of the classical risk-assessment approach for evaluating the commercial use of transgenic crops is due to the observation that the resulting technological risk can exist on several different levels of decision-making simultaneously: environmental, human health, agricultural management, economical, societal, and ethical. In the regulatory appraisal of risk, it is therefore difficult to adequately deal with the many levels of risk especially when no single units of performance are available. This is because the classical risk-assessment methodologies were developed to assess closed systems where there was a direct hazard, such as toxicity, pathogenicity, or flammability.

Most regulatory agencies require a full risk assessment of transgenic crops, case by case, before commercial release. For risk assessment both the genotype and phenotype of a transgenic organism has to be evaluated. For the genotype, this means the molecular characterization of the gene cassette introduced, consisting of promoter, structural gene, and terminator, as well as the number of copies actually integrated into the plant genome. But it is only when the plant characters are actually expressed that the plant interacts with the environment and can affect human health (Auberson-Huang, 2002).

Annex II of the Cartagena Protocol lays down the stepwise procedure of risk assessment; it consists of hazard identification in association with the phenotype, estimation of the likelihood that the hazard will occur, description of the consequences should the hazards occur, and determination of how far the overall risks are acceptable.

In the case of closed technological applications, information obtained during a risk assessment are highly useful for the design of suitable containment measures to prevent the inadvertent release of harmful chemicals or dangerous organisms into the environment. The risk-assessment methodology for the release of GMOs follows a stepwise procedure similar to that for closed systems, yet a very different approach is actually needed. During the risk assessment for the deliberate release of GMOs, we need to ask the question the other way around: once transgenic crops are growing in the environment, what would be the estimated damage potential from the spread of the genes from the transformed plant for wild flora, soil composition and microbial community, ecology, and human health? Some very urgent issues in the context of environmental and agricultural biotechnology are exemplified by the overuse of antibiotics for prophylaxis in animal husbandry or the effects of pesticide use on insect communities. The manure of livestock treated with antibiotics harbours large numbers of resistant bacteria. When this manure is recycled on cropland, farmers inadvertently inoculate their soil with antibiotic resistant bacteria. This warrants the need for examining the damage potential from the transfer of antibiotic marker genes, from transgenic plants to soil microorganisms; it becomes desirable to understand what gene transfers are already taking place and hence to re-examine the desirability of other agricultural impacts on the environment (Auberson-Huang, 2002).

The precautionary approach is an ideal policy as it leaves the door open to making decisions regarding technological applications as a matter of "choice," in the absence of scientific certainty. Environmental processes being highly complex, ecological awareness generates the ability to perceive the interdependence of all forms of life and life processes. Yet its vast dimensions extend beyond the level of an individual, beyond science, and beyond economics.

Risk assessment makes up the core in the regulation of transgenic crops. Its procedure and scientific methods are rigorous. The call for precaution neither contradicts the science of risk assessment nor does it set a requirement for zero risk. Instead, we can decide between precaution and risk assessment by considering variations that could occur in measured characteristics or properties; possible flaws in the methodology used to assess effects; and the position of the baseline level of significance or insignificance (Auberson-Huang, 2002).

There is a difference between decisions based on reason (risk assessment) and those based on choice (precautionary approach); only the latter is amenable to judicial review. According to Christoforous (2000), the task of the members of the dispute-settlement system should be to decide whether a contested measure is based on a risk assessment, not whether the scientific theory upon which the conclusions of the risk assessment are based is scientifically correct and acceptable.

The Cartagena Protocol on Biosafety will be the first legally binding international agreement to honour the precautionary approach as a policy instrument for obeying collective ethical values, such as the conservation and sustainable use of biological diversity and the protection of human health-both relevant to the transboundary movement of GMOs. Risk assessment is the primary tool under the Protocol for concerned authorities to evaluate the potential adverse effects of living modified organisms on the conservation and sustainable use of biological diversity in the likely potential receiving environment, while duly considering risks to human health (see Auberson-Huang, 2000).

Scientific, Political and Ethical Dimensions of Genetically Modified Organisms

In the United States the decade of the 1980s was marked by a laissez-faire regulatory environment that coincided with novel technologies and new investment emerging from the private sector resulting in prototype GMO and GM food products (Flavr Savr tomatoes and virus-, herbicide-, and insect-resistant crops). These new products were in a sense thrust into existing regulations at the EPA, the USDA, or both, with only perfunctory overview from the FDA. It is ironical that whatever regulatory oversight these early products attracted was actually demanded by industry!

Genetically modified (GM) foods are those that contain ingredients from GMO crop varieties, though there is continuing debate as to whether there should or should not be a statistical dimension (regarding "tolerance" of the maximum permissible amounts of one or another GM ingredient) in defining what are and are not GM foods. Also the definition-even the usage-of GMO is debatable. Buttel and Goodman (2001) prefer the terminology "genetically engineered crop".

In the 1980s and 1990s, several major chemical companies became "life sciences" companies. Most of these companies were the commercial proponents of GMO crops. They decided to reject labeling of their products and failed to educate the public about the merits of the products. They also ignored the early signs from Europe of a developing controversy.

By 1999, agricultural biotechnology in the United States had acquired much momentum. As of the 1998 growing season, about 36 percent of U.S. soybean acreage, 20% of U.S. cotton acreage, and 22% of U.S. corn acreage was devoted to genetically engineered varieties, and about 60% of Canadian canola acreage was devoted to genetically engineered varieties (James, 1998). The adoption rates for American GMO soybeans, corn, and cotton and for Canadian canola were very impressive. The controversy over recombinant bovine somatotropin (rBST; also known as recombinant bovine growth hormone, or rBGH), which had been the United States's most contested new biotechnology product, had largely fizzled out by late 1998, and in retrospect nothing that could have been learned about corporate approaches to public concerns over agricultural technology was in fact learned (Buttel and Goodman, 2001).

By 1999, the European Union (EU) had approved three *Bacillus thuringiensis* (*Bt*) corn varieties, one herbicide-tolerant corn variety, and at least six additional *Bt* and herbicide-resistant crop varieties. Two "stacked" (both *Bt* and herbicide-resistant) varieties were under regulatory review. The authority of the World Trade Organization (WTO), which included a number of provisions on intellectual property, nontariff barriers to trade, and the harmonization of national standards of food regulations that were favorable to commercial biotechnology, was respected and the WTO became increasingly well institutionalized. WTO rules required the EU countries to not only approve these new agricultural input products but also to accept imports of GM grains and oilseeds products.

Later in the same year, however, a transnational social conflict erupted over GMOs. The EU began to restrict imports of GM corn and soybeans and imposed de facto moratorium on approval of new GM input products. The Seattle protests at the 1999 WTO ministerial meeting were galvanized around consensus among environmental, labour, consumer, sustainable agriculture, and human rights groups that there should be resistance against GM foods and more importantly, against WTO rules that limit the ability of nations and consumers to choose not to consume GM food ingredients (Buttel and Goodman, 2001). By 2000, the resistance to GM foods had spread to Japan, Korea, Thailand, Australia, and India. The precise definition of "GMO" and "GMO food" has generated some debate. Some observers-including even many of the most active proponents and opponents of molecular biological technologies used in agriculture-regard GMOs as being synonymous with "agricultural biotechnology." Biotechnology happens to be a broad term that covers a suite of conventional methods-including tissue culture-as also more modern techniques based on molecular biology used for enhanced management of plant-breeding programs and in diagnosis of crop diseases and stresses. These biotechnology methods are not what the GMO controversy is about. Rather, the focus of the controversy is on crop varieties and the foods derived therefrom that have been developed with the use of genetic engineering. Genetic engineering refers to the construction of genes engineered from recombinant DNA made in the laboratory and introduced into the chromosomes of a crop plant. Such genes, collectively called transgenes, when expressed in the recipient plant confer a new trait or property on the plant (Buttel and Goodman, 2001).

Currently, most GMO crop plants contain single-gene (or a small number of) transgenes that confer two major types of traits: There are the *Bt* crops (chiefly corn, cotton, and potatoes) that, as a result of expression of a gene taken from *Bacillus thuringiensis*, have become resistant to insect pests, and herbicide-resistant (HR) crops (chiefly soybean, corn, and canola) engineered using bacterial or modified plant genes. Virus-resistant crops also come within the above definition of GMOs-they involve one or a few transgenes that encode proteins that affect input traits. Virus-resistant crops have not been so controversial, partly because they were not adopted rapidly and have not been the commercial blockbuster products that *Bt* cotton and corn and HR soybeans have been. Also, most environmental and related organizations consider virus-resistant crops to be fairly environmentally benign.

One unique aspect of the GMO controversy has been that it was essentially prompted by international trade and by the WTO's rules governing trade. European resistance to GMOs was spear-headed by the realization in Europe and EU that adherence to WTO rules would result in a widespread presence of GM foods in the European food supply. The genesis and momentum of the GMO controversy lay in WTO rules and European reactions to these trade rules.

Another distinctive aspect of the GMO controversy is that it is global and has become a North-South and international development controversy. As the prospects for food-production increase in the developing world, the voices of the contending parties tend to become louder when they discuss whether GMOs-or biotechnology in general-will be beneficial for the developing world.

Much of the North-South GMO debate has centered on the "golden rice" issue. GMO proponents tout the potentials of golden rice-a transgenic rice containing one daffodil gene and two bacterial genes that together encode an increased level of provitamin A-for its being able to reduce the incidence of night blindness and other disorders that lead vitamin A deficiency to be associated with elevated rates of childhood mortality. In contrast, GMO opponents suggest that golden rice will probably never be deployed in a widespread manner because it is covered by numerous patents, many of which may preclude its commercialization. Also, according to opponents, golden rice cannot be a good solution to the problem of poverty and homogenization of the food supply. Poor villagers do not need golden rice as much as they need social arrangements that enable them to diversify their production systems and to have access to balanced diets containing sufficient vitamin A.

While a large number of agricultural scientists and researchers support GMOs, a smaller but nevertheless impressive-sized group of agricultural scientists and ecologists has expressed concerns about GMO technology. This minority of scientists is concerned that the methods and regulatory procedures for determining the environmental risks of these technologies are inadequate, and that these technologies may already be showing grave environmental (as well as socioeconomic) problems such as weed resistance to herbicides, genetic drift to wild and weedy relatives, and insect resistance to *Bt*. A third group includes many other scientists whose views about GMOs are ambivalent-they do accept the importance of molecular tools in agricultural research and see some advantages to GMOs, but they also feel that the current generation of GMO products has shortcomings and that public opposition to GMOs carries the risk of antagonising the public on agricultural research as a whole.

According to Borlaug (2000), ultimately the central issues in evaluating the matter of biotechnology and the future of global agriculture relate to the relative safety of GM crops for humans and the environment, and the fact that the future food security status of the poor, the peasants and the urban slum dwellers of low-income countries depend on pursuing biotechnological research. Borlaug points out that GM crop varieties do not differ significantly from conventional or nontransgenic ones, and that the new GMOs are as safe as, and in some ways even better than, conventional varieties on human health and ecological grounds. But Borlaug's most direct comments are on the role that biotechnological research and GMO technology will have to play in winning the race against population growth in the developing world, and on the related topic of the environmental and other activist groups that, according to him, are impeding the pursuit of food production innovations needed by the poor.

In contrast to Borlaug's confidence in biotechnology and his conviction that the future well-being of billions of the world's poor people depends on rapid development of these novel technologies, Kirschenmann (2001) has raised several concerns about GMOs and biotechnology. GMO technology, according to him, is rooted in an ideology of biological determinism, which sees agricultural problems and their necessary solutions primarily in genetic terms, and in terms of "quick fixes." Not only does this ideology tend to lead to de-emphasis on ecological and social risks; biological determinism tends to underplay such promising alternatives as agroecological approaches that use, rather than suppress, biological and habitat diversity (Buttel and Goodman, 2001).

According to Burkhardt (2001), although the scientific and legal-political dimensions of GMOs have become quite important and are widely discussed, the GMO issue is ultimately an ethical one–*i.e.*, whether GMOs and GM foods are ultimately morally and ethically acceptable.

Borlaug (2000), Kirschenmann (2001), and Burkhardt (2001) all refer frequently to the question of whether GMOs and GM foods are critical to economic development and food security in the low-income countries of the South. Ruttan (2001) addresses this issue proactively. It has become quite typical that those actively involved in the early stages of the Green Revolution usually support GMOs as well as biotechnology research and development more generally, because of biotechnology's promise in generating sustained agricultural productivity improvement in developing countries. Ruttan (2001) discusses how important it will be to achieve new trajectories of productivity and output improvement in agriculture in the South. According to him, however, it is an open question as to whether GMO-type technology will have adequate potential to remove the current physiological constraints to yield increase that are now emerging in crop agriculture across the developing world (see Buttel and Goodman, 2001).

The implications of GMOs and GM foods for international development are some of the most forceful, social and ethical issues in evaluating these new technologies, but there are also crucial domestic policy dimensions of GMOs. In some ways, GMO technology resembles that of rBST in the dairy sector, both technologies being apparently scale-neutral because the input product can be purchased in either small or large lots and can be used on farms ranging from very small to very large. At the same time, it appears that both technologies are much more applicable to large-farm operations, suggesting the likelihood that GMOs as well as other agricultural biotechnology products will benefit larger farmers more than smaller ones.

It is generally agreed now that the GMO/GM foods issue should ultimately be resolved on ethical grounds, or on grounds of the public good, even though different experts have varying views about how ethical and public good considerations should be weighed. Also, while some countries address GMO/GM foods policy issues in terms of domestic considerations, in reality these issues are, by their nature, global. The GMO/GM foods issues cannot be approached solely in terms of group or national interests, since the welfare of much of the rest of the world depends on the quality of the judgement the developed nations such as USA and the international community will make.

Transgenic Crops and Small Farmers

Two-acre peasant farmers are the primary producers of staple foods, accounting for very high percentages of national production in most Third World countries. So important for food production, this sector itself is characterised by poverty and hunger, and in some cases lagging agricultural productivity. Solving these problems requires a clear understanding of their causes. If the causes lie in inadequate technology, then a technological solution (such as genetic engineering) is a distinct possibility, otherwise not.

Historical Background

The history of the third world since the advent of colonialism has been a history of un-sustainable development. Colonial land grabs displaced rural food producing societies away from the fertile farming lands - lands that were converted to production for export in the new global economy dominated by the colonial powers. Instead of producing staple foods for local populations, they became extensive cattle ranches or plantations of rubber, sugar, cotton and other highly valued products (Rossett, 2001).

Farming people-accustomed to the continuous production of annual crops on fertile, well-drained soils with good access to water-were driven into marginal areas. Forests were cut and many fragile habitats were subject to unsustainable production practices, in this case by poor, newly destitute and displaced farmers. The favoured lands were, simultaneously, being degraded by continuous export cropping at the hands of Europeans.

National liberation from colonialism did not help to alleviate the above environmental and social problems. Post-colonial national elite came to power with strong linkages to the global export-oriented economy, often connected to former colonial powers. The period of national liberation, corresponded with the rise of capitalist market and production relations on a global scale, and in particular, with their penetration of developing countries' economies and villages. This era of modernisation had its dominant ideology of "bigger is better". In villages small farms were consolidated into larger farms that could be mechanised. It generated the notion that the "backward and inefficient" peasantry should abandon farming and migrate to the cities where they would provide the labour force for industrialisation. This ushered in a new era of land concentration in the hands of the wealthy, and drove the growing problem of landlessness in rural areas. The landless rapidly became the poorest of the poor, subsisting as part-time seasonal agricultural or day labourers, share croppers or migrating to the agricultural frontier to fell forests (Rossett, 2001).

Thus rural areas in the Third World are today characterised by extreme inequalities in access to land, in security of land tenure and in the quality of the land farmed. Ironically, food and other farm products flow from areas of hunger and need to areas where money is concentrated, in the North.

The situation is no better in the more favourable lands where even the better soils have been concentrated into large holdings used for mechanised, pesticide and chemical fertiliser-intensive, monocultural production for export. Many of our planet's best soils are being rapidly degraded, and in some cases abandoned completely, in the short-term pursuit of export profits and competitiveness. The productive capacity of these soils is declining rapidly due to soil compaction, erosion, waterlogging, and fertility loss, together with growing resistance of pests to pesticides and the loss of in-soil and above-ground functional biodiversity. The threat of the growing problem of "yield decline" in these areas looms large, adversely affecting global food production.

Changes in Macro-economic Policies

The past half a century has witnessed several changes in national and global governance mechanisms - changes that were made within a paradigm that saw international trade as the key resource of promoting economic growth in national economies, and growth as the solution to all ills. The balance of governance over national economies has tended to shift away from governments toward market mechanisms and global regulatory bodies like the WTO. Southern governments have progressively lost the majority of the management tools in their macro-economic policy toolboxes. The ability of Southern nations to ensure the social welfare of poor and vulnerable people, achieve social justice, guarantee human rights, and protect and sustainably manage their natural resources, has been severely eroded (Rossett, 2001).

Lagging Productivity

Third world food producers have suffered lagging productivity not because they lack 'miracle' seeds that contain their own insecticide or tolerate massive doses of herbicide, but because they have been displaced onto marginal, rain-fed lands, and have to contend with macroeconomic policies that are more and more unfavourable to food production by small farmers. In fact, in many parts of the third world, farmers today produce far less than they could with presently available know-how and technology, because there is no incentive for them to do otherwise-there are only low prices and few buyers. No new seed, good or bad, can change that, and thus it is very unlikely that, in the absence of urgently needed structural changes in access to land and in agricultural and trade policies, genetic engineering could enhance food production by the world's poorer farmers (Rossett, 2001). This is because genetic engineering does not address the primary constraints they face.

Diverse, Risk-prone Agriculture

Peasant farmers have historically been displaced into marginal zones characterised by broken terrain, slopes, irregular rainfall, little irrigation, and/or low soil fertility. Their agriculture is best characterised as complex, diverse and risk-prone. To survive under such circumstances, and to improve their standard of living, they should be able to tailor agricultural technologies to their variable but unique circumstances, in terms of local climate, topography, soils, biodiversity, cropping systems, market insertion, and resources. For this reason such farmers have over millennia evolved complex farming and livelihood systems which balance risks. Typically, their cropping systems involve multiple annual and perennial crops, animals, fodder, even fish, and a variety of foraged wild products (Rossett, 2001).

The Erroneous Top-down Research

Such farmers have rarely benefited from 'top down' formal institutional research and 'green revolution' technologies. Any new strategy to truly address productivity and poverty concerns will have to meet their needs for multiple suitable varieties. Formal research which values monocultural 'yield' above all else cannot handle the vast complexity of physical and socio-economic conditions in most third world agriculture. Seeds have multiple characteristics that cannot be captured by a single yield measure, and farmers have multiple site-specific requirements for their seeds, not just controlled-condition high yields. These interconnections are in sharp contrast to formal breeding procedures. In the face of such conditions, a different approach-participatory breeding by organised farmers themselves-which takes into account the multiple characteristics of both seed varieties and farmers, is essential. Genetic engineering stands in direct contrast to participatory, farmer-led research.

Proponents of genetically engineered varieties are repeating the very 'top down' errors which led first generation green revolution crop varieties to have low adoption rates among poorer farmers. Yet, the biotech movement is racing ahead. What then, are the risks associated with 'forcing' genetically engineered varieties into complex, diverse and risk-prone circumstances?

The commonest transgenic varieties currently available are those that tolerate proprietary brands of herbicides, and those that contain insecticide genes. Herbicide tolerant crops do not mean much to peasant farmers who plant diverse mixtures of crop and fodder species; chemicals would only destroy key components of their cropping systems.

Transgenic plants, which produce their own insecticides-usually using the '*Bt*' gene-are rapidly failing as pests build up resistance to insecticides. Instead of the failed "one pest-one chemical" model, genetic engineering adopts a "one pest-one gene" approach, shown over and over again in laboratory trials to fail, as pest species rapidly adapt and develop resistance to the insecticide present in the plant. *Bt* crops violate the basic principle of integrated pest management (IPM), *viz.*, that reliance on any single pest management technology tends to trigger shifts in pest species or the evolution of resistance. In general, the greater the selection pressure across time and space, the quicker and more profound the pests' evolutionary response. Thus IPM approaches employ multiple pest control mechanisms, and use pesticides minimally, only as a last resort. Adopting this principle reduces pest exposure to pesticides, slowing the evolution of resistance. But when the product is engineered into the plant itself, pest exposure jumps from minimal and occasional to massive and continuous exposure, dramatically accelerating resistance. Many entomologists feel that *Bt* will rapidly become useless, both as a feature of the new seeds and as an old standby natural insecticide sprayed when needed by farmers that want to exit from the pesticide treadmill.

The use of *Bt* crops also affects non-target organisms and ecological processes. The *Bt* toxin can affect beneficial insect predators that feed on insect pests present on *Bt* crops, and windblown pollen from *Bt* crops found on natural vegetation surrounding transgenic fields can kill non-target insects. Small farmers rely for insect pest control on the rich community of predators and parasites associated with their mixed cropping systems (Altieri *et al.*, 1998; Rossett, 2001).

In fact *Bt* retains its insecticidal properties after crop residues have been plowed into the soil, and is protected against microbial degradation by being bound to soil particles, persisting in various soils for at least 230 days. This is a matter of serious concern for poor farmers who cannot purchase expensive chemical fertilisers, and who instead rely on local residues, organic matter and soil microorganisms (key invertebrate, fungal or bacterial species) for soil fertility, which can be negatively affected by the soil bound toxin.

When the *Bt* genes fail, what would poor farmers be left with? They would possibly face the serious rebound of pest populations freed of natural control by the impact *Bt* had on predators and parasites, and reduced soil fertility because of the impacts of *Bt* crop residues plowed into the ground.

In the Third World there will typically be more sexually compatible wild relatives of crops present, making pollen transfer to weed populations of insecticidal properties, virus resistance, and other genetically engineered traits more likely, with possible food chain and superweed consequences. With massive releases of transgenic crops, these impacts are expected to scale up in those developing countries which constitute centres of genetic diversity.

Thus, the various risks outweigh any potential benefits for peasant farmers, especially when we consider the factors that currently limit their ability to improve their livelihoods, and the proven agroecological, participatory and empowering alternatives available to them (Rossett, 2001; Altieri and Rossett, 1999).

It is not a lack of technology which holds 2-acre farmers back, but rather injustices and inequities in access to resources, including land, credit, market access and other anti-poor policy biases. Two approaches make the most sense under such conditions: 1. technologies, which have pro-poor diseconomies of scale, like agroecology; and 2. organisation into social movements capable of exerting sufficient political pressure to reverse policy biases. There is little useful role that genetically engineered crops can play.

Liabilities and Economics

Two major threats to current crop production practices in intensive agriculture are the cross-pollination of GM crop varieties with conventional varieties, and the germination of volunteer GM seeds (seeds dropped, blown, or inadvertently planted). Cross-pollination of GM crops with crop varieties destined for non-GM, GM-free, or "organic" niche markets is generating concern to the corresponding producers (Smyth *et al.*, 2002).

The first generation of GM crop products derived from canola, corn and soybean have been available in Canada, USA and some other countries for some seven years. These products were marketed when regulatory control was minimal. But the imminent introduction of next-generation GM crops with novel uses is likely to face restructured regulatory systems, radically altered marketplaces, and new technology options (Phillips and Khachatourians, 2001).

Currently, the agrochemical industry has to incur heavy expenditure on development and commercialization costs and GM-related technologies. Corporations and regulators have also to ensure that the new traits and varieties created do not pose risks or liabilities that offset the value generated. At the farm level there is significant risk of profit reduction and for co-mingling of transgenic plants with other crops, creating potential new liabilities.

In general, first-generation, "input-trait" GM crops (*i.e.* traits conferring agronomic benefits) have been regarded by regulators as substantially equivalent to existing varieties and so were allowed to be introduced into many of the existing commodity food systems without segregation. But some of these GM crops can potentially cross-pollinate with other compatible crops of the same species or with weedy relatives, or to become volunteers in other crops, creating potential new environmental or crop-weed management risks which may offset the crops' benefits. Second-generation crops (nutraceuticals traits with health and nutritional benefits), will sell only if their purity or quality can be guaranteed-which is problematic due to the difficulty of ensuring gene containment. Third-generation crops with new industrial, nutraceutical, or pharmaceutical properties will require effective gene control systems, or even may not be permitted to be released.

According to Smyth *et al.*, the current intellectual property (IP) protection does not fully control the use of transgenes once they are expressed in seed. Many GM crops can be propagated in subsequent years with seed from previous years. Although regulations and market pressure may help to prevent some gene transfers, several genes may continue to be dispersed and survive beyond a season. Regardless of how effective regulations or contracts are, some producers (either intentionally or inadvertently) can misappropriate these new technologies and thereby create new risks and liabilities. Further, even if patent violation could be controlled, many plant species are promiscuous sexually and create natural gene flow to related species (Smyth *et al.*, 2002).

The liability cost of genes from GM crops escaping and becoming rogue, or co-mingling and harming the quality of other plant-based products, is significant.

Smyth *et al.* identified four important liability issues that can at least be partially quantified: first, the potential for volunteer GM seed to germinate the following year; second, the potential for pollen flow from GM crops to non-GM crops; third, the potential for co-mingling of GM and non-GM crops, and fourth, the potential for environmental risks associated with uncontrolled gene flow from GM varieties into related plants to block export of GM varieties to countries that do or do not adopt the new technologies.

There are two ways in which gene flow from GM plants can create liabilities. First, through conventional harvesting which leave behind some seeds which germinate in the next season and often create a tolerance-level liability. Second, the pollen of a GM plant may fertilize a non-transgenic plant and the resulting hybrid seed might possess the trait for that transgene.

In Canada many canola farmers have had to contend with the problem of managing volunteer canola (CCC, 2001). Most farmers planting transgenic herbicide-tolerant canola in 2000, said that the difficulty of managing volunteer transgenic herbicide-tolerant canola was about the same as that of volunteer conventional canola.

While about 90% of corn pollen is deposited within five meters of the edge of cornfields, canola pollen is dispersed as far as 20 km. Canola, particularly the *Brassica rapa* variety, is an open-pollinating crop; this has aroused considerable concern about canola genes escaping into related wild species. Even in the self-pollinating *Brassica napus* (also canola), frequently up to 50% of its pollen is dispersed. It has been estimated that any infestation of herbicide-resistant wild mustard above four plants per square meter would reduce the benefits of transgenic herbicide-tolerant canola to below zero (see Smyth *et al.*, 2002).

Though there are several uncertainties, it seems certain that volunteer plants and cross-pollinated varieties will be co-mingled in the commodity food system. Several surveys have shown substantial and rising consumer preferences for organic and non-GM crops or plant products. Consumers wish to know what they are eating and they wish to discriminate between organic, GM-free, and GM foods and ingredients but precise discrimination is not yet scientifically possible. Control of GM cross-pollination and volunteer GM seed will be essential to build consumer trust in products labeled non-GM and organic. Co-mingling of GM and non-GM seeds has already imposed significant costs on the food industry. There are several examples where contaminations have jeopardized entire product lines. The introduction of transgenic herbicide-tolerant canola in western Canada destroyed the market for organic canola.

The likelihood of outcrossing and pollen flow has deterred buyers buying organically produced western Canadian canola because it might contain transgenes.

The European Union has already banned import of Canadian honey because it may contain some pollen from GM canola plants.

Finally, in many countries IP protection mechanisms are ineffective. This makes these countries unattractive as markets for new technologies, and they lag in the adoption of new traits and varieties. In the canola sector, few companies export new cultivars to China or India because of lack of effective IP protections there.

Despite subsidies for irrigation and fertilizer, India and China have average canola crop yields almost 40% and 3% lower than Canada's, respectively (Smyth *et al.*, 2002).

Institutional control, biological control or a combination of both is an essential pre-requisite for managing the risks of GM crops. The public sector evaluates new GM crops for safety considerations by examining the new products against existing products to determine whether they bring any new risks related to human consumption, the environment, or livestock.

Most regulators can intervene if an unexpected risk is detected. The private sector is generally responsible for managing the risks of new GM products once they enter the market; it uses a combination of contracts, testing, and auditing to ensure compliance.

But these instruments cannot keep all the risks in check-in the field, genes can be transferred and/or transported. Regardless of how effective regulations or contracts are, some actors do either deliberately or inadvertently misuse new technologies, thereby creating potential new risks and liabilities.

Crops and weeds have been exchanging some genes for centuries. Genetic engineering now raises concerns because it not only enables introduction into ecosystems of genes that confer novel fitness-related traits, but also allows novel genes to be introduced into many diverse types of crops each with its own specific potential to outcross (Snow, 2002). Most cultivated plants mate with one or more wild relatives in some area of their geographic range. Several crops naturalize and persist as feral weed populations (Holm *et al.*, 1997). Hence, newly introduced genes may potentially disperse into neighbouring populations, bringing along new phenotypic traits such as resistance to insects, diseases, herbicides, or harsh growing conditions. Better knowledge of the impact of crop gene introgression into populations growing on roadsides, field margins, or uncultivated areas is needed.

A gene being a gene, transgenes disperse and become incorporated into the genomes of other species in the same manner as other crop genes. Gene flow can be surprisingly widespread. New cases of crop-to-wild gene flow are still being discovered and crop alleles can persist in weed populations for decades. Certain crops (such as oilseed rape) can pass genes to a wild relative even when those genes are carried on unshared (non-homologous) chromosomes. Some commercially important grass species can hybridize with nearby congeners and then switch to asexual seed production (apomixis), allowing crop genes to spread widely, even when F1 hybrids are sterile (Snow, 2002).

In some cases, novel genes spread to free-living plant populations and can potentially create or exacerbate weed problems by providing novel traits that allow these plants to compete better, produce more seeds, and become more abundant.

It has so far proved impossible to prevent gene flow between sexually compatible species in the same area. Pollen and seeds disperse too easily and too far to make containment feasible.

It appears that current political and societal pressures are likely to lead to more stringent regulation of future GM varieties. Regulators will have to choose between either outright rejection or imposition of regulations for detailed production and market segregation.

The first wave of GM products remained largely unnoticed by consumers and much of the food industry. This pattern will not be repeated for second- or third-generation products. The costs of denying the risks are potentially enormous. Similarly, control mechanisms are not cheap.

Many of the risks and potential liabilities of GM crops are only partially manageable. Although institutional costs to manage risks are quite high, the cost of failure is even greater. Inability to contain the risks and control the liabilities will reduce net returns on investments so much that GM technology may become infeasible. This issue may be addressed by governments improving the regulatory oversight of GM crops, aggressively pursuing the use of refugia, contract registration, regional regulation, and mandatory crop rotations and audits (Smyth *et al.*, 2002; Khachatourians *et al.*, 2002).

Containing the GM Revolution

Roundup Ready (RR) canola was patented by Monsanto in 1993 as a herbicide resistant crop-resistant to Monsanto's very successful herbicide Roundup. Used in combination with Roundup, RR canola was claimed to increase farmers' yields and (over time) lower their input costs by eliminating weeds and plants that might inhibit the plants' growth. This canola was introduced commercially into Canada in 1997. But the plant's *in situ* behaviour has created difficulties for farmers growing canola from their own selected and saved seed and for organic farmers.

Monsanto places strict constraints on farmers who decide to use the RR product. These constraints are contained in Technology User Agreements (TUAs) which farmers have to sign before obtaining and using the product. Farmers must agree to pay Monsanto a royalty fee of Can $15 per acre. They must not save and replant any of the Roundup Ready seeds, only use Roundup on their canola crop and allow Monsanto free access to verify compliance with the terms of the TUA for a period of three years (Phillipson, 2001).

Corporate Control

For millennia, farmers have been saving seed from one year's crop to plant the following year. The proliferation of patents over genetically modified seeds now threatens that practice. Under the terms of a TUA, farmers cannot save seed and must return to the patentee each year for a new supply, thus becoming fully dependent on these corporations for their seeds.

As more and more conventional plant varieties are genetically modified and patented, these corporations (as opposed to farmers) will exert an ever-increasing level of control over the overall seed supply and on the variety of crops grown. Corporations will choose to market seeds of GM crops that they deem to be commercially viable, rather than allowing farmers to determine the types of crops they wish to grow. All this will lead to higher seed prices-a disaster scenario for the developing world. Higher seed prices are unthinkable in a context where food security is already precarious. In the developed world also increased input costs in the guise of higher seed prices will drive small farmers off the land, leaving farming to large-scale corporate farmers. Such an industrial model of rural development will only increase the control of large multinationals over rural economies and communities.

A huge power shift is now taking place in global agriculture. Traditional farming practices are being eclipsed by the extension of patent monopolies to the fundamentals of food production. A relatively small number of large corporations are gaining increasing control over certain fundamentals of agricultural production.

The problem of "volunteer" GM plants (*i.e.*, the invasion of some GM plant materials/pollen in a crop field that was not planted with seed of the GM variety) is of particular concern to organic farmers. GM crops are incompatible with organic certification systems and if an organic farm is contaminated with "volunteer" GM crops, certification will be lost. Loss of certification spells commercial ruin as the price premiums enjoyed by certified organic farmers depend on their being able to conform to stringent conditions of production. Such premiums are necessary compensation for the increased labour costs and lower yields associated with organic production.

Initially, the legal debate must address the issue of whether the manufacturers of GM crop systems owe any obligation for damage caused by the unwanted proliferation of these crops. Future debate should pinpoint the obligations of the governments that grant patents over (and approve the sale of) GM crop systems. Some governments (such as those in the USA, Canada and Argentina) are enthusiastic supporters of GM technology. The extent to which this support manifests itself in the facilitation of market access (for GM crops) via the elimination of regulatory hurdles should be examined in the context of potential future liability (Phillipson, 2001). According to Phillipson, the emergence of GM agricultural technology represents a major challenge to traditional agricultural practices and to the nature of farming itself. But civil society also should have a role in determining the manner in which such technology is introduced and adopted, if at all.

Bt Maize

Maize (*Zea mays*) has been a primary focus of agricultural research on the African continent since the Green Revolution. In Kenya, maize production system relies primarily on smallholder agriculture that uses minimal inputs and open-pollinated varieties (OPVs) of seed. Large-scale maize production, characterized by monocultures of hybrid seed purchased each season, is only common in parts of the Rift Valley, Western Province, and a few other areas.

Besides drought, stresses facing maize farmers, both in Africa and Asia, can include pests and diseases, poor soil fertility, high costs of inputs and low maize prices. The significance of these factors varies across agro-ecological zones. Attack by stemborers and other insect pests is a major constraint on maize production. In Kenya, stemborers, including *Chilo partellus, C. orichalcociliellus, Busseola fusca, Eldana saccharina* and *Sesamia calamistis,* are estimated to cause losses of around 15% and in some areas are recognized as the most severe pest problem facing maize production. The insects are most destructive in the larval stage when they tunnel inside the stalk after hatching and are therefore very difficult to control. Once inside, their feeding may lead to reductions in the number of ears or structural damage increasing the likelihood of the plant or the cob toppling down when a strong wind blows. In some cases the pests also attack maize ears making the cob vulnerable to cob rots, such as *Aspergillus* (fungi), which produce harmful aflatoxins.

Conventional methods of stemborer control that employ chemicals or biopesticide sprays, including those based on *Bt*, have not always been effective owing to the challenge of correctly timing these applications and the resulting difficulties in eradicating the pest once it has infested the crops. Very few farmers use these techniques, even in nations practising industrial maize production.

Formulations consisting of the spores of *Bacillus thuringiensis (Bt)* and isolated toxins have been developed as natural pesticides. *Bt* sprays have become a popular biopesticide mainly due to their extremely specific spectrum of action, relative safety towards humans and low environmental persistence. *Bt* formulations are therefore being recognized as insecticides approved for use on organically certified crops. In Kenya, local *Bt* strains have proved effective against *Chilo partellus, Busseola fusca* and *Sesamia calamistis* (Mwangi and Ely, 2001).

Recombinant DNA technologies have made possible the insertion of the *Bt* genes responsible for production of insecticidal toxins into the maize genome. The resulting *Bt* maize plant produces the toxins throughout the various tissues over its life cycle. In the

USA, *Bt* genes that are effective against the European cornborer have been identified and successfully incorporated into maize. European cornborer larvae that penetrate the plant tissues are killed when they ingest the toxin produced in the *Bt* maize cells. The technology results in increased yields and a reduced need for insecticide sprays. By 2000, *Bt* maize constituted a significant proportion of overall maize plantings in the USA, Argentina and Canada. In Europe and elsewhere, adoption of this technology has been slower due to uncertainty over the risks that it may pose to the environment or to consumer health. Austria has blocked imports of *Bt* maize from the time of its initial acceptance by the European Commission in 1996; it is one of the few countries in the European Union where no GM crops have yet been cleared for release.

France and Portugal approved *Bt* maize for environmental release, but later moved towards a more precautionary approach. The primary adverse effects to be borne in mind are the following:

1. Effects on non-target organisms (typified by conclusions drawn from initial studies on the monarch butterfly), on natural enemies of stemborers such as the green lacewing *Chrysoperla carnea* and on soil biota.
2. The possibility that spread of the transgene to neighbouring conventional crops might lead to economic losses to farmers wishing to preserve non-GM identity.
3. Rapid development of resistance among insect pests toward the *Bt* toxins used, potentially leading to ineffectiveness of the *Bt* maize and the related *Bt* biopesticide.
4. Effects of the presence of antibiotic resistance genes used in the production of some *Bt* maize varieties, which were feared to cause increases in antibiotic resistance among clinically significant bacteria.

Some recent European Directives have aimed to ensure labelling and traceability of all GM ingredients throughout the market introduction process and to establish a minimum 1% threshold for transgenic DNA and protein, below which products need not be labelled (Mwangi and Ely, 2001).

Based on mammalian toxicity tests and the concept of substantial equivalence, several strains of *Bt* maize have been declared safe for human consumption by the United States Food and Drug Administration. Unlike food safety, however, environmental safety is region-specific and requires region-by-region evaluation and monitoring. The Cartagena Biosafety protocol and national regulations require their assessment prior to release.

Whereas environmental release in the developed world can usually be recalled (as illustrated by Starlink *Bt* maize in the USA), in areas where smallholder agriculture, open pollinated varieties and seed-saving and exchange are more common, releases are more likely to be irreversible, so that local ecological risk assessments are even more critical.

Maize is a highly domesticated crop whose wild relatives occur only in Central America. Gene-flow leading to weediness is not therefore a significant risk for maize in either Africa or Europe. However, there are general concerns over contamination of well-adapted landraces of maize that have evolved since the species introduction in East Africa. These landraces are a vital source of genetic diversity for breeding locally adapted varieties and should be conserved (see Mwangi and Ely, 2001). Potential loss of landraces can be addressed by conservation of seeds at a national gene bank and the use of these landraces in the development of improved varieties. *In situ* conservation without contamination poses a strong challenge, since maize is wind pollinated with long dispersion distances. This, in the context of predominantly smallholder agriculture, makes it practically impossible to ensure noncontamination of landraces with the *Bt* gene.

Prolonged exposure of target insects to *Bt* toxins can potentially enhance the development of resistance because of increased selection pressure. The high dose/refugia strategy has been the preferred method of insect resistance management in large-farm areas such as the USA, where no significant insect resistance has been reported since licenses for growing *Bt* maize were first awarded in 1995. The strategy adopted in each case depends on the crop in question and on the various regulatory and socioeconomic conditions prevailing in the area. Options include licensing, labelling, central control of seed and continual monitoring of use of *Bt* seed and resistance development. Informal seed distribution systems and the possibility of introgression of *Bt* genes into local OPVs can reduce the effectiveness of labelling options and make enforcement of refuges very difficult.

Despite these difficulties, traditional mixed farming with non-*Bt* crops such as sorghum and pearl millet (also hosts to maize stemborers) may provide some refugia. Furthermore, many indigenous grasses (*e.g. Cenchrus ciliaris, Panicum maximum, Pennisetum purpureum* and *Echinochloa* spp.), which occur as weeds in maize fields are alternative hosts to many stemborers and could also act as natural refugia. It is important to note however, that in order for these to be effective, maize farmers must adjust their cultivation practices to ensure that stemborers can emerge from these hosts before the crop is harvested. Incorporation of natural pest resistance through conventional breeding into the *Bt* plant (termed pyramiding) on the other hand, is an approach to insect resistance management that requires little adaptation of farming systems.

As both the improved maize varieties and the local ones are being planted, depending on the efficiency of artificial selection (preference by farmers) and natural selection for the *Bt* trait including that resulting from pollen competition, it seems likely that the toxin genes may quickly spread to the fields and affect even the saved seed stores of farmers who use these varieties and who had never purchased the genetically modified seeds. In this instance, *Bt* maize would therefore be a possible approach to stemborer control that requires no adaptation of farming methods or extra external inputs by farmers. However, these benefits will obviously come at the expense of individual choice on the part of both farmers and consumers and may lead to further loss of maize biodiversity.

Socioeconomic assessments surrounding intellectual property, farmers' rights and equitable benefits of the technology also should not be excluded in preference of environmental issues. Both socioeconomic and environmental risks need to be considered critically to assesses the GM technology against existing alternatives.

The U.S. Environmental Protection Agency (EPA) has granted approval for genetically engineered *Bt* corn for seven more years, despite serious questions about the dangers the GE crops pose to human health or the environment.

Bt plants produce an insecticidal or *Bt* toxin. To develop these *Bt* crops, a company clones the insecticidal gene from the bacterium and inserts it into a crop plant. The plant then produces the toxin in most, if not all, parts of the plant through all or most of a growing season.

The five varieties of *Bt* corn still on the market are made by Monsanto, Pioneer/DuPont, Dow and Syngenta. At least three types of *Bt* corn were previously taken off the market. Most *Bt* corn varieties had received initial approval from EPA in 1995.

Allergenicity Concerns

Human allergenicity is a key issue related to *Bt* crops; however, EPA failed to take into account several recent studies showing that *Bt* toxins could be possible allergens. The Agency approved the corn for planting before appropriate tests were carried out.

Environmental Impacts

Studies on potential environmental impacts of *Bt* corn have been inadequate. EPA ignored the concerns of several University researchers about impacts of *Bt* corn on monarch butterflies. According to Hickey (2001), it is possible that exposure to *Bt* corn may have long-term, harmful effects on the butterflies.

Bt sprays constitute an important pest management tool for many organic and some conventional farmers, but continued use of *Bt* crops may reduce their effectiveness. Toxins in *Bt* sprays break down rapidly in the environment as opposed to the *Bt* in genetically engineered crops, which breaks down more slowly. With widespread use of *Bt* crops, there is increased insect exposure to the toxin, and insect resistance is much more likely to develop, resulting in the loss of *Bt* sprays as a valuable tool.

In 2000, it was found that StarLink corn, a type of *Bt* corn approved only for animal feed, had contaminated the human food supply. Many farmers had not followed the guidelines for growing StarLink, and had not segregated the corn after harvest. Regulations and guidelines agreed to by Aventis, the corporation that produced the StarLink seed, were not passed on to farmers, and little if any follow up by the corporation or EPA was done to see if the plans were being implemented (Hickey, 2001).

Alternatives to *Bt* Maize

Traditional small-scale, low external input systems of maize farming based on multi-cropping, integration with livestock husbandry and seed saving are the norm in several developing countries. Despite its limitations this system of agriculture still provides sustenance and food security to millions of 2-acre farmers.

Cultural control techniques involve the removal of the crop residue for fodder (or burning) after the maize is harvested. This prevents repopulation of the fields by the progeny of any stemborers remaining in the stalks. This method may have some negative effects on soil conservation (as it can reduce soil fertility and increase the risk of soil erosion).

Neem extract applied at the time when the stemborers' eggs hatch, can almost completely control maize stemborer problems. Practices based on indigenous knowledge such as the application of soil, ashes or chilli powder to clusters of maize plants have also been recorded.

While *Bt* maize is a significant step in the control of stemborers of maize internationally, regulatory approval of the crop has been stifled in some parts of the world, due to some concerns of environmental and consumer health. The crop should be grown in a way that ensures that the development of resistance in target insects is avoided.

IPM and LEISA-based alternatives to the protection afforded by *Bt* maize do exist, but also have their limitations. The appropriate choice will vary from farmer to farmer, depending on their own particular circumstances, and the different options are not all mutually exclusive. Habitat management and other options may be strategically combined with *Bt* maize both for added stemborer control and also to provide the refuges necessary for insect resistance management. Indeed, if *Bt* maize is approved for commercial release, farmers will not be faced with the choice of biotechnology against alternatives to biotechnology but will have to decide what combination of strategies to use in combination with the *Bt* maize that is likely to accumulate in their seed stocks (Mwangi and Ely, 2001).

References

Altieri, M., Rossett, P. Ten reasons why biotechnology will not ensure food security, protect the environment and reduce poverty in the developing world. *AgBioForum* 2(3&4): 155-162 (1999).

Altieri, M., Rossett, P., Thrupp, L. The potential of agroecology to combat hunger in the developing world. Institute for Food and Development Policy. *Food First Policy Brief* No. 2 (1998).

Auberson-Huang, L. The dialogue between precaution and risk. *Nature Biotechnol.* 20: 1076-1078 (2002).

Beachy, R. plus 17 others. Divergent perspectives on GM food. *Nature Biotechnol.* 20: 1195-1196 (2002).

Betz, F.S., Hammond, B.G., Fuchs, R.L. Safety and advantages of *Bacillus thuringiensis*-protected plants to control insect pests. *Regulatory Toxicol. Pharmacol.* 32 (2): 156-173 (2000).

Borlaug, N.E. Ending world hunger: the promise of biotechnology and the threat of antiscience zealotry. *Plant Physiol.* 124: 487-490 (2000).

Burkhardt, J. The genetically modified organism and genetically modified food debates: why ethics matters. *Trans. Wiscousin Acad.* 89: 63-82 (2001).

Buttel, F.H., Goodman, R.M. Introduction to the scientific, political, and ethical dialogue on genetically modified organisms. *Trans. Wisconsin Acad.* 89: 1-13 (2001).

CCC. (Canola Council of Canada). An agronomic and economic assessment of transgenic canola. Canola Council of Canada, Winnipeg (Jan., 2001).

Choudhary, B. Legislating for a genetic heritage. *Biotech. Develop. Monitor* Issue No. 48: 19-21 (Dec., 2001).

Christoforous, T. *Environ. Law J.* 8: 622-648 (2000).

Commoner, B. Failure of the Watson Crick theory as a chemical explanation of inheritance. *Nature* 220: 334-340 (1968).

Crick, F.H.C. On protein synthesis. In *Sympos. Soc. Exp. Biol. XII.* p. 153 Academic Press, New York (1958).

Crick, F.H.C. The Central Dogma of molecular biology, *Nature* 227: 561-563 (1970).

Ellis, R.J., Hemmingsen, S.M. Molecular chaperones. *Trends in Bioch Sci. (TIBS)* 14(8): 339-342 (1989).

Firbank, L.G., Forcella, F. Genetically modified crops and farmland biodiversity. *Science* 289: 1481-1482 (2000).

Goklany, I.M. From precautionary principle to risk-risk analysis. *Nature Biotechnol.* 20: 1075 (2002).

Goklany, I.M. *The Precautionary Principle: A Critical Appraisal of Environmental Risk Assessment*. Cato Institute, Washington, D.C. (2001).

GRAIN. Blinded by the gene (Editorial). *Seedling* (GRAIN, Barcelona) pp. 1-5 (July, 2003).

Hickey, E. U.S. EPA approves *Bt* corn despite lack of testing. *Global Pesticide Campaigner* 11(3): 17 (Dec. 2001).

Holm, L.G. *et al. World Weeds: Natural Histories and Distributions.* Wiley, New York (1997).

IHGSC (International Human Genome Sequencing Consortium). Initial sequencing and analysis of the human genome. *Nature* 409: 860-921 (2001).

Ismael, Y., Bennett, R., Morse, S. Farm level impact of *Bt* cotton in South Africa. *Biotech. Develop. Monitor* Issue No. 48: 15-19 (Dec., 2001).

James, C.A. *Global Review of Commercialized Transgenic Crops: 1998.* No. 9-98. International Service for the Acquisition of AgBiotech Applications, Ithaca, N.Y. (1998).

Khachatourians, G.G., McHughen, A., Scorza, R., Nip, W., Hui, Y. *Transgenic Plants and Crops.* Marcel Dekker, New York (2002).

Kirschenmann, F. Questioning biotechnology's claims and imaging alternatives. *Tarns. Wisconsin Acad.* 89: 35-61 (2001).

Miller, H., Conko, G. *Nature Biotechnol.* 18: 697 (2000).

Mwangi, P.N. Ely, Assessing risks and benefits: *Bt* maize in Kenya. *Biotech. Develop. Monitor* Issue No. 48: 6-9 (Dec., 2001).

Phillips, P.W.B., Khachatourians, G.G. *The Biotechnology Revolution in Global Agriculture: Invention, Innovation and Investment in the Canola Sector.* CABI, Wallingford (2001).

Phillipson, M. Agricultural law: containing the GM revolution. *Biotech. Develop. Monitor* Issue No. 48: 3-5 (2001).

Raffensperger, C., Tickner, J. (eds.). p. 8. *Protecting Public Health & the Environment. Implementing the Precautionary Principle* (Island Press, Washington, D.C., 1999).

Rossett, P. Genetically engineered crops. *LEISA* Magazine 17(4): 6-8 (Dec., 2001).

Ruttan, V.W. Biotechnology and agriculture: a skeptical perspective. *Trans-Wisconsin Acad.* 89: 83-92 (2001).

Schmucker, D. *et al. Drosophila* Dscam is an axon guidance receptor exhibiting extraordinary molecular diversity. *Cell* 101: 671-684 (2000).

Schubert, D. A different perspective on GM food. *Nature Biotechnol.* 20: 969 (2002).

Smyth, S., Khachatourians, G.G., Phillips, P.W.B. Liabilities and economics of transgenic crops. *Nature Biotechnol.* 20: 537-541 (2002).

Snow, A.A. Transgenic crops-why gene flow matters. *Nature Biotechnol.* 20: 542 (2002).

Stirling, A. *et al. On Science and Precaution in the Management of Technological Risk: A Synthesis Report of Case Studies.* Vol. 1. Inst. Prospective Technol. Studies, Sevilla (1999).

Svitashev, K., Somers, D.A. *Genome* 44: 691-697 (2002).

Venter, C. *et al.* The sequence of the human genome. *Science* 291: 1304-1351 (2001).

Visser, B., van der Meer, I., Louwaars, N., Beekwilder, J. The impact of 'terminator' technology. *Biotech. Develop. Monitor* Issue No. 48: 9-12 (Dec., 2001).

Watkinson, A.R., Freckleton, R.P., Robinson, R.A., Sutherland, W.J. Predictions of biodiversity response to genetically modified herbicide-tolerant crops. *Science* 289: 1554-1556 (2000).

Windels, P. *et al.* Characterisation of the roundup ready soybean insert. *Eur. Food Res. Technol.* 213: 107-112 (2001).

Chapter 5

Natural Selection, Plant Breeding and Improvement of Selected Crops

Introduction

There has been some slowing down of progress in genetic improvement of crop yield potential since the mid-1980s. While world grain yields per unit area increased 26% between 1965 and 1975 and 31% between 1975 and 1985, they increased only 9% between 1985 and 1995 (Brown *et al.*, 1998). Also, grain production per capita fell more than 10% between 1985 and 1995. This has warranted the need for more sustainable management of agricultural ecosystems. According to Denison *et al.* (2003), a better understanding of natural selection may help solve both problems. They illustrated this point with two specific examples. First, the genetic legacy of crop plants has been refined through the millennia by natural selection, often driven by competition among plants. This prompted Denison *et al.*, to suggest that most simple, tradeoff-free options to increase competitiveness (*e.g.*, increased gene expression, or minor modifications of existing plant genes) have already been tested by natural selection. Further genetic improvement of crop yield potential in the coming few years will mainly involve tradeoffs, either between fitness in past versus present environments, or between individual competitiveness and the collective performance of plant communities. Eventually, agronomists may be able to predict the consequences of genetic alterations so radical that they have not yet been tested by natural selection. Second, natural selection acts chiefly at the level of genes, individuals, and family groups, rather than ecosystems as a whole so we should not expect the structure of natural ecosystems (diversity, spatial, or temporal patterns) to be a reliable model for agricultural ecosystems. Natural ecosystems are nonetheless an important source of information that should be gainfully used to improve agriculture.

What Denison *et al.* (2003) have critically dicusssed is the implications of past natural selection for crop genetic improvement and for the overall design of agricultural ecosystems. They have presented two main hypotheses. First, that natural selection had ample opportunity, before the wild ancestors of our crops were domesticated, to test alternative solutions to problems that limited individual fitness under preagricultural conditions. Such plant characters as efficient enzymes, competitiveness, stress tolerance, and broad-spectrum defenses against pests would have consistently conferred enhanced individual survival and reproduction relative to alternative genotypes in preagricultural environments, so further improvement of these traits may be very unlikely. Instead, opportunities for further genetic improvement of crop yield will mainly involve tradeoffs between plant adaptation to agricultural versus natural conditions, or between the competitiveness of individual plants and the overall performance of plant communities. Breeding of crops resistant to evolving pests should continue to be an important ongoing activity, as also breeding to suit changing consumer preferences.

The second main hypothesis emanates from the fact that natural selection is the only reliable source of (agricultural) improvement in natural ecosystems that operates on a time scale longer than the lifetime of individual plants. Natural selection operates at the level of genes, individuals, and family groups, but not communities and ecosystems. This makes the second hypothesis inconsistent with the suggestion of Soule and Piper (1992) and Lefroy *et al.* (1999) that agricultural ecosystems whose species composition, spatial and temporal patterns are based on natural ecosystems will be consistently more efficient, sustainable, and productive. No doubt this hypothesis rejects mindless mimicry; but still, if properly understood, natural ecosystems remain a valuable reservoir of ideas for agriculture (Denison *et al.*, 2003).

Sometimes, past natural selection and present human goals conflict; in such cases, it may be possible to improve crops by deliberately reversing the negative effects of past natural selection on the collective performance of plant communities. Likewise, it should be possible to design agricultural ecosystems that maximize resource-use efficiency or control pests in ways not observed in natural ecosystems. A consideration of natural selection can explain past successes and predict which approaches are likely to succeed in the future (Denison *et al.*, 2003).

If natural selection did have enough opportunity, before the domestication of wild ancestors of our crops, to explore alternative solutions to problems which limited individual fitness under preagricultural conditions, then plant breeders may not be able to beat natural selection. The problem is not really a lack of genetic variability. The problem, for traits that would enhance both crop yield and individual plant fitness, is that rare variants happen to be rare precisely because they are less fit. The goal of natural selection was never to maximize the oil or protein content of seed; that is why there has been prolonged response to directional selection by humans for these traits.

Natural selection usually operates in such away as to maximize transmission of some alleles relative to alternative alleles usually by maximizing the reproductive success of individuals carrying those alleles (Dawkins, 1976). Some maternally inherited genes in plants that increase seed production at the expense of pollen production facilitate their own transmission at the expense of the plant's overall fitness (Dominguez, 1995).

Similarly, selection by humans can change the balance between vegetative and reproductive growth away from that favoured by natural selection, in order to favour seed production in grain crops or biomass production in forage crops or trees grown for biomass (Denison *et al.*, 2003). Many of the plant improvements suggested by molecular biologists concern traits that have already been subject to long-term improvement by natural selection and are currently under stabilizing selection–even molecular tools may not deliver much (Simmonds, 1998). However, we still can resort to the following alternative avenues for genetic improvement of crop yield:

1. Red Queen breeding (see Dawkins and Krebs, 1979; Cao *et al.*, 1998) *i.e.*, keep ahead in evolutionary arms races with pests;
2. Accelerating adaptation to new physical or chemical conditions such as different climates or greater soil fertility;
3. Reversing the effects of past natural selection for traits that enhance individual competitiveness but limit community-level performance; this may be achieved either through deliberate selection of less competitive plants with specific traits, or through human-mediated group selection; and
4. Possibly designing truly novel phenotypes not previously tested by natural selection (Denison *et al.*, 2003).

Agroecosystem Design

The concept that agricultural ecosystems would be more productive or sustainable if they were more similar to natural ecosystems has attracted considerable interest. Jackson and Piper (1989) suggested that agroecosystems should be modelled on nature's standards with agriculture being based on natural communities. Similarly, Altieri (1987) suggested that agricultural pest management strategies should be modelled on natural ecosystems. However, while natural ecosystems may well have withstood the test of evolutionary time (Ewel 1999), the test criteria have not been stated clearly. Some have assumed that natural selection operates, if not for the benefit of an entire ecosystem, at least for the benefit of a species as a whole. According to Jackson and Koch (1997), natural selection may conflict with the goals of plant breeders but it does operate to continue the survival of species.

According to the second hypothesis of Denison *et al.* (2003), natural selection is the only process that consistently improves the performance of anything in nature. Should this hypothesis be true, then nature's wisdom should express only in those entities that have been subject to natural selection. Individual plants and animals contain genes that have repeatedly passed through the screen of natural selection; but the important issue is: has anything analogous to natural selection also improved the structure and function of the ecosystems where those species live.

As the productivity of natural ecosystems is usually affected adversely by human disturbance, undisturbed ecosystems should somehow be expected to be structured in ways that enhance productivity. Since human-induced changes in ecosystem structure tend to adversely affect specific ecosystem functions, it seems to prove that the structure of undisturbed natural ecosystems is well-suited to those functions. Here the question is whether undisturbed natural ecosystems actually excel in those functions relevant to agriculture, such as the ability to export food sustainably? Although few, if any, comparisons between natural and agricultural ecosystems have strictly controlled for the effects of soil, climate, or external inputs, it seems unlikely that natural ecosystems will consistently outperform agricultural systems by most agricultural criteria (Denison *et al.*, 2003). Denison *et al.* concluded that "Nature's wisdom resides mainly in the sophisticated adaptatations of individual plants and animals, not in ecosystem structure. Mindless mimicry of natural ecosystems reveals ignorance of, not respect for, ecological principles."

Crop Diversity

That a Darwinian perspective can potentially contribute to the design of agricultural ecosystems is illustrated by the use of crop diversity in controlling specialist agricultural pests. Many natural ecosystems contain interacting groups of plant species. Likewise it can be advantageous to grow mixtures of crops. Some pests are less common with intercropping (see Risch *et al.*, 1983; Fukai, 1993).

Mimicking natural ecosystems cannot give us benefit that we can have from deploying the available crop diversity. Deploying crop diversity on the landscape scale also seems more compatible with crop rotation which deploys diversity over time. When we consider species mixtures in nature, the effects on pests of diversity in time may commonly be different from those of diversity in space. Pests adapted to considerable spatial diversity (as found in natural ecosystems) tend to be poorly adapted to major changes in vegetation from one year to the next, because this sort of change is relatively uncommon in many natural ecosystems (Denison *et al.*, 2003). A three-year rotation, growing each crop in sequence, could perhaps result in better pest control than growing the same three-crop mixture each year.

The Contrasting Attitudes of Plant Breeders and Molecular Biologists Towards Rice

Rice is an important crop not only for Asia but also is the staple food for nearly 3 billion people in the world. By 2025 rice consumers will increase to over 4 billion. In 1999 the International Rice Research Institute, Manila estimated that the rice yield in India (2.9 tonnes/hectare) is much lower than in China (6.3 t/ha) or Japan (6.4 t/ha). As production based on green revolution technologies has reached a plateau, there is a strong need to explore the potential of biotechnology for further increasing rice production. Scientists have been approaching the problem from different angles.

Rice can be grown all over the world under diverse ecosystems as rainfed uplands, rainfed shallow, semideep and deepwater lowlands, irrigated lands and hills. In India, it is cultivated on 42 million hectares (mha) under four major ecosystems - irrigated (19 mha), rainfed lowland (14 mha), flood prone (3 mha) and rainfed upland (6 mha). Rice ecosystems in India represent 24% of irrigated areas, 34% of rainfed lowlands, 26% of flood-prone areas and 37% of rainfed uplands cultivated to rice in the entire world (see IRRI, 1995). No other country in the world has such a diversity of rice ecosystems.

Concerted efforts have been made to improve yielding ability, increase efficiency in the use of external inputs and incorporate resistance to biotic and abiotic stresses. The multilocational testing of breeding stock developed at different research centres is being organized.

Indica rice varieties produce high biomass and show many variations in panicle weight and grain number per panicle, but are prone to lodging and consequent yield losses. In the past three decades, several improved varieties of rice have been produced and grown, and have also been distributed to other countries (Table 5.1).

Several Indian varieties have produced stable high yields in the international tests, and those showing resistance to stresses or good quality traits have found wide acceptance in several countries around the world (Prasad *et al.*, 2001).

The demand for rice is expected to grow rapidly over the next two decades, especially in some of the Asian countries-65% in the Philippines, 56% in Malaysia, 51% in Bangladesh, 46% in India,

Table 5.1: Some Rice Varieties Developed in India and Released in Countries Around the World (after Prasad *et al.*, 2001)

Country Where Released	*Number*	*Name/Designation*	*Released Region*	*Year Released*	*Ecosystem*
Afghanistan	IET 1415	CR 44-11	South Asia	1975	Irrigated
Afghanistan	IET 355	Cauvery	South Asia	1975	Upland
Afghanistan	IET 953	Padma	South Asia	1975	Irrigated
Bhutan	—	Barkat	South Asia	1992	Irrigated
Brazil	IET 2881	Seshu	Latin America and The Caribbean islands	1984	Upland
Burkina Faso	IET 2885	Vikram	Sub-Saharan Africa	1979	Irrigated
Burkina Faso	IET 1879	Vijaya	Sub-Saharan Africa	1997	Rainfed lowland
Burundi	—	Savithri	Sub-Saharan Africa	—	Irrigated
Cambodia	IET 7435	OR 142-99	South East Asia	1992	Rainfed lowland
China P.R.	—	M 114	EastAsia	1981	Irrigated
Ivory Coast	IET 723	Jaya	Sub-Saharan Africa	—	Irrigated
Ghana	IET 2885	Vikram	Sub-Saharan Africa	1982	Irrigated
Iran	IET 1990	Sona	West Asia & North Africa	1982	Irrigated
Iraq	IET 8113	RP 2095-5-8-31	Sub-Saharan Africa	—	Rainfed lowland
Kenya	—	Basmati 217	Sub-Saharan Africa	—	Irrigated

Contd...

Table 5.1–Contd...

Country Where Released	*Number*	*Name/Designation*	*Released Region*	*Year Released*	*Ecosystem*
Malawai	IET 4094	Kitish	Sub-Saharan Africa	1993	Irrigated
Mali	IET 1444	Rasi	Sub-Saharan Africa	1984	Rainfed lowland
Mali	IET 1879	Vijaya	Sub-Saharan Africa	1978	Irrigated
Myanmar	—	Mahsuri mutant	South—east Asia	1977	Irrigated
Nepal	IET 1444	Rasi	South Asia	1981	Upland
Nepal	—	K39-96-1-1-1-2	South Asia	—	Irrigated
Nepal	—	IR 3941-4-PLP2B	South Asia	1982	Irrigated
Paraguay	IET 4094	Kitish	Latin America and The Caribbean islands	1989	Irrigated
Paraguay	IET 5612	R 22-2-10-1	Latin America and The Caribbean islands	1989	Irrigated
Senegal	IET 1444	Rasi	Sub-Saharan Africa	1981	Upland
Tanzania	IET 4790	BIET 360	Sub-Saharan Africa	1986	Irrigated
Tanzania	IET 2397	RP 143-4	Sub-Saharan Africa	1984	Rainfed lowland
Venezuela	IET 7288	PR 106	Latin America and The Caribbean islands	1984	Irrigated
Vietnam	IET 723	Jaya	South—east Asia	—	Irrigated
Zambia	IET 7543	RTN 500-5-1	Sub-Saharan Africa	—	Irrigated

45% in Vietnam, 42% in Myanmar and 38% in Indonesia (Prasad *et al.*, 2001). With a rich and diverse germplasm resource, India can lead in rice improvement programme to sustain self-sufficiency in the future.

There have been many claims that biotechnology can play a significant role in India's economic development and technological self-reliance. According to Swaminathan (1987), while new technologies based on high-yielding varieties are scale-neutral in respect of their suitability for being grown by farmers irrespective of the size of their holdings, they are not resource-neutral. This necessitates the addition of a dimension of resource-neutrality to scale-neutrality in technology development.

For over a decade now, scientists have been attempting to use molecular biology tools to better understand the rice plant with the goal of developing varieties that are resistant to biotic stresses caused by pests and insects and abiotic stresses caused by drought or salinity. A related goal is to overcome yield barriers.

Whereas molecular biologists tend to consider rice as a generic plant that can be understood at the molecular level and manipulated genetically in a laboratory setting, breeders view rice as a crop to be improved via selection in the field. The scale of operation and impact of plant breeders is much larger than that of molecular biologists. Plant breeders take a holistic view of the rice plant as a crop whereas molecular biologists have a narrow view. Plant breeders also usually incorporate environment-specific factors (such as climate) in their research. Whereas the benchmark of excellence for molecular biologists is publications, for plant breeders the yardstick is the development of improved plant varieties. Rice breeders are guided by operating principles that are more inclusive than those that guide rice molecular biologists. Breeders take into account the material interests of end-users. To some extent plant breeders adopt the role of spokespersons for the relevant end-user groups in different contexts (Haribabu, 2000). This is not so with molecular biologists. While most plant breeders have never seen DNA, most molecular biologists have never seen a rice field. The field view of rice as a crop, held by the breeders, and the laboratory view of rice as a plant, held by molecular biologists, needs to be integrated. One strategy to this end is for the breeders to become familiar with the techniques of molecular biology.

Marker-assisted Selection versus Transgenic Approaches

Breeders have traditionally relied on selection based on phenotypes. It is only recently that they are exploring the possibilities of selection based on genotypes. Even so, rice breeders and molecular biologists in India have different perceptions as regards the most appropriate strategy for the improvement of rice. The majority of rice breeders advocate marker-assisted selection (MAS) as the appropriate strategy. MAS is based on the intra-species transfer of genetic material and has the possibility of 'gene pyramiding' by accumulating genetic markers for desired multiple traits such as higher yield and resistance to diseases and drought. Intra-species transfer of genetic material, as used in MAS-based technology, may not be all that precise, at least not at the level of defining what exactly is being transferred. Some rice breeders feel that MAS may be more acceptable socially, at least in the short-term. Molecular biologists emphasise genetic engineering methods that might also involve the inter-species transfer of genetic material. They feel that a transgenic approach is inherently more precise and more efficient than conventional crossing and selection.

What rice breeders want is routine techniques and tools that can be handled easily in different field conditions and at affordable cost. The transition of biotechnology to such a routine phase requires active collaboration between molecular biologists, plant breeders, entomologists, pathologists, and agronomists. Rice breeders believe that MAS is more appropriate at this juncture from the point of view of the feasibility of integrating it with breeding practices, biosafety restrictions and current social acceptability. However, a social cost-benefit analysis should be undertaken to assess the relative advantages of MAS *vis-a-vis* transgenic methods. The differing types of incentives regarding Intellectual Property Rights (IPRs) that plant breeders and molecular biologists are exposed to, point to the likely generation of new tensions and conflicts. Molecular biologists view the patent system as an unmixed incentive whereas plant breeders tend to accept the utility of Plant Breeders' Rights (PBR) or Plant Varietal Protection (PVP) rights (Spillane, 2000). Breeders' Rights or Varietal Protection Rights are a set of rights that provide patent-like protection to breeders over plant varieties.

The System of Rice Intensification

The system of rice intensification (SRI) was developed in Madagascar in the 1980s. It is now emerging elsewhere also that it can be more productive than other methods for growing irrigated rice. SRI is conducive to healthy plants and higher yields.

Agroecological Management

Agroecological management requires attuning practices to crops and conditions, rather than applying a fixed set of practices (Uphoff, 2001). The latter "technological" management assumes that the crucial factors are genetic, rather than the management of interactions between genetic potential and environmental conditions which guides an agroecological approach to crop production.

The SRI has revealed for rice-and possibly also for other crops-that there is considerable genetic potential that can be exploited by adjusting the agronomic management practices. SRI practices can double production for practically all of the rice varieties, local or improved. Indeed, the highest yields with SRI methods have come from "improved" varieties: a World Bank 1996 report on rice in Madagascar, noted that four farmers in the Andapa region who used SRI methods with the high-yielding variety IR-46, averaged 13.7 t/ha, with one farmer reaching 16.5 t/ha.

Thus the SRI experience leads one to two very different conclusions:

1. There appears to exist substantial genetic potential in existing rice varieties that can still be tapped through agroecologically sound practices, implying that genetic modification work is not necessary if increasing food production and lowering costs of production are the main objectives. Some farmers in Sri Lanka have found not only that SRI methods raised their yields from 2.9 t/ha with conventional methods to 8.5 t/ha, but also reduced the costs of production, from 6 rupees/kg to 3 rupees/kg.
2. Agroecological methods of crop production seem to give better dividends with genetically-improved varieties. Some varieties are more suitable for tillering, root growth and grain filling than others in response to wide spacing, aerated soil, and other SRI practices. Quite likely, some

varieties may also have better pest and disease resistance, or greater drought tolerance, when grown with SRI practices. Thus, there may still be considerable potential for conventional breeding and selection to identify varieties that are specifically adapted and responsive to SRI practices (Uphoff, 2001).

Soil Health

One criticism of GMOs is that a preoccupation with genetic changes deflects efforts at studying and improving what may be the most important factor in increasing yield: the management of natural resources or soil health.

There has been much concern about the conservation of biological diversity in recent decades but much of it has focused on above-ground flora and fauna. Soil biodiversity-the vast and complex communities of bacteria, fungi, mycorrhiza, actinomycetes, protozoa and nematodes as well as earthworms and other soil 'megafauna'-may hold the key to high productivity with SRI methods. Increases in the variety and number of micro-organisms playing different roles in plant nutrition such as biological nitrogen fixation and phosphorus solubilisation are very beneficial.

SRI is not necessarily an 'organic' methodology; it can also be used with agrochemicals. But continuous use of compost gives higher yields than does NPK fertilizer. Also, pest and disease problems under SRI are usually not serious enough to warrant the use of biocides, and thus a 'healthy' soil can be maintained, which in turn further reduces the need for chemicals.

Optimal Use of Resources

SRI may not be appropriate or feasible in all rice-growing areas. For instance, there needs to be sufficient irrigation and drainage infrastructure to control water applications and maintain well-drained soils. Although the SRI practices may potentially double or triple the production of rice in the world, this is not a reasonable goal or use of the methodology. Rather it should allow small farmers to raise the productivity of their land, labour and water resources, while trying to meet their staple food requirements.

Ultimately, productivity is what is most important to farmers: how to get the most from their limited land, labour, capital and water.

SRI attests to the general value of an agroecological perspective, and should increasingly guide agricultural research, not only for rice but also for other crops (Uphoff, 2001).

Shuttle Breeding for Rainfed Rice

Although rainfed lowland rice accounts for about 40% of the rice area in India, it has received less research attention as compared to irrigated rice.

The DRR (Directorate of Rice Research) programme has been instrumental in the development and release of 632 rice varieties in India so far of which 257 (41%) are for rainfed eco-system and the remaining 375 (59%) for irrigated ecosystem (Table 5.2). The number of released varieties for rainfed lowland ecosystem is only 167 (26%). Some reasons for the slow progress in varietal development for rainfed lowland ecosystem are the following (see Mallik, 1995; Mallik *et al.*, 2002):

1. Harsh, heterogeneous and unpredictable environment.
2. Wide variability in intra- and inter-location;
3. Fewer researchers involved compared to irrigated ecosystem.
4. Lower priority given to rainfed lowland programme.
5. Lack of proper testing facilities or methodologies.
6. Inadequate donors for desirable traits, required for this ecosystem.
7. Inadequate generation of new material with wide genetic base.
8. Advancing the generations (F2 onwards) but lack of follow-up of *in situ* testing in many of the locations.
9. Lack of proper environmental characterization such as water depth, duration and depth of submergence during trials.
10. Crop generally grown once a year, as mostly it is long-duration/photoperiod sensitive.
11. Lesser involvement of NGOs in varietal improvement than in hybrid rice development programmes.

Table 5.2: Varieties Released in India for Different Ecosystems up to 2000 (after Mallik *et al.*, 2002)

Ecosystem	*No. of Varieties*	*Percentage*
Rainfed		
Upland	84	13
Shallow*	123	19.5
Semi-deep	30	5
Deep	14	2.2
Hill	6	1
Total	**257**	**40.7**
Irrigated		
Early	123	19.5
Medium-E	17	3
Medium	173	27
Saline-alkaline	15	2.4
Scented	20	3.1
Hill	27	4.3
Total	**375**	**59.3**
Grand total	**632**	**100**

Attempts at rice varietal development for rainfed lowland, which occupies 17.2 m ha in India, were intensified only during the early nineties. Shuttle breeding programme, a collaborative project between the Indian Council of Agricultural Research and the International Rice Research Institute, Philippines has provided an opportunity for exchange of breeding materials of diverse origin among the eastern Indian states to strengthen the breeding programme. The materials developed through this project have served as input to the other breeding programmes for this ecosystem. Many promising selections are now in advanced stage (F6/F7) (Mallik *et al.*, 2002).

In this programme, the parents are selected either through artificial or field screening. Natural calamities such as drought and floods have also helped to identify donors. The F1s are grown in shallow water. From F2 onwards, the materials are grown in field condition where dry seeds are sown from April to early May prior to monsoons. Water starts accumulating in the field from mid-July and increases gradually, depending upon rainfall and surface run-off from neighbouring fields; some water remains till harvest. The breeding materials during generation advancing are naturally exposed to such abiotic stresses as drought at early vegetative stage, flooding of varying depths and durations at different growth stages, and biotic stresses like aquatic weeds, insects and diseases. Preliminary yield trials are conducted using F6 under transplanted condition before the replicated yield trial in F7 (Mallik *et al.*, 2002).

Exchanges of breeding materials (F3/F4) among breeders are encouraged at different sites; this helps them in selecting breeding materials suited to their condition. Any potentially promising breeding lines, identified after replicated yield trials are nominated to national trials.

Through the shuttle-breeding programme several rice varieties have already been released. Also, several hundreds of farmers in each state have been given the seeds of new varieties to expedite inter-state flow of promising rice varieties.

Research for developing superior genotype for rainfed lowland ecosystem through biotechnology is underway at several institutions. Selection efficiency in breeding population can be enhanced by use of molecular markers. Identification of the RFLP/AFLP markers flanking Sub (1) t locus is an important step in this direction. Researchers at CSIRO, Australia are engaged in genetically altering levels of pdc (pyruvate decarboxylase) and adh (alcohol dehydrogenase) in rice (Mohanty *et al.*, 2000). Three different rice *pdc* genes (*pdc1*, *pdc2* and *pdc3*) have been cloned and sequenced, and their plasmid constructions have been introduced into rice to yield transgenic lines. The 'Rice Genome Project' is providing nucleotide sequence of the complete rice genome-to pinpoint function and location of additional genes important for flooding tolerance, needed for rainfed lowland rice varieties (Mallik *et al.*, 2002).

Dwarf Rice

Dwarf rice that produces normal grain has been successfully developed by targeting increased gibberellin turnover to the sites of synthesis in vegetative tissues (see Hedden, 2003).

Some have rightly opined that the introduction into cereals of genes to decrease stem height has saved more lives than any other scientific development. In the past half a century the ability to feed the growing human population became possible by the development of high-yielding, semi-dwarf varieties of wheat and rice in the 'green revolution' (Evans, 1998). These dwarf varieties are more resistant to wind and rain damage than their tall parents, and they also give higher grain yields because more of the plant's resources go into grain formation rather than the vegetative tissues. Unfortunately, it is not possible to grow semi-dwarf varieties of many crop species because suitable dwarfing genes are not available; this necessitates the application of chemical growth retardants to achieve growth control. Sakamoto *et al.* (2003) have produced dwarf plants of rice without detrimental consequences for grain development; their method can, in principle, be adopted for several other species of plants (Hedden, 2003). What Sakamoto *et al.* did was to ectopically express a gene encoding gibberellin (GA) 2-oxidase, an enzyme that deactivates the GA plant growth hormones. These latter control several physiological processes, including elongation of the stem and development of reproductive organs. The GAs are major targets for control of vegetative growth in crop plants. Indeed, growth retardants retard growth by inhibiting their biosynthesis. Furthermore, the dwarfing genes that drove the green revolution interfere with either GA biosynthesis or signal transduction (Hedden, 2003).

Sakamoto *et al.* have shown the potential for manipulating crop architecture by genetic modification of GA signaling (biosynthesis and signal transduction). In this, targeting the intervention to the appropriate tissues is crucial. The complexity of the GA-biosynthetic pathway offers many possible points for manipulation and different approaches may be required for different species. Besides their agronomic utility, modification of GA metabolism in transgenic plants should provide further information on the regulation of GA biosynthesis and on the role of GAs in plant development (Hedden, 2003).

High-sugar Rice?

Both associative nitrogen fixers and free-living nitrogen fixers share a common enzymology-participation of Fe, Mo, and V, and use of ATP as energy source. But the systems are labile to oxygen. The issue is how to get the high levels of ATP needed without interference from oxygen? Anaerobic operation requires quantities of substrate that are prohibitive in a productive system. The symbiotic systems have developed the use of haemoglobin with an affinity for oxygen high enough to prevent it from inactivating the O_2-sensitive nitrogenase but still providing enough O_2 to support the ATP-generating respiratory system. Cyanobacteria form heterocysts that shield the nitrogenase from the adjacent photosynthesizing O_2-generating cells. *Azotobacter* cells apparently respire at such a high rate that they keep the pO_2 under control, but this is very wasteful of energy. The roots of rice growing in water may be exposed to an optimum pO_2 compatible with N_2 fixation, but, if the N_2 fixers (*e.g.*, cyanobacteria) are outside the roots, the products of N_2 they fix are washed away, so some way has to be devised to get the fixing organisms to thrive inside the roots while the plant provides them with adequate substrates to support fixation, provides them with a pO_2 adequate to support fixation but one that is not too high to inactivate their nitrogenase system. The fact that sugarcane plus *Acetobacter diazotrophicus* can fix adequate N is an encouraging pointer to the feasibility of the above approach and warrants the possibility of evolving a high-sugar rice. This will have to be done at the expense of the starch content of the rice grain. Even notwithstanding such a trade-off, however, the solution to the problem of rice fixing its own nitrogen will not come easily.

Golden Rice

The golden rice was developed by Peter Beyer and Ingo Potrykus. The two inventers entered into a unique arrangement with the agribusiness company Syngenta; the inventers assigned all commercial rights to Syngenta with the condition that subsistence farmers in any developing country can cultivate Golden Rice varieties, once available, licence-free (Subsistence farmers are defined as having annual earnings below $10,000 per year).

Rice plants do not normally produce carotenoids (vitamin A precursors) in the grain. Syngenta is the world's largest agribusiness. Early in 2001 it announced the completion of the entire rice genome, and certain other companies which hold patents relating to different steps in the development of Golden Rice, have agreed to waive licence fees for the project.

The first free samples of this new type of rice - genetically engineered to contain vitamin A precursors - have already been sent to the International Rice Research Institute (IRRI) in Los Baños, Philippines and similar samples of the rice are planned to be sent *gratis* to non-commercial research institutes in China, India, Africa and Latin America. The institutes will conduct biosafety studies, and then use traditional breeding techniques to confer the beneficial traits of the transgenic rice to locally adapted strains.

Some of the debates over genetic modification are politically inspired. There are important social, religious, political, and economic reasons for care in the application of genetic modification-especially for food and agriculture.

One recent example of this has been the outcry over Golden Rice, a strain of rice modified to be rich in vitamin A (*Science* 287: 303-305, 2000). In theory, Golden Rice could help prevent blindness, caused by a deficiency in the vitamin, in children in developing countries. Greenpeace calculated that-at the concentrations produced by current strains of Golden Rice-a child would need to eat seven kilograms of cooked rice a day to get the recommended daily dose of vitamin A. But Gordon Conway, president of the Rockefeller Foundation, which funded the project, rightly pointed out to Greenpeace that vitamin A deficiencies can arise when children lack "10%, 20% or 50% of their daily requirements, not 100%". An average daily rice quota could, therefore, be beneficial. Indeed, excess intake of vitamin A can prove very injurious because, except in low doses, vitamin A is toxic.

Golden rice has been touted as the miracle cure for relieving twin scourges of malnutrition and hunger which afflict over 800 million people in the world. But according to Manicka Velu *et al.* (2001), vitamin A rice cannot remove Vitamin A Deficiency (VAD). Currently, it is not even known how much vitamin A the genetically engineered rice will actually have. The goal is being set at about 33 μg/100 gm of rice. Even if this goal can be achieved it will still be ineffective in removing VAD. The daily average requirement of vitamin A being about 750 micrograms, the golden rice would only provide 9.9 micrograms which is 1.32% of the requirement, considering the fact that one serving contains 30 gms of rice on a dry weight basis.

Out of the three new genes introduced into rice, two have come from daffodils and one from a microorganism. The transgenic rice is yellow in colour and produces beta-carotene, a precursor to vitamin A. However, the production of beta carotene will be only 2.8% of that obtained from amaranth leaves and 2.4% of that obtained from coriander or drumstick leaves.

Even the World Bank has stated that the use of local plants, particularly green leafy vegetables can dramatically and efficiently reduce VAD in malnourished children at very low cost. Given the diversity of plants and crops in the Third World, farmers, especially women, have bred and used such vitamin A rich food plants as coriander, amaranth, carrot, mango and drumstick. Thus a far better route to alleviating vitamin A deficiency is biodiversity conservation, propagation and use of plants that are naturally rich in vitamin A.

Corn

According to Harlan (1995), modern domesticated maize (*Zea mays*) provides about 21% of global human nutrition. The major product of maize (except for forage) is the seeds which are produced on thick, long cobs. The similarity of the wild female maize spike to the spike structure of wheat, rye, and barley prompted Lev-Yadun *et al.* (2002) to propose that these major crops could be engineered with cobs resembling those of modern maize.

The wild progenitors of domesticated maize, the annual teosintes (*Zea mays* spp. *mexicana* and spp. *parviglumis*), have a thin female inflorescence having several nodes that carry female flowers resembling the hermaphroditic spikes of wheat, rye, and barley. While the male inflorescence has not changed greatly during domestication, the modest female inflorescence of teosinte has evolved into a very different structure, the large, thick maize cob. The domesticated plant also has become less branched and more robust.

Three of the major genetic factors possibly involved in the above changes (*teosinte branched, teosinte glume architexture,* and the *two ranked* gene) have been characterized. The branched gene changes the architecture of the plant from branched and grasslike to the single-stalk form of cultivated maize. The glume architecture gene eliminates the large, hard casing on teosinte kernels. The two ranked gene leads to the production of kernels around the entire circumference of the ear (see Lev-Yadun *et al.*, 2002).

Some wild diploid wheats of the section Sitopsis (such as *Aegilops speltoides*) that were the progenitors of the wild tetraploid emmer wheat (in turn a progenitor of bread wheat) have an ear morphology like that of wild maize ears, even more than do current domesticated wheats, barley, or rye. In fact, today's wheats, barley, and rye have already undergone some of the changes that can lead to a maize cob-like morphology (Lev-Yadun *et al.*, 2002).

Maize is a C4 plant adapted to warm regions with high solar irradiation; it cannot be used as a crop everywhere. Wheat, rye, and barley, however, grow in cooler climates under different soil conditions. Barley can tolerate more arid conditions and also highly saline water unlike wheat or rye (Harlan, 1995). Wheat yields were improved genetically largely through changes in the harvest index (grain biomass/total biomass), rather than through increased biomass production. It has only emerged in recent years that increased wheat biomass production is achievable.

In maize, the morphological changes that occurred in the course of domestication resulted in both higher productivity and superior grain quality. A similar gain resulting from analogous changes in wheat, rye, and barley could be very useful.

Although cereal genomes have reshuffled during evolution, some blocks of genes have remained in the same order, even in new chromosomal locations. This raises the possibility to clone those genes in wheat, rye, and barley that are homologous to maize genes, or to transform these species with maize genes and produce new types of GM crops (Lev-Yadun *et al.*, 2002).

Bt Maize - Risks and Benefits

Bt corn is maize (*Zea mays* I.,) that has been genetically modified to express the *cry1Ab* gene from *Bacillus thuringiensis* and produce an insecticidal toxin to kill lepidopteran pests, especially the European corn borer (*Ostrinia nubilalis*). In the USA, 20 million acres of transgenic *Bt* corn were planted in 1999. This aroused concerns that the Cry1Ab protein may pose some risk to natural and agricultural ecosystems and that non-*Bt* crops grown in soil previously used to grow *Bt* corn may take up the pesticide.

Recent researches in which toxin was purified and exogenously added or naturally released into soil from root exudates or biomass of *Bt* corn, indicate that plants do not take up the pesticide. Tissue samples (leaves, stems, and roots) of non-*Bt* corn, carrot, radish, and turnip were both immunologically negative for the presence of the toxin and nonlethal to the larvae of *Manduca sexta* after 4-6 months of growth in soil previously used to grow *Bt* corn to maturity, soil to which *Bt* corn biomass had been added, or soil to which purified Cry1Ab protein had been added (Saxena and Stotzky, 2001).

In contrast to the above, samples of soil from the rhizosphere of *Bt* corn plants or amended with biomass of *Bt* corn or with 3.2 µg/g of soil, oven-dry equivalent, of purified Cry1Ab protein were both immunologically positive and lethal to the larvae. No toxin could be detected in the tissue of non-*Bt* corn, carrot, radish, and turnip grown for 3 months in nonsterile soil amended with different amounts of clay minerals and into which purified *Bt* toxin had been incorporated.

These researches of Saxena *et al.* (1999) and Saxena and Stotzky (2001) demonstrate that non-*Bt* corn and other species do not take up toxin released to soil in root exudates of *Bt* corn, from the degradation of the biomass of *Bt* corn or even as purified *Bt* toxin. It appears that *Bt* toxin released to soil is not taken up by crops subsequently grown in soils in which *Bt* corn has been grown. The persistence of the toxin in soil for 6 months after its release also suggests that the pesticide remains bound to surface-active soil particles, which protect the toxin from biodegradation.

Benefits

Bacillus thuringiensis (Bt) is an insecticidal bacterium. Formulations consisting of the bacterial spores and isolated toxins have been used as natural pesticides for half a century and *Bt* sprays have become a popular biopesticide mainly due to their specific spectrum of action, relative safety towards humans, and low environmental persistence. This environmental benefits make *Bt* sprays acceptable for use on organically certified crops.

Recombinant DNA technologies have made possible the insertion of the *Bt* genes involved in production of insecticidal toxins into the maize genome. The resulting transgenic maize plant produces the toxins in its tissues. In the USA, *Bt* genes effective against the European cornborer were identified and successfully introduced into maize. European corn-borer larvae that penetrate the plant tissues die upon ingesting the toxin produced in the *Bt* maize cells. The technology helps in increasing maize yields and decreases the need for insecticide sprays. By 2000, *Bt* maize constituted a significant proportion of overall maize plantings in the USA, Argentina and Canada.

Risks

Many other countries, particularly in Europe, have been more cautious to adopt the new technology in view of uncertainty over the risks that it may pose to the environment or to consumer health.

The primary "adverse effects" to be borne in mind are the following:

1. Effects on non-target organisms (*e.g.* monarch butterfly), on natural enemies of stemborers such as the green lacewing *Chrysoperla carnea*, and on soil biota.
2. Whether spread of the transgene to neighbouring conventional crops can spill economic losses to farmers who wish to preserve non-GM identity.
3. Rapid development of resistance among insect pests toward the *Bt* toxins used.

Despite these risks traditional mixed farming with other crops (*i.e.*, not engineered with *Bt*) such as sorghum and pearl millet that are also hosts to maize stemborers may probably provide some refugia. Furthermore, some native grasses (*e.g. Cenchrus ciliaris, Panicum maximum, Pennisetum purpureum* and *Echinochloa haploclada*), which occur as weeds in maize fields serve as alternate hosts to many stemborers and could also act as natural refugia. Another option is incorporation of natural pest resistance through conventional breeding into the *Bt* plant (termed pyramiding); this approach to insect resistance management requires little adaptation of farming systems (see Mwangi and Ely, 2001).

Alternatives to the Use of *Bt* Maize

In Kenya, traditional, small-scale, low external input systems of maize farming have long been based on multi-cropping, integration with livestock husbandry and seed saving; over 95%

farms covering less than 8 hectares each have adopted this practice. Even though this traditional system has some limitations, it still provides sustenance and food security to the vast majority of Kenyan population. After harvesting the maize, the crop residues are removed for use as fodder or are burnt for fodder. This makes it difficult for the progeny of any stemborers surviving in the stalks to repopulate the fields. This method has some negative effects on soil conservation and can reduce soil fertility while increasing soil erosion (see Mwangi and Ely, 2001).

Many farmers place neem extract or some pyrethrum extracts in the heart of the plants at the time of hatching of the stemborers so as to control maize stemborer problems.

Certain traditional multi-cropping or intercropping practices can be adopted with wild grasses that repel stemborer females, or planting *Sorghum vulgare sudanense* and *Pennisetum purpureum* at the periphery of maize fields which, as highly susceptible trap plants, attract the stemborers away from the crop. Also a fodder legume *e.g.* silverleaf desmodium (*Desmodium uncinatum*), which suppresses parasitism by witchweed *(Striga hermonthica)*, is planted among the crop. Such "push-pull" strategies have been shown to result in substantial yield increases over maize monocropping.

Quality Protein Maize

Globally, about 200 million children younger than five years are undernourished for protein, and suffer from stunted growth, weakened resistance to infection and impaired mental development. Many of these children eat maize-based diets-maize (*Zea mays*) is a major cereal crop worldwide. Being rich in carbohydrates, fats, proteins and some of the important vitamins and minerals, maize is reputed as a 'poor man's nutricereal'. Several million people, particularly in developing countries, get their protein and calorie requirements from maize. The maize grain accounts for about 15 to 56% of the total daily calories in diets of people in about 25 developing countries, particularly in Africa and Latin America. But cereal proteins have poor nutritional value for monogastric animals (including humans) because of reduced content of essential amino acids such as lysine, tryptophan and threonine. Cereal proteins contain about 2% lysine, which is less than one-half of the concentration recommended for human nutrition by the Food and

Agriculture Organization (FAO) of the United Nations. Therefore, healthy diets must include alternate sources of lysine and tryptophan. From the human nutrition viewpoint, lysine is the most critical limiting amino acid in the maize endosperm protein, followed by tryptophan. The problem has been mainly addressed by supplementing grains with essential amino acids produced by bacterial fermentation. This approach works well for feeding animals but is expensive. For this and other reasons, it is wiser to adopt a genetic enhancement strategy in which essential amino acids are either incorporated or increased in grain proteins. Substantial advances have been made in genetic enhancement of several cereal crops (Prasanna *et al.*, 2001), particularly in developing quality protein maize (QPM).

Breeding for improved protein quality in maize began in the mid-1960s with the discovery of mutants, such as *opaque*-2, that produce enhanced levels of lysine and tryptophan (see Olson and Frey, 1987; Chopra, 2001). Concerned research efforts led to amelioration of negative features of the opaque phenotype, and to some development of QPM which has superior nutritional and biological value and is essentially interchangeable with normal maize in cultivation and kernel phenotype.

The guiding principle in developing competitive QPM genotypes was combining the nutritional advantages offered by the *o2* mutation with the *o2* modifiers that contribute to the genetically complex endosperm modification trait. Since the *o2* gene increases lysine levels two-fold, efforts were devoted to maintaining rather than further enhancing the levels of lysine at protein levels of 9-10% in the whole grain. This approach greatly facilitated breeding of agronomically superior QPM genotypes, with emphasis on alleviation of key problems related to grain texture of *o2* genotypes (see Hallauer, 1994; Mertz, 1990).

Considerable progress has been made in the identification of the number or location of major genetic loci involved in endosperm modification. Analysis of segregating progenies and Recombinant Inbred Lines (RILs) derived from crosses between *o2* and modified *o2* genotypes indicated two independent loci affecting seed opacity and density. Endosperm modification was found to be associated with enhanced accumulation of g-zeins, suggesting that either the g-zeins are directly involved in the process of seed modification or the modifier gene(s) could be tightly linked to those responsible for g-zein synthesis. Two major loci involved in *o2* modification have been revealed by RFLP analysis; one locus maps near the centromere of chromosome 7 and the second maps near the telomere on the long arm of chromosome 7; *opaque*-15, a mutation that maps near the telomere of chromosome 7L, appears to have the properties of a defective *o2* modifier.

It appears that two or three major loci influence endosperm modification from the opaque to vitreous phenotype. Availability of a combination of molecular probes that would allow selection of endosperm modifiers in *o2* genotypes, prior to selection for agronomic characteristics, should facilitate rapid and efficient conversion of non-QPM inbred lines into QPM counterparts. Some DNA-based markers have been identified that could be of value in selection of endosperm modifiers contributing to the QPM phenotype.

Nutritional genomics has the potential to further enhance the nutritional value of maize through elucidation and effective manipulation of biochemical pathways and molecular mechanisms controlling kernel quality (see Hallauer, 1994; Singh, 2001; Chopra, 2001). Genomic techniques are being employed to investigate the patterns of gene expression in mutants influencing maize kernal texture (see Prasanna *et al.*, 2001).

Modification of genes encoding zeins and genetic engineering of key enzymes involved in the lysine biosynthetic pathway, namely, aspartate kinase and dihydropicolinate synthase, are some possible options to enhance the nutritional value of maize grain by increasing the lysine content. Deregulation of lysine biosynthetic pathway via genetic engineering may prove effective, provided there is no impairment of normal metabolic functions in the vegetative tissues, and the increased lysine is confined to the kernel (see Prasanna *et al.*, 2001).

Contamination of Corn in Mexico

Genetically engineered material has contaminated native corn varieties in Mexico. Out of 22 communities tested by government agencies in the state of Oaxaca, contamination of corn by transgenics was found in 15 (Cryan, 2001). This represents the first proven case of transgenic contamination affecting a crop at its center of origin-in the region where it evolved, where numerous landraces and wild ancestors still exist.

Mexico placed a moratorium on the planting of genetically engineered corn in 1998 so as to protect the grain's biodiversity base. The government did not, however, eliminate or regulate the import of U.S. corn to be used for human food or animal-feed corn. At least 25% of corn produced in the U.S. is transgenic (James, 2000). Some of the imported U.S. corn may have been planted by Mexican farmers, who could not have known that the corn they purchased might be genetically engineered.

There is now growing concern that besides endangering biodiversity, the introduction of transgenic crops allows some multinational corporations to strengthen their control of the seed supply and expand pesticide sales. Farmers who wish to grow local varieties or who own only a few acres of land are unable to compete with bigger farms that buy seeds and pesticides from the agri-biotech giants-especially now that the genetic material of the small farmers' local corn varieties cannot be protected from contamination by the agri-biotech companies' products (Cryan, 2001).

Resistance to Rootworms

Corn rootworms (an adaptive complex of beetles) feed on corn roots, resulting in yield loss and harvesting problems. The search for corn plants resistant to rootworms has been largely fruitless; corn growers rely on two strategies for managing rootworm damage: crop rotation (which prevents larvae from feeding on corn roots during the next season, thereby interrupting the rootworm life cycle) or insecticides (which directly kill rootworms infesting a crop). For several reasons, both strategies are becoming less effective, and growers are now uncertain about the most effective management strategies for these pests. Moellenbeck *et al.* (2001) provided one solution: successfully developing transgenic corn expressing two novel proteins from *Bacillus thuringiensis (Bt)* strain PS149B1 that show toxicity to corn rootworms.

Corn rootworms belong to several species of *Diabrotica, e.g., virgifera virgifera, barberi, virgifera zea,* and *undecimpunctata howardii.* These species join the ever-growing list of insect pests targeted by *Bt* technology, including the Colorado potato beetle (which has been targeted by Cry1A(b), Cry1A(c), Cry9C, Cry1F, or Cry2A(b)), the cotton boll feeders and defoliators (targeted by Cry1A(c) or Cry2A(b)), and the soybean defoliators (targeted by Cry1A(c)). Two distinct commercial efforts have focused on producing transgenic corn resistant to rootworm infestations: Dow AgroSciences' (Huxley, IA) and Pioneer HiBred International's (Johnston, IA) collaborative effort to develop the *Bt* PS149B1 toxin (described by Moellenbeck *et al.*); and Monsanto's (St. Louis, MO) *Bt* Cry3B(b) toxin (see Ostlie, 2001).

In terms of yield losses, harvesting delays, and preventative insecticide expenses, the combined impacts of the principal indigenous rootworm species exceed over a billion U.S. dollars per year. Corn rootworms are the predominant reason why corn has the second largest insecticide use among field crops in the United States. Although crop rotation is used on large acreages of corn, approximately 25% of the US corn crop receives a soil insecticide application annually. This insecticide use has aroused such environmental concerns as bioaccumulation of chlorinated hydrocarbons, avian toxicity, ground and surface water contamination, other non-target effects, also health concerns, *e.g.*, toxicity to workers who handle the chemicals. By delaying egg hatching, some species have become resistant to crop rotation, whereas others have shifted egg laying to other rotational crops; this threatens to increase insecticide use. Faced with these problems, growers need new management options.

Transgenic corn resistant to corn rootworms may simplify corn rootworm management while avoiding the potential environmental and health concerns. Moellenbeck *et al.* have demonstrated that the novel binary proteins in *Bt* PS149B1 successfully control *D. virgifera virgifera* rootworms in transgenic corn as effectively as soil insecticides.

Transgenic technology is easy to use, does not delay planting, and requires no application equipment. Besides improved root protection, corn rootworm-resistant corn also eliminates applicator, handler, and farm worker exposure to soil or foliar insecticides. With a narrow spectrum of activity, transgenic corn greatly reduces the nontarget concerns generated by broad-spectrum soil insecticides.

But more research is needed to assess the actual dose effects of *Bt* corn, corn rootworm biology and ecology, the need for integrating corn rootworm and soil insect management, and the impacts of different resistance management options.

Transgenic technology, such as the *Bt* PS149B1 toxin plants reported by Moellenbeck *et al.* can potentially reshape corn rootworm management. Resistance management plans developed for other pests in corn, cotton, and potatoes are based on a high dose/refuge model that may simply not be appropriate.

Soybean

Roundup Ready (RR) soybean was commercially released in the USA in 1995, after successful field results revealed yield increases and reductions in chemical inputs. American farmers adopted the GM crop without much hesitation. Soon thereafter, Monsanto's GM crops heralded a rapid but silent revolution that brought millions of hectares of arable land in the USA under GM crops. Later Argentina, Canada, China and South Africa also adopted GM crops. But, GM crops are also being grown in other countries where levels of political and social stability are not high and where it is usually very difficult to implement any biosafety legislation.

Four major transgenic crops have so far dominated world markets: RR soybean accounts for 58% or 25.8 million ha of the total area under GM crops (as in 2000-2001); transgenic corn for 10.3 million ha; transgenic cotton for 5.3 million ha; and GM canola for 2.8 million ha (see Jones, 2001; Schenkelaars, 2001). Argentina and the USA lead in GM crops-in Argentina 95% of all soybean is transgenic whereas in the USA this figure is 54%.

Monsanto launched its biotechnology programme in the mid-1980s, but faced serious problems in exporting its transgenic crops to foreign countries. In its attempts to sell its products outside North America, Monsanto encountered traditional communities having philosophies and regulatory systems quite different from those prevailing in USA. While the company's initial promotion of GM crops in the USA emphasized yield increases and cost reduction, more recent arguments in support of its GM product have been prompted in response to its European experience, being derived from the lexicons of development agencies and the environmental movement-the key words frequently used now are environmental friendliness, sustainability, poverty alleviation, and food security. Monsanto seems to be moving towards the notion of the "greening of the multinationals" with a view to improving its public image. It pledged to pursuing a policy of environmental responsibility for all its products. The company's latest pledge, in 2001, commits Monsanto to "openness" and "transparency" in its communications with the public. It has pledged to post summaries of the "safety evaluation of agricultural biotechnology products" on its website.

Roundup Ready (RR) canola was first patented by Monsanto Corporation in 1993. It entered the commercialization phase in Canada in 1997, but the behaviour of the GM crop in the field created difficulties for farmers growing canola from their own saved seed as well as for organic farmers.

RR canola shows resistance to the company's highly successful herbicide Roundup. The company claimed that in combination with Roundup, RR canola increases farmers' yields and (over time) lowers their input costs by eliminating weeds and plants that otherwise would inhibit the crop growth.

The patenting system concerning genetically modified seeds threatens traditional farming practices because farmers are expressly forbidden from saving seed and have to return to the patentee each year for a fresh supply. This practice makes farmers dependent on seed corporations for their seeds.

With more and more conventional plant varieties being currently modified genetically and patented, these corporations will undoubtedly exercise increasing control over the overall seed supply, and hence will determine the variety of crops grown.

This ongoing power shift in global agriculture is rendering traditional farming practices illegal by the extension of patent monopolies to the fundamentals of food production. A fairly small number of big corporations are exerting greater and greater control over these fundamentals.

The problem of "volunteer" GM plants poses serious concern to organic farmers because GM crops are incompatible with organic certification systems.

Governments will have to address the issue of whether the developers of GM crop seeds owe any obligation for damage caused by the unwanted proliferation of these crops. Also, there is need to focus on the obligations of the governments that grant patents over GM crop systems. In Argentina, organic farmers are planning to sue the government over its large-scale approval of GM corn and soybean crops. As Argentina has enthusiastically welcomed GM crop systems, their proliferation threatens the existence of Argentina's organic farming sector.

Whilst nobody can argue against some potential of gene technology as applied to agriculture, the speed with which companies such as Monsanto have introduced a complex, new and innovative science into the public domain has naturally generated some distrust (see Jones, 2001).

The hasty commercialization of GM crops has often dangerously involved oversimplifications, for the complexities of hunger and poverty have to be considered in their socio-economic dimensions, not science alone. The promotional activities of most transnational companies have erected serious barriers to the development of informed public awareness of the realities of the transgenic revolution (see Jones, 2001).

It has emerged from recent judgements in litigation surrounding patenting systems that multinational corporate patent holders successfully defend their intellectual property rights in the courts, with the result that corporate control over the fundamentals of agricultural production is going to increase in the coming years.

Today, multinational corporate patent holders possess a strong arsenal of legal rights. But sooner or later they will also have to reckon with their obligations. Governments should immediately define the obligations that need to be discharged along with the legal right to sell and exploit GM agricultural technology. Undoubtedly, emergence of GM agricultural technology constitutes a strong challenge to traditional agricultural practices. To address this challenge, a global debate over the rights and obligations of the purveyors of such technology is the need of the hour (see Phillipson, 2001).

GM Soybean in Brazil

The United States, Brazil and Argentina account for about 90% of soybean exports with Brazil occupying 26% of grain, 25% of soybean meal and 16% of soybean oil exports worldwide. The demand for non-biotech soybean has grown to 25% of the EU market, 44% of which is supplied by Brazil (von der Weid and Tardin, 2001).

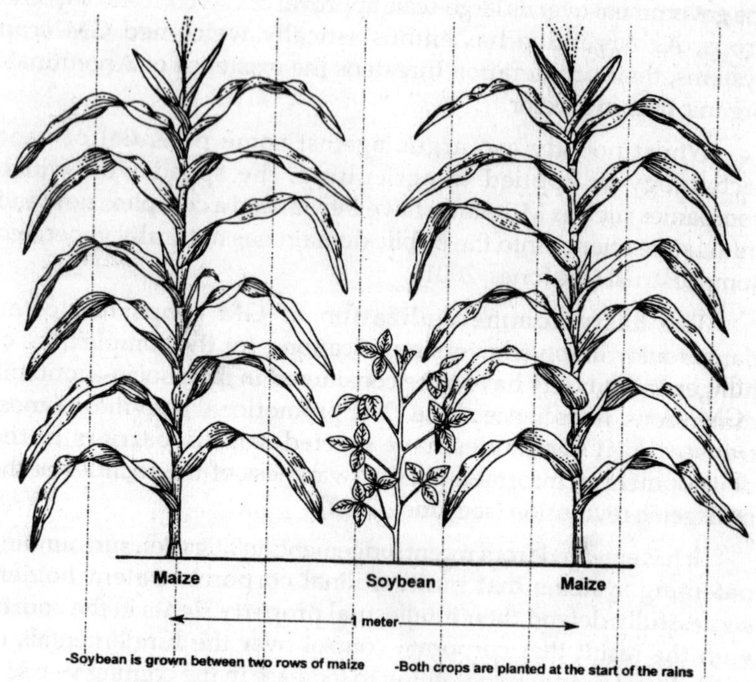

Fig. 5.1: Intercropping as a Feature of the Agroecological System of Soya Production

According to the Brazilian government, transnational corporations like Monsanto and Syngenta and the big soybean producers, the country will lose markets if it lags behind in biotechnological improvements in soybean production. Nevertheless, more careful assessments of agricultural performances, costs and market opportunities point to this position as being unreal; Brazil is doing well without GM soybean, and what's more, agroecological soybean production is a viable and competitive option for many small-scale farmers.

Conventional (non-GM) Soybean Production

Over the past 3 decades soybean production has developed enormously with increases in both yields and area planted.

Conventional genetic improvement has permitted farmers in various ecosystems to choose from around 170 varieties. The widespread adoption of nitrogen-fixing bacteria (NFB) has eliminated the use of expensive nitrogen fertilizers. The major pest threat to soybean production in Brazil, the *Anticarsia gemmatalis* worm, is controlled with a cheap biological agent, *Baculovirus anticarsia*. Much reduction of soybean production costs has resulted from the fact that at least 30% of the seeds used are produced by the farmers themselves.

Comparison of GM to non-GM Soybean

Does GM soybean have enough advantages to compensate for the risks of adverse impacts? If we compare yields of Brazilian non-GM soybean to USA yields (50% planted to GM soybean), the latter attained 2560 kg/ha in 2000/01, and the former, 2710 kg/ha. In the 5-year period since the introduction of GM soybean in the USA the average yield was 2,520 kg/ha. compared to 2,400 kg/ha. in Brazil (von der Weid and Tardin, 2001). The yields of non-GM soybean in Brazil have increased faster than the GM soybean yields in the USA, largely because production costs are higher for GM soybeans.

With the increasing demand for non-GM soybeans from Europe, Brazilian exports soared from 11 million tons in 1999 to 14 million tons in 2000, whilst US exports stagnated. Moreoever, non-GM soybeans have got a premium of 11 US$/ton, whereas prices for biotech products have dropped.

The Agroecological Alternative

In southern Brazil, agroecological soybean production is being promoted. It is based on direct sowing, no-till system using green manure varieties as cover crops and weed control based on mechanical hand weeding. It suffers minimum soil losses, minimal losses from leaching, and no soil, food or water contamination.

If we compare agroecological soybean production with conventional systems in Brazil the advantages stand out with a yield of 2,677 kg/ha from the soybean plot for the agricultural year 2000/01, whilst production costs were 240.95 US$/ha. Prices for conventional and organic soybeans also differed significantly, 17.20 US$ compared to 24.60 US$ per 60kg bag of grain. In the southeastern Parana region agroecological soybean production is already a major economic alternative for family farmers, with nearly 400 of them involved in this activity since 1995.

How can such extreme cost differences between GM (USA), conventional (Brazil) and agroecological (Brazil) soybean productions be explained?

In the agroecology case, weed infestations dropped by 50% in 4 years with continuous agroecological practices (see Fig. 4.1) in bean, maize and soybean plots, further reducing production costs in comparison to chemical systems. No chemicals are used in agroecology system, but crop rotations, green manuring and biofertilizers produced on the farm are used. No pesticides or fungicides are used. Low doses of lime are used to neutralize soil acidity.

Thus it emerges that agroecological soybean production is competitive with both conventional and GM cropping systems, without the harmful impacts of the former and the apparent risks of the latter. Wider application of these experiments depends more on enabling public policies, mostly related to credit, rural extension and participatory research (von der Weid and Tardin, 2001).

Bt Cotton

In South Africa, *Bt* cotton has benefited farmers by giving higher yields as compared to non-*Bt* varieties, and also reduced pesticide costs. The farmers' profit margin has substantially improved.

Some major agronomic problems faced by cotton farmers in South Africa relate to pests, particularly the bollworm, and weeds.

It appears that the advantages of *Bt* cotton become particularly apparent in times of environmental stress-an important consideration for risk-prone, resource-poor farmers. Also smaller holdings benefit more from growing the *Bt* variety.

After considering the issue of commercial use of *Bt* cotton for a few years the environment ministry has finally cleared the cultivation of the insect protected variety of cotton in India. This will benefit many farmers from India's cotton-growing states who can now buy and sow the genetically-modified seeds. Gujarat, Maharashtra, Andhra and Punjab are the major cotton growing states and a fourth of the world's cotton acreage is in India.

Field tests have been conducted on *Bt* cotton for four years. *Bt* cotton has shown proven resistance to the bollworm, the cotton crop's incurable pest. The issue attracted headlines in 2001 when yet another bollworm attack wiped out much of the standing crop, save the *Bt* variety, which a Gujarat-based firm had illegally got and sold. A government order to uproot the standing crop of *Bt* cotton was vigorously resisted by the farmers; in addition, those who hadn't planted it demanded they too have access to it. Whereas Indian government has been testing for years, China has, over the last few years, already approved *Bt* cotton seeds for commercial use. India and China both got into biotechnology testing at the same time.

Canola and Mustard

According to Ginns (2002), some field tests conducted in South Australia have shown that 0.05% to 0.01% of pollen from canola can travel beyond 400 metres (which is the recommended planting buffer zone).

This implies that proponents and activists of GM crops need to be particularly careful in properly segregating GM and non-GM crops and to ensure that GM seeds do not end up where they are not intended to. The rather lax attitude of users of the technology is a clear warning sign to patent holders and licensed users.

One of the more debatable aspects of the emerging GM technology is the 'terminator' gene, originally designed to be incorporated into varieties such as Roundup tolerant canola, to prevent the problem of the unwanted appearance of volunteer plants and their unwitting or intentional reproduction. This technology has been strongly criticised by consumer and farming groups. One of the major intentions of the terminator gene was to protect farmers from passive patent infringement

Concerns have again emerged about the possible emergence of 'super- weeds' after it was noticed in Canada that crossing between GM and non-GM canola varieties is actually more frequent than expected (see Ginns, 2002).

According to Ginns, we can protect farmers and the environment by adopting a mechanism specifically designed to 'self regulate' reproduction of GM varieties. Many hybrid crop, fruit and vegetable varieties (produced by conventional plant breeding) are sterile.

Applying the terminator and similar genes to GM varieties may be the logical way of avoiding the problems of patent infringement.

Herbicides are currently being used extensively for weed control in crops of wheat and rice. In view of the increasing concern about pollution of the environment, toxicity to animals and persistence of the herbicides or their residues in the soil and water, herbicides like glyphosate and L-phosphinothricin (PPT) which are highly effective at low dosage, safe for animals and rapidly degraded in soil might prove more useful for weed control. Unfortunately these herbicides are non-selective and therefore have to be used in conjunction with transgenic crops that are resistant to these herbicides. Globally, crops with resistance to non-selective herbicides occupy about 20 million hectares out of the total of 28 million hectares under transgenic crops. In India, a major benefit of herbicides could be in no-till or low-till agriculture for moisture conservation in the rain-fed areas and for multiple cropping in the irrigated, intensively cultivated areas.

DL-phosphinothricin (glufosinate) is a broad spectrum, post-emergence, contact herbicide (with some systemic activity), which is marketed as Basta (Agrevo). Glufosinate is a modified version of bialaphos, a naturally occurring herbicidal tripeptide antibiotic 'L-phosphino thricyl-L-alanyl-L-alanine' produced by *Streptomyces hygroscopicus*. Glufosinate lacks the terminal alanine and contains PPT which is the active herbicidal ingredient.

Transgenic lines resistant to herbicide PPT have been developed in mustard (*Brassica juncea*), a major oilseed crop grown on 6 million hectares of land in North India. Seedling-derived hypocotyl explants were transformed with a disarmed *Agrobacterium tumefaciens* strain GV3101. The developed constructs contained the *bar* gene encoding the enzyme phosphinothricin-acetyl-transferase (PAT) which acetylates phosphinothricin (PPT), thereby inactivating it. The expression of the *bar* gene was controlled either by the double enhancer version of CaMV35S promoter (35Sde*bar*) or a CaMV35S promoter with a leader sequence from RNA4 of alfalfa mosaic virus introduced at the 5' end of the *bar* gene (35SAMVL*bar*) or without (35S*bar*) it. Plant viral leader sequences have been shown to be translational enhancers. Mehra *et al.* (2000) carried out *in-vitro* selections for transformed plants on a medium containing PPT, and recovered transgenic shoots at a frequency of 16%-23%. Single copy transgenics were selfed to develop homozygous lines which could be used for the study of resistance to herbicide PPT at the field level and to correlate this protection with expression levels observed through molecular analysis. Herbicide-tolerant lines could be used for testing the possibility of low-till or no-till cultivation of mustard in the rain-fed areas where it is extensively grown (Mehra *et al.*, 2000).

Effect of Climate on Yields of Corn and Soybean

There is no doubt that substantial increases in crop yields will be needed to satisfy the future demand for food worldwide. Changes in climate and diminishing returns from technological advances could limit the ability of many regions to achieve the required gains (Cassman, 1999). Rosenzweig and Parry (1994) had predicted the effect of future climate changes on crop production using a both field studies and models, but there is no evidence that relates decadal-scale climate change to large-scale crop production. Now Lobell and Asner (2003) have reported that recent trends in temperature have increased the productivity of the two major U.S. crops and that accounting for climate significantly reduces the perceived gains due to various factors, especially management.

A surprisingly high percentage of the improvement in crop yields recorded in the USA since the 1940s seems to have been due not to farm management but to climate change. Food production in the United States seems be more vulnerable to changes in climate than was previously suspected, a fact that could affect global food security. Lobell and Asner (2003) analyzed the role of climate and other factors in American agriculture. They studied the interactions among temperature, rainfall, amount of sunshine, and bushels of corn and soybeans per acre from 1982 to 1998. During this time, summers in a large area of the Midwest became slightly cooler. Lower temperatures in the region increased the yields of corn and soybeans by about 30% over the study period.

The cooling climate was found to be responsible for about 20% of the gains (see Stokstad, 2003). Some gains also accrued from management and other factors, such as increased carbon dioxide in the atmosphere. It is emerging that food production may be more vulnerable to shifts in climate than assumed previously; calculations made by Lobell and Asner indicate that yields would drop by 17%

for each degree that the growing season warms. Most climate models predict that the Corn Belt of the U.S. Midwest will warm over the next few decades (Stokstad, 2003).

These yield trends in the U.S. also have global implications. For example, drop in U.S. production might stimulate more planting of soybeans in environmentally sensitive areas such as Amazonian watersheds in Brazil.

Should climate in the United States turn unfavourable, farmers would be hard pressed to compensate, in view of the fact that U.S. agriculture is already near its maximum efficiency. But some feel that if the temperature rises, the crop breeders could develop hybrids that can tolerate it. More vulnerable to climate change are developing countries, where temperatures are already high, soils are often poor, and management is not satisfactory. Climate change could really affect them greatly.

Besides temperature, climate, variability and the number of extreme events, such as floods, also strongly affect yields. Besides, temperature shifts might influence harvests in other ways, such as by bringing more or less land into production, according to Mark Rosegrant of the International Food Policy Research Institute in Washington, D.C. (see also Dyson, 1999).

Genetic Manipulation of Gibberellin Metabolism

Dwarfing is a very valuable trait in crop breeding, because it results in plants that are more resistant to lodging and give higher yields (Evans, 1993). It was the introduction of the semi-dwarf trait into cereal crop cultivars that fueled the 'green revolution' (see Khush, 1999; Peng *et al.*, 1999). The endogenous phytohormone gibberellin GA is one of several factors associated with the dwarf phenotype.

The GA biosynthesis pathway has been established, and most genes encoding GA biosynthesis enzymes have been identified. By using the genes for these GA biosynthetic or catabolic enzymes, one can control the level of bioactive GAs to produce dwarf crop plants by genetic engineering (Hedden and Phillips, 2000, 2000a). Sakamoto *et al.* (2003) adopted a biotechnological approach to produce dwarf rice plants by modifying the GA level by overproduction of a GA catabolic enzyme, GA 2-oxidase. When the gene encoding GA 2-

oxidase, *OsGA2ox1*, was constitutively expressed by the actin promoter, transgenic rice showed severe dwarfism but failed to set grain because GA is involved in both shoot elongation and reproductive development. In contrast, *OsGA2ox1* ectopic expression at the site of bioactive GA synthesis in shoots under the control of the promoter of a GA biosynthesis gene, *OsGA3ox2 (D18)*, resulted in a semi-dwarf phenotype that is normal in flowering and grain development. The stability and inheritance of these traits point to the feasibility of genetic improvement of cereal crops by modulation of GA catabolism and bioactive GA content (Sakamoto *et al.*, 2003).

The green revolution genes, wheat *Rht* and rice *sd1*, are involved in gibberellin signaling and biosynthesis, respectively (Sasaki *et al.*, 2002; Peng *et al.*, 1999). This makes genetic manipulation of the bioactive GA level a practical strategy to develop varieties having suitable dwarf architecture for high grain yield.

Two possible approaches to reducing the endogenous GA content are (1) suppression of GA biosynthesis and (2) enhancement of GA catabolism. Antisense expression of GA biosynthesis genes exemplifies the first approach. However, only a few transformants showed a semi-dwarf phenotype with reduced bioactive GA content.

The second approach is exemplified by overproduction of a GA catabolic enzyme, GA 2-oxidase. In contrast to the first approach, overproduced GA 2-oxidase strongly affects bioactive GA content and plant height. The second approach therefore seems to be better than the first approach for the breeding of dwarf plants (Hedden and Phillips, 2000, 2000a).

Genetic Engineering of Disease-Resistant Grapes

The possibility looms large that wine makers will face massive consumer rejection if they accept genetically engineered grapes. Several requests for field trials of engineered grapes have already been granted in the USA including research into genetically engineered grapes resistant to Pierce's disease. A Greenpeace survey of British retailers representing 80% of the UK's wine sales found that all would refuse to carry wine made from genetically engineered grapes. Britain is the largest export market for California wine, consuming 30% of the state's wine exports.

The Greenpeace report, *GM Grapevines* also found widespread rejection of genetic engineering elsewhere in Europe. In France, wine makers have called for a moratorium on genetically engineered vines and for increased government research into alternative solutions. The institute that supervises specifications of French wines currently forbids any genetically modified grape varieties. In Italy, the Association of Wine Producing Towns and the regional councils of Tuscany and Valle d'Aosta have opposed genetically engineered wine experiments.

Although new reports of field trials conducted by the University of Florida to engineer a wine grape resistant to Pierce's disease have projected the grape as a potential panacea for California's grape disease problem, researchers do acknowledge that the gene inserted into wine grapes makes a protein similar to an ingredient in bee venom that can cause anaphylactic shock in some people. They also admit that it may not be possible to alter the grapes in such a way to keep the potentially deadly substance out of the human food chain.

Genetically engineered grapes pose a tremendous risk to the California wine industry and to organic winegrape growers. Organic winemakers and importers are very concerned about the contamination of organic vineyards by wind- and insect-borne pollen from genetically engineered grapes. If there is accidental contamination, the damage can be irreversible.

References

Altieri, M.A. *Agroecology: The Scientific Basis of Alternative Agriculture.* Westview Press, Boulder (1987).

Ammann, K., Jacot, Y., Simonsen, V., Kjellsson, G. (eds.). *Methods of Risk Assessment of Transgenic Plants.* Birkäuser Verlag, Basel (1999).

Betz, F.S., Hammond, B.G., Fuchs, R.L. Safety and advantages of *Bacillus thuringiensis*-protected plants to control insect pests. *Regulatory Toxicol. Pharmacol.* 32 (2): 156-173 (2000).

Brown, I.R., Renner, M., Flavin C. *Vital Signs 1998: The Environmental Trends That Are Shaping our Future.* W.W. Norton, New York (1998).

Cao H, Li, X., Dong, X. Generation of broad-spectrum disease resistance by overexpression of an essential regulatory gene in systemic acquired resistance. *PNAS* (USA) 95: 6531-6536 (1998).

Cassman, K.G. *PNAS* (USA) 96: 5952 (1999).

Chopra, V.L. (Ed.). *Breeding Field Crops.* Oxford & IBH, New Delhi (2001).

Cryan, P. GE pollution in Mexico: Native corn contaminated. *Global Pesticide Campaigner* 11(3): 16 (Dec. 2001).

Dawkins, R. *The Selfish Gene.* Oxford University Press, Oxford (1976).

Dawkins, R., Krebs, J.R. Arms races between and within species. *Proc. Royal Soc. London* Series B 205: 489-511 (1979).

Denison, R.F., Kiers, E.T., West, S.A. Darwinian agriculture: When can humans find solutions beyond the reach of natural selection? *Quart Rev. Biology* 78: 145-168 (2003).

Dominguez, C.A. Genetic conflicts of interest in plants. *Trends in Ecology & Evolution* 10: 412-416 (1995).

DRR. High yielding rice varieties of India - 2000, Bulletin 2001-1, Directorate of Rice Research, Hyderabad, pp. 1-102 (2001).

Dyson, T. *PNAS* (USA) 96: 5929 (1999).

Evans, L.T. *Crop Evolution, Adaptation and Yield.* Cambridge Univ. Press, Cambridge (1993).

Evans, L.T. *Feeding the Ten Billion. Plant and Population Growth.* Cambridge Univ. Press, Cambridge, (1998).

Ewel, J.J. Natural systems as models for the design of sustainable systems of land use. pp. 1-21 in *Agriculture As a Mimic of Natural Ecosystems,* Lefroy, E.C. *et al.* (eds.) Kluwer, Dordrecht (1999).

Fukai, S. Intercropping-bases of productivity. *Field Crops Research* 34: 239-245 (1993).

Ginns, D. Timely lesson in genetically modified canola case. *Austral. Farm J.* 11(12): 48-49 (2002).

Halluer, A.R. (Ed.). *Specialty Corns.* CRC Press, Boca Raton, Fl pp. 80-121 (1994).

Haribabu, E. Cognitive empathy in inter-disciplinary research: the contrasting attitudes of plant breeders and molecular biologists towards rice. *J. Biosci.* 25: 323-330 (2000).

Harlan, J.R. *The Living Fields. Our Agricultural Heritage.* Cambridge Univ. Press, Cambridge (1995).

Hedden, P. Constructing dwarf rice. *Nature Biotechnol.* 21: 873-874 (2003).

Hedden, P., Phillips, A.L. Gibberellin metabolism: new insights revealed by the genes. *Trends Plant Sci.* 5: 523-530 (2000).

Hedden, P., Phillips, A.L. Manipulation of hormone biosynthetic genes in transgenic plants. *Curr. Opin. Biotechnol.* 11: 130-137 (2000a).

IRRI (Internat. Rice Res. Inst.). *Fragile Lives in Fragile Environments.* Int. Rice Res. Inst. Manila pp. 369-394 (1995).

Jackson, *I.E.*, Koch, G.W. The ecophysiology of crops and their wild relatives, pp. 3-37 in *Ecology in Agriculture,* Jackson, *I.E.* (Ed.) Academic Press, San Diego (1997).

Jackson, W., Piper, J. The necessary marriage between ecology and agriculture. *Ecology* 70: 1591-1593 (1989).

James, C. Global status of commercialized transgenic crops: 2000. International Service for the Acquisition of Agri-Biotech Applications (ISAAA) Briefs No. 21: Preview. ISAAA: Ithaca, NY (2000).

Jones, M.M. Monsanto: Rewriting the script. *Biotech. Develop. Monitor* 48: 13-14 (2001).

Khush, G.S. Green revolution: preparing for the 21st century. *Genome* 42: 646-655 (1999).

Lefroy, E.C., Hobbs, R.J., O'Connor, M.H., Pate, J.S. (eds.). *Agriculture as a Mimic of Natural Ecosystems.* Kluwer, Dordrecht (1999).

Lev-Yadun, S., Abbo, S., Doebley, J. Wheat, rye, and barley on the cob? *Nature Biotech.* 20: 337-338 (2002).

Lobell, D.B., Asner, G.P. Climate and management contributions to recent trends in U.S. agricultural yields. *Science* 299: 1032 (2003).

Mallik, S. in *Sustaining Crop and Animal Productivity - The Challenge of the Decade* (ed. Deb, D.L.), Associated Publishing Co., New Delhi, pp. 37-46 (1995).

Mallik, S., Mandal, B.K., Sen, S.N., Sarkarung, S. Shuttle-breeding: An effective tool for rice varietal improvement in rainfed lowland ecosystem in eastern India. *Curr. Sci.* 83: 1097-1102 (2002).

Manicka Velu, A., Durai, R.S.R., Malar, R.P.G. Golden rice. *LEISA* India 3: 4 (Dec., 2001).

Mehra, S., Pareek, A., Bandypadhyay, P., Sharma, P., Burma, P.K., Pental, D. Development of transgenics in Indian oilseed mustard (*Brassica juncea*) resistant to herbicide phosphinothricin. *Current Science* 78: 1358 (2000).

Mertz, E.T. (Ed.). *Quality Protein Maize.* Am. Assoc. Cereal Chemists, St. Paul, MN (1990).

Moellenbeck, D.J. *et al. Nature Biotechnol.* 19: 668-672 (2001).

Mohanty, H.K., Mallik, S., Grover, A. *Curr. Sci.* 78: 132-137 (2000).

Mwangi, P.N., Ely, A. Assessing risks and benefits: *Bt* maize in Kenya. *Biotech. Develop. Monitor* 48: 6-9 (2001).

Olson, R.A., Frey, O.J. *Nutritional Quality of Cereal Grains: Genetic and Agronomic Improvement.* pp. 183-236 Am. Soc. Agron., Madison, WI (1987).

Ostlie, K. Crafting crop resistance to corn rootworms. *Nature Biotechnol.* 19: 624-625 (2001).

Peng, J. *et al.* 'Green revolution' genes encode mutant gibberellin response modulators. *Nature* 400: 256-261 (1999).

Phillipson, M. Agricultural law: containing the GM revolution. *Biotech. Develop. Monitor* 48: 3-5 (2001).

Prasad, G.S.V., Prasadarao, U., Rani, N.S., Rao, L.V.S., Pasalu, I.C., Muralidharan, K. Indian rice varieties released in countries around the world. *Current Science* 80: 1508-1511 (2001).

Prasad, G.S.V., Prasadarao, U., Sobharani, N., Rao, L.V.S., Pasalu, I.C., Muralidharan, K. *Curr. Sci.* 80: 1508-1511 (2001).

Prasanna, B.M., Vasal, S.K., Kassahun, B., Singh, N.N. Quality protein maize. *Current Science* 81: 1308-1319 (2001).

Risch, S.J., Andow, D., Altieri, M.A. Agroecosystem diversity and pest control: data, tentative conclusions, and new research directions. *Environ. Entomol.* 12: 625-629 (1983).

Rosenzweig, C., Parry, M.L. *Nature* 367: 133 (1994).

Sakamoto, T. *et al. Nature Biotechnol.* 21: 909-913 (2003).

Sakamoto, T., Morinaka, Y., Ishiyama, K., Kobayashi, M., Itoh, H., Kayano, T., Iwahori, S., Matsuoka, M., Tanaka, H. Genetic manipulation of gibberellin metabolism in transgenic rice. *Nature Biotech.* 21: 909-913 (2003).

Sasaki, A. *et al.* A mutant of gibberellin-synthesis gene in rice. *Nature* 416: 701-702 (2002).

Saxena, D., Flores, S., Stotzky, G. Transgenic plants: Insecticidal toxin in root exudates from *Bt* corn. *Nature* 402: 480 (1999).

Saxena, D., Stotzky, G. *Bt* toxin uptake from soil by plants. *Nature Biotechnol.* 19: 199 (2001).

Schenkelaars, P. Agronomic and environmental effects of growing glyphosate soybean, Centrum voor Landbouw en Milieu and Schenkelaars Biotechnology Consultancy (June 2001).

Simmonds, N.W. "Shades of Green." Review of *The Doubly Green Revolution: Food For All in the Twenty-First Century,* by Conway, G. *Nature* 391: 139 (1998).

Singh, U. *Quality Protein Maize Products for Human Nutrition.* Directorate of Maize Research, New Delhi (2001).

Soule, J.D., Piper, J.K. *Farming in Nature's Image. An Ecological Approach to Agriculture.* Island Press, Corelo, Calif. (1992).

Spillane, C. Could agricultural biotechnology contribute to poverty alleviation? *AgBiotechNet* 2: ABN042 (2000).

Stokstad, E. Study shows richer harvests owe much to climate. *Science* 299: 997 (2003).

Swaminathan, M.S. Biotechnology and agricultural betterment in the developing countries; in *Biotechnology in Agriculture.* Natesh, S., Chopra, V.L., Ramachandran, S. (eds.) Oxford and IBH (New Delhi) pp. 3-11 (1987).

Uphoff, N. The system of rice intensification: agro-ecological opportunities for small farmers? *LEISA Magazine* 17(4): 15-16 (2001).

von der Weid, J.M., Tardin, J.M. Genetically modified soybeans. *LEISA* Magazine 17(4): 19-20 (Dec., 2001).

Chapter 6

Food Security, Organic Farming and Alternatives to Botechnology

Introduction

Proponents of genetic engineering have sometimes justified it as a humane technology–one that can feed more people with better food. True. But in reality, with very few exceptions, the primary agenda of companies engaged in genetic engineering seems to be to increase the sales of chemicals and bio-engineered products feeding the billions is only secondary.

The new genetic engineering revolution has made it possible to break through natural species' barriers, systematically moving genes from one species to another that do not normally combine in nature. Genetic material can be transferred, for instance, from bacteria to plants. Proponents of genetic engineering (GE) believe that it can provide improved plants and animals, crops that produce their own pesticide so reducing the use of chemical pesticides; crops that produce drugs; plants tolerant to salt and drought, and enriched food to restore micronutrient deficiencies. But the following disturbing questions lurk behind the above claims:

1. Who benefits from genetic engineering and who loses?
2. What are the risks and who will bear them?
3. Are there alternatives to genetic engineering?

In fact, genetic engineering differs radically from previous technologies because it allows for the transfer of genes between different species across natural boundaries, thereby making the risks unpredictable.

There are several concerns about the implications of GM crops-these relate to the consequences for the ecological systems into which the GM crops are introduced. This is best exemplified by a report from the USA where genes from the bacterium *Xanthomonas* were transferred to another soil bacterium, *Klebsiella planticola*. The new organism was intended to ferment stubble into alcohol, so providing extra income to farmers who were otherwise burning the stubble. It later turned out that wheat planted in the soil containing the new organism was killed by it!

The development of GE virtually uniquely benefits companies or transnational corporations (TNCs). It is these corporations which also hold the patents to the so-called. "Terminator" technology in which genes are manipulated to be able to switch seeds on and off by treatment with chemicals provided by the same GM seed company, effectively preventing farmers from saving their seeds for replanting. Strong public opposition has compelled the companies not to enforce this technology but they still hold the patents. The companies chiefly promote high-input, highly industrialized monoculture systems, which force farmers to buy packages of inputs from a single company.

Organic farmers cannot make sure that their crops are totally free from GM crops, considering that seeds and pollen spread by wind, water, birds and insects and large areas can become contaminated by the introduction of GM crops by a single farmer. In the US, contamination by GM crops has become so troublesome that organic farmers find it very difficult, if not impossible, to obtain GM-free seeds.

With the advent of agricultural genetic engineering, the costs of contamination and of falling market shares are being imposed on farmers, consumers and the general environment everywhere. The risk of an unintended introduction of GM crops is greatest in those countries where either no legal framework exists at all or where the legal enforcement is weak.

Impact of Modern Biotechnology on Global Agri-food Industry

Increasing commercial application of biotechnology in the agri-food sector has aroused considerable criticism, raising doubts about how extensively it will be used in the future. Nevertheless, the rapid adoption of this new transformative technology points to a number of crucial developments that are expected to fructify or materialize over the coming few decades (Phillips, 2002).

The scientific base for agri-food production has expanded greatly. Until very recently, virtually all food used to be developed using such techniques as selection by the farmer, advanced phenotype-based breeding, wide crosses by embryo-rescue, or by mutagenesis. But all these have now been overtaken or even eclipsed by new gene splicing techniques and rapid advancement in molecular breeding. A wide array of new factors and objectives have been incorporated into agricultural research concerning the manner of use of this new technology. Many people interested in biotechnology and agri-food policies believe in a positive future in which technology will overcome food shortages, improve the environment, heal or eliminate disease and will lead to a prosperous and healthy society. In contrast, some policy makers, environmentalists and consumers fear that the technology can exacerbate food insecurity, threaten the environment, endanger human health and ultimately even impoverish society. Considerable debate on the pros and cons of biotechnology has led to the general consensus that while biotechnology does have the technical potential to overcome some food production constraints, mitigate some environmental stresses and develop enriched foods and new medicines, it is hard to say whether our social institutions will be able to adapt, adopt and use the technology in a manner that will satisfy society and improve social welfare (Phillips, 2002).

The Biotech Revolution

The first successful plant transformation in the year 1986 heralded the revolution in the agri-food sector. Within about a decade since then (by 1997), biotech was increasingly exploited by the private sector in the agri-food sector to genetically modify about 60 crops for 10 different traits. Most of the attributes commercialized in the first generation of products relate to input and yield performance.

Two of these, *viz.* insect resistance (*e.g. Bt* cotton and corn) and herbicide tolerance (*e.g.* Roundup Ready™ soybeans), accounted for about 99% of the GM world acreage in 2001 by which year plants were also engineered to resist viruses and to be sterile. Several second-generation output trait crops have also been developed and commercialized, including delayed ripening tomatoes (*i.e.* Flavr Savr™), and new or modified oils. While most of the effort focused on modifying crops, some research has been done to develop non-crop related agri-food applications, such as animal vaccines, hormones, fish, food animals, microbes and active food ingredients (Phillips, 2002).

The first GM crop, a herbicide tolerant tobacco, was released in China around 1990, and full commercialization began in 1994, with approval for the Flavr Savr™ tomato in the USA. By 2002, several more crops, modified for one or more of about 50 phenotypic traits, had been commercialized. But only four crops *viz.*, soybean, maize, cotton and canola constituted around 99% of the total acreage planted to GM crops in 2001.

Currently, corporate and public researchers are working hard to improve crop tolerance to abiotic stresses, to enhance nutritional or functional attributes, to express for nutraceutical or pharmaceutical proteins and enzymes and to be used for phytoremediation of soils contaminated with heavy metals. Some researchers look to the new fields of genomics and proteomics to solve many of the constraints in current research efforts.

One fact that emerged from the wide adoption of GM crops is that both science and the markets need to grow in tandem. If the science and markets advance as per expectation, the value of GM seed sales could (possibly) double by 2005 and reach US$25 billion by 2010 (Phillips, 2002).

According to Phillips, the following five factors will influence the future use of biotechnology in the agri-food sector.

Research and Development Policy

A major long term threat lies in the demand for land, labour and capital in more productive local sectors. Policy makers are increasingly looking to research as a means to increase productivity in domestic agriculture.

One debatable issue is whether intellectual property rights (IPRs) create a 'freedom to operate' (FTO) problem for research and commercialization. If there are only a few core technologies (*e.g.* transformation systems), then a company with a patent could block competitors. As IPRs create transaction costs, this could restrict FTO, reducing the spillovers that create increasing returns to scale and thereby limiting commercialization of new products. For instance, the development of the much acclaimed Golden rice™ involved as many as 70 patented technologies, so its inventors decided to tender the product to any company that could assemble the FTO for commercialization (Zeneca won the tender). One remedy is to just wait for property right to lapse (*i.e.* patents past 20 years and PBR (plant breeder rights) past 18 years, and most key biotechnology patents have only 10 or so years left). Recent years have seen an increasing trend to pair public funds with private capital, which appears to be shifting research priorities towards commercially attractive applications.

Location

Location is in the limelight for discussion of the dynamics and benefits of knowledge-based sectors such as biotechnology-and how to capture benefits from this new technology when only a few multinational companies in advanced countries own the technology. The aim is to attract research and commercialization locally, so that the local public and consumers benefit. Commercialization of research seems to be linked to specific sites through the location of key research 'stars' in the networks (see Wolf and Zilberman, 2001). There are only about 25 leading centres around the world in which biotechnology research and development is clustered. According to Phillips and Khachatourians (2001), studies of one such cluster (the Saskatoon-based canola cluster in Canada) suggest that research and commercialization is tied up locally by a combination of open-platform research institutions that pool and create new pre-commercial and non-competitive knowledge, a corresponding specialized labour market for researchers and sophisticated, progressive farmers. Especially for developing countries, clusters could play a crucial role in the development and commercialization of many smaller agri-food applications.

Society

The manner in which biotechnology is being applied to the agri-food sector has prompted several socio-economic and ethical issues. Much current debate relates to how society should manage biotechnology gainfully and adequately manage risks. One notable fact is that, for whatever reason, GM products have often been approved and produced in countries that were already major producers and exporters, rather than in importing countries; this may be due to diverging regulatory systems-major exporting countries know about the benefits of production and adopt the technology aggressively whereas traditional importers consider the technology as a threat to their already uncompetitive agri-food sectors and raise barriers (Buckingham and Phillips, 2002). That is why GM foods approved and generally accepted in the USA, Canada and Japan are rigorously scrutinized, or stigmatised with labels in Europe and Australia. This often puts small firms, small markets and public research institutions at a disadvantage because they cannot reap the full economies of scale and market potential for their products.

Marketing

The fragmenting trade consensus and changing consumer preferences are challenging markets to better establish and control quality parameters for agri-food products. Regardless of how the regulators sort out their differences, markets are establishing their own private regimes for labelling food quality attributes and are developing private identity-preserved production and marketing (IPPM) systems for differentiable products (Phillips, 2002). These systems involve close non-market relationships in at least part of the supply chains, including packages of inputs (seed, chemicals and financing), marketing contracts with growers and investment in the supply chain. But all these systems are mostly purpose-built for their products, incur fairly high costs and face high risks of failure (Phillips, 2002). Except for a few systems involving differentiated GM or GM-free produce, none of the systems is cost competitive. Two general problems have emerged. First, even if IPPM systems could be created, without adequate detection and enforcement systems, differentiated markets may not be sustainable. Second, the opportunity cost of not differentiating products may become too high when second and third generation GM foods enter the market because they require segregation to capture the value inherent in their development. To tackle these two problems agri-food sector will have to develop suitable standards, reduce costs, create more accurate and timely detection methods and develop a reliable system of external quality assurance (see Knoppers and Mathios, 1998).

Winners and Losers

The future of agricultural biotechnology depends on the returns it can offer. If these are high enough and are distributed to match with opportunity costs, then use of the technology may increase. But if the returns are low and inequitably distributed, then the technology may fail.

Recent studies on GM soybeans and *Bt* cotton have shown that innovators usually gain the major share of the benefits, that consumers gain some, and that producers can gain or lose (*e.g.* early adopters win, non-adopters lose) (Kalaitzandonakes, 2003). Most of the reported farmer benefits (*e.g.* ease of use and reduced risk) have not yet been quantified in monetary terms. No doubt, gross annual returns of biotechnology applications are often quite high, but the cost of developing the innovations and any externalities of adopting the technology are also high. For herbicide tolerant canola, the gross investment by the industry could only be recovered in the seventh year of adoption and, even by then, industry had not recovered the value of the pre-existing chemical markets that were cannibalised by the new technology (Kalaitzandonakes, 2003).

The new agri-food biotechnology does have a strong potential to transform the agri-food sector provided that the social, economic and institutional challenges can be addressed. According to Phillips (2002), "the prospects for a better bioengineered world will not be determined by the science but by the capacity for public and private institutions to adapt, adopt and use the new technology in commercially and socially beneficial ways."

Detection of GMOs in Foods

The production of genetically modified varieties of, for instance, corn, cotton, canola and soybean has necessitated enactment of laws to regulate the presence of GMOs in foods. At least 50% of food products in American supermarkets contain GMOs (Beachy, 1999;

Ahmed, 2002). Although in North America consumers have generally accepted the newly developed food biotechnology and there has been much less public resistance to the introduction of genetically modified foods, these modified foods have not gained general worldwide acceptance particularly in many European countries because of consumer concerns resulting from earlier food and environmental concerns, transparent regulatory oversight and mistrust in government bureaucracies-factors that generated debates about the environmental and public health safety issues of introduced genes (*e.g.* potential gene flow to other organisms, the destruction of agricultural diversity, allerginicity, antibiotic resistance and gastrointestinal problems) (Haslberger, 2000).

Also certain economical and ethical issues relating to intellectual property rights were raised (see Serageldin, 1999) when it was realized that inadvertent contamination of non-GM seeds with GMOs can occur. These concerns prompted the European Union (EU) to either restrict the import of bioengineered foods or to enact legislation requiring mandatory labeling of GM foods and food ingredients that are no longer equivalent to their conventional counterparts (Anonymous, 1997; 2000), or foods or ingredients containing additives and flavourings that have been produced from GMOs (Anonymous, 2000a). EU regulations require labeling of foods containing GMOs, and fix a 1% threshold for contamination of unmodified foods with GM food products. In the USA, labeling of GM foods is not mandatory but a voluntary labeling of bioengineered foods is recommended (see Ahmed, 2002).

An agreement called the 'Cartagena Biosafety Protocol', has laid down rules governing the trade and transfer of GMOs across international borders (*e.g.* labeling of shipments of GM commodities); it allows governments to prohibit the import of GM food when there is concern over its safety (Gupta, 2000).

As universal legislation requires that governments and others concerned, *e.g.*, crop producers should develop ways to accurately quantitate GMOs in crops, foods and food ingredients to assure compliance with threshold levels of GM products, various test methods, their potential and their limitations have been reported (Ahmed, 2002).

These methods are generally either protein-based or DNA-based and employ western blots, enzyme-linked immunosorbent assay, lateral flow strips, Southern blots, qualitative-, quantitative-, real-time- and limiting dilution-PCR techniques. In cases where information on modified gene sequences is lacking, novel approaches, *e.g.*, near-infrared spectrometry, might facilitate detection of non-approved genetically modified foods (Ahmed, 2002).

GM products usually contain some additional trait coded for by an introduced gene(s), which generally produce an additional protein(s) responsible for expressing the trait. Either raw grains or processed foods derived from GM crops may in principle be identified by looking for the presence of introduced DNA, or by detecting expressed novel proteins encoded by the genetic material. Both qualitative and quantitative methods are available (Schreiber, 1999).

GeneScan Europe (2001) has developed a test kit that detects GMOs in food products, by allowing a multiplex PCR for the specific detection of DNA sequences from plant species and GM traits. The procedure involves isolation and purification of DNA from the sample, followed by amplification of specific DNA sequences from both plant species and GM traits (Table 6.1) in two separate multiplex PCR reactions; the products of both reactions are mixed, and single-stranded (ss) DNA appears following digestion with an exonuclease. After mixing with hybridization buffer, the sample is spread on the chip and amplified sequences that will hybridize with cDNA probes covalently bound in the chip are stained with the fluorescent dye Cy5 and analyzed by a biochip reader (Ahmed, 2002).

Table 6.1: Some Plant Species and Genetic Elements Included in the GeneScan™ GMO Chip[a,b]

DNA	*For the Detection of*
from plants	Corn, canola, soybean, rice, CaMV
from GMOs	CaMV 35S-promoter; *bar* gene (BASTA resistance for Sh); Gene encoding phosphinothricine-acetyltransferase from Sv; Nos-Terminator; *Bt* 11 corn; Roundup ReadyTM soy

[a] Abbreviations: *Bt* , *Bacillus thuringiensis;* CaMV, cauliflower mosaic virus; GMO, genetically modified organism; Sh, *Streptomyces hygroscopicus;* Sv, *Streptomyces viridochromogens* (after GeneScan Europe, 2001).

Table 6.2 summarizes some of the methods commonly used for detecting recombinant DNA products of GM foods.

Table 6.2: Some Methods that Specifically Detect rDNA Products Produced by GM foods[a-c] (after Magn *et al.*, 2000; Ahmed, 2002)

Parameter	*Protein-based*			*DNA-based*			
	Western Blot	*ELISA*	*Lateral Flow Strip*	*Southern Blot*	*Qualitative PCR[d]*	*QC-PCR and Limiting Dilution*	*Real-time PCR*
Ease of use	Difficult	Moderate	Simple	Difficult	Difficult	Difficult	Difficult
Sensitivity	High	High	High	Moderate	Very high	High	High
Duration[e]	2 d	30-90 min	10 min	6 h[f]	1.5 d	2 d	1 d
Cost/sample	Moderate	Low	Low	Moderate	High	High	High
Quantitation of results	No	Yes[g]	No	No	No	Yes	Yeh[h]
Suitable for field test	No	Yes[g]	Yes	No	No	No	No
Employed mainly in	Labs	Test facility	Field	Labs	Test facility	Test facility	Test facility

[a]Abbreviations: ELISA, enzyme-linked immunosorbent assay; GM, genetically modified; QC-PCR, quantitative-competitive PCR; rDNA, recombinant deoxyribonucleic acid. [b]Modified from Magn *et al.* (2000); [c]Near infra-red detects structural changes (not DNA or protein), is fast (<1 min) and inexpensive (~US$1); [d]including nested PCR and GMO Chip; [e]Excluding time alloted for sample preparation; [f]When nonradioactive probes are used; otherwise 30 h with ^{32}P-labeled probes; [g]As in the antibody-coated tube format; [h]With high precision. All methods, except lateral flow strip require special equipments.

Alternatives to GM Technology

The real issue is whether we really need GM technology to combat malnutrition or to enhance local and agriculture production. Have GM crops in any way helped to remove or reduce poverty? In fact, the problem is not one of food production but of purchasing power. Any residual hunger problems are due to poverty, rather than food availability. This means that achieving the objective of food security for all needs to be based on a premise other than GE. Alternative approaches to agricultural production are, therefore needed.

There exist several agroecological, low-external-input alternatives to agricultural production. There are many plants in nature which provide us with valuable clues to better pest management. In India, agricultural research is only exploiting some 3% of the total of 3000 known rice varieties that are known. Indeed, many of the problems faced by farmers are amenable to fairly cheap, controllable and low external input solutions by paying greater attention to hitherto neglected ecological principles. Many such principles are still being overlooked, underestimated or sidelined; they deserve more attention.

Some notable examples of ecologically-sound alternatives that already exist are the push-pull system in Kenya, organic cotton production in Senegal and zero-tillage no-herbicide soybean cultivation in Brazil. They pose no danger to the environment, nor do they make the farmers dependent on agricultural supply companies.

It is the small or peasant farmers that produce a high percentage of staple foods in most Third World countries. But this important sector suffers greatly from poverty and hunger. In fact, rural areas in the Third World typically suffer from extreme inequalities in access to land, in security of land tenure and in the quality of the land farmed (Altieri and Rossett, 1999). Wages and living standards are low, so that healthy domestic markets can never emerge, reinforcing export orientation. There is a downward spiral into deeper poverty and marginalization, even as national exports become more "competitive" in the global economy. An unfortunate irony of our world is that food and other farm products flow from areas of hunger and need to weathy and affluent people in the North (Rossett, 2001).

Another similar irony is that the situation is no better in the developed nations where the better soils are converted into large holdings and used for mechanized, pesticide and chemical fertilizer-intensive, monocultural production for export. The productive capacity of the soils is declining through fertility loss, growing resistance of pests to pesticides, and the loss of edaptic and above-ground functional biodiversity.

Lagging Productivity

Third world farmers have to face lagging productivity largely because they have been displaced onto marginal, rain-fed lands, and have to deal with structures and macroeconomic policies that are becoming more and more unfavourable to food production by small farmers. The marginal zones to which they are displaced are characterized by broken terrain, irregular slopes, erratic rainfall, little irrigation, and/or low soil fertility. Faced with such unfavourable circumstances, these peasants can only survive if they can tailor agricultural technologies to match their variable but unique circumstances, in terms of local climate, topography, soils, biodiversity, cropping systems, market system and resources. Farmers have over millennia evolved complex farming and livelihood systems which balance risks. The result is that their cropping system are fairly diversified and include multiple annual and perennial crops, animals, fodder, and a variety of foraged wild products (Rossett, 2001).

By and large, such farmers have not significantly benefited from any 'green revolution' technologies. Addressing their productivity and poverty concerns must be based on meeting their needs for multiple suitable varieties, and should deal with the vast complexity of physical and socio-economic conditions in most third world agriculture. The obsession with monocultural 'yield' above all else must go and complex rural realities should be considered. The fact is that farmers have multiple site-specific requirements for their seeds, not just controlled-condition high yields. These interlinkages are at variance with and incompatible with, formal breeding procedures. A different approach involving participatory breeding by organized farmers themselves, based on the multiple characteristics of both seed varieties and farmers, is essential. And, genetic engineering is in direct conflict with participatory, farmer-led research (Rossett, 2001).

Risks for Small Farmers

The currently common transgenic varieties fall into two classes, those that tolerate proprietary brands of herbicides, and those that contain insecticide genes. The former are irrelevant to peasant farmers who plant diverse mixtures of crop and fodder species; also, chemical herbicides would only destroy key components of their cropping systems.

Transgenic plants that make their own insecticides–commonly using the '*Bt*' gene-are rapidly failing as pests build up resistance to insecticides. Pest species rapidly adapt and develop resistance to the insecticide present in the plant. Indeed, *Bt* crops violate the principle of "integrated pest management" (IPM), which postulates that reliance on any single pest management technology tends to trigger shifts in pest species or the evolution of resistance. Thus IPM approaches are based on multiple pest-control mechanisms, and use pesticides minimally, only as a last resort. Most entomologists agree that *Bt* is destined to rapidly become useless.

Further, the use of *Bt* crops affects non-target organisms and ecological processes. The *Bt* toxin can affect beneficial insect predators that feed on natural insect pests of *Bt* crops, and windblown pollen from *Bt* crops found on natural vegetation surrounding transgenic fields can kill non-target insects. Most small farmers depend for insect pest control on the rich community of predators and parasites associated with their diversified cropping systems.

Citizens' Juries on GMOs

Developers of GM technology should ensure that their work addresses the requirements of small farmers in developing countries. The system of patents and licenses should be modified suitably so as to facilitate sharing of GM technology with scientists in developing countries.

Many people have expressed doubts about the ability of golden rice to eliminate vitamin A deficiency. They feel that a better approach to fight vitamin A deficiency and malnutrition is to go in for various inexpensive and nutritious foods already available (*e.g.* green leafy vegetables, *Moringa* drumstick fruits, etc.) and in diversifying food production systems in the fields and at household level.

Ecological soil management is a key to sustainable agriculture. Many problems in agriculture relate to soil and soil fertility management. Examples are soil erosion, low efficiency of fertilizers, nutrient imbalances, soil borne diseases, pests and weeds, soil compaction, farmer induced drought and water pollution. Productivity and ecological sustainability in agriculture may be successfully raised by improved soil management. We can learn much from the traditional ecological management practices adopted by farmers including organic agriculture, agroecology, natural farming, integrated soil fertility management, zero tillage and integrated pest management. Perhaps a judicious combination of traditional and organic technologies with chemical fertilisers may yield rich dividends. Organic soil management technologies being highly labour demanding, appropriate mechanization is important. Ecological soil management being site and crop rotation specific it is desirable to fine-tune ecological soil management for specific conditions and rotations.

Over the past two decades, several 'participatory' methods have been developed with a view to democratising policy-making (ODI, 1999; Campolina, 2001). These include citizens' juries, neighbourhood forums, consensus conferences, scenario workshops, multi-criteria mapping, participatory rural appraisal, visioning exercises and polling. These methods can potentially empower people to go beyond being passive recipients of development policies or users of externally-imposed technologies.

Karnataka

A citizens' jury was organised by Action Aid India and took place on a farm in a small village in the state of Karnataka. This dryland area of the Chitradurga district has a high proportion of marginal farmers and landless peasants.

As it is the lives of small farmers that would be primarily affected by the introduction of GMOs, the jury was composed of fourteen small and marginal farmers (six men and eight women) who represented the variety of farming traditions, income levels and social groupings. The jury also included expert witnesses who presented evidence for and gainst GMOs and other participants and observers. Scientific Institutes, commercial biotechnology corporations (Monsanto), development NGOs, Farmers Unions and Government agencies were represented among them.

The jury spent four days hearing information from 'witnesses' on the merits and limitations of GMOs. The subject under discussion was the possible future role of GMOs in the context of reducing rural poverty and promoting sustainable agriculture (Pimbert *et al.*, 2001).

The jury deliberated on the question: Would you sow the new commercial (GMO) seeds proposed by the Indian Department of Biotechnology & Monsanto on your fields? The verdict was: 4 yes, 9 no, 1 invalid ballot paper (by secret ballot). The jury's rejection of the GMO seeds was not simply a negative response. It was supplemented by the following suggestive actions that should be taken by the government and transnational corporations to get better acceptance for their new seeds.

1. Microbes and beneficial insects should not be damaged. New seeds should not harm animal populations.
2. They should be lawfully released only after extensive field trials for at least 5 years in which farmers shall be involved not only in yield assessments but also in safety, environmental and other aspects.
3. They should not damage the succeeding crop that is grown on the same field or adjoining fields.
4. The success of the new seeds should be judged not just under lab conditions, but also, on farmers fields.
5. The technology must be easy to adapt.

The jury also considered the issue of increasing farmer confidence in multinational corporations (MNCs) and biotechnology:

1. Some members of the jury were afraid of any contact with MNCs, having heard about them in the context of WTO and patents. They felt that the powerful MNCs, which develop their seeds in laboratory conditions, could ultimately gain control over seeds and farmers' sovereignty.
2. If the seeds fail for any reason, whether to do with the technology itself, or weather conditions, the MNCs should not only compensate for the losses, but also buy up the whole crop at double the price.

Andhra Pradesh

Prajateerpu, the citizens' jury on food and farming futures in Andhra Pradesh was another exercise in involving rural people in decisions having a strong impact on their livelihoods. This participatory process was jointly organised and facilitated by the British-based International Institute for Environment and Development (IIED) and the Institute of Development Studies (IDS) and the India-based Andhra Pradesh Coalition in Defence of Diversity, the University of Hyderabad and the all-India National Biodiversity Strategy and Action Plan (NBSAP). This jury was made up of small and marginal farmers, food processors and consumers, dalit (untouchable caste) and indigenous ('adivasi') people. The jury members were presented with three different scenarios. For four days they listened to and cross-questioned twelve witnesses including representatives of the Government of A.P., the Indian branch of the International Federation of Organic Agriculture Movements (IFOAM) and Syngenta, the major biotechnology corporation. It was up to the jury to decide which of the three scenarios, or combination of elements from each, was most likely to provide them with the best opportunities to enhance their livelihoods, food security and the environment, by the year 2020.

Three Visions

Vision 1: Vision 2020

This scenario was suggested by Andhra Pradesh's Chief Minister backed by a loan from the World Bank. It proposes to consolidate small farms and to rapidly increase mechanisation and modernisation. Production enhancing technologies such as genetic modification will be introduced in farming and food processing, reducing the number of people on the land from 70% to 40% by 2020.

Vision 2

An export-based cash crop model of future organic production is based on certain proposals for environmentally friendly farming linked to national and international markets. It is also increasingly driven by the demand of supermarkets in the North to have a cheap supply of organic produce and comply with new eco-labelling standards.

Vision 3

This is a future scenario of localised food systems based on increased self-reliance for rural communities, low external input agriculture, the re-localisation of food production, markets and local economies, -with long distance trade in goods that are surplus to production or not produced locally.

The key conclusions reached by the jury - their 'vision' - included a desire for:

1. Food and farming for self reliance and community control over resources.
2. To maintain healthy soils, diverse crops, trees and livestock, and to build on our indigenous knowledge, practical skills and local institutions.

The jury was opposed to:

1. The proposed reduction of those making their livelihood from the land from 70%-40%.
2. Land consolidation and displacement of rural people.
3. GM crops - including vitamin A rice and *Bt* cotton.
4. Labour-displacing mechanisation
5. Contract farming
6. Loss of control over medicinal plants including their export.

Rejection of GMOs by Brazilian Small-scale Farmers

A citizen's jury on GMOs met in Belem do Para, capital of the Amazonian State of Para, Brazil, in September 2001. Hundreds of small-scale farmers, landless people and poor urban consumers attended the event.

Before the event, the organisers chose 6 community-based associations (2 landless settlement associations, 2 rural workers' labour unions and 2 urban associations). These organisations provided the organisers with complete lists of membership. Four members (2 men and 2 women), were picked at random from these lists. Thus, 24 potential members for the jury were identified. The first task of the jury was to select, at random, 7 members from this list for the jury (4 women and 3 men). After that, the judge, head of the Law faculty of the Federal University of Pará, read the case, which had been agreed upon by the prosecution and defence, prior to the hearing. This case presented a definition of GMOs, the scope of the jury-GM agricultural varieties tolerant to herbicides, insects, plant diseases and new nutritional qualities, and the questions that the members of the jury were supposed to answer: (1) Can GMOs address the problem of hunger? (2) Can they improve the food security of small-scale farmers? (3) Is there enough evidence that they do not threaten the environment? (4) Is there enough evidence that they do not threaten food safety? (5) Is the process of liberalisation of tests and commercial use of GMOs democratic, transparent and careful enough?

After the presentation of the case, the prosecution and the defence lawyers made their first speeches, presenting the main arguments against and in favour of GMOs. The lawyers then invited their witnesses, three each. Each witness gave a 20-minute long presentation, and was then cross-examined by the defending and prosecuting lawyers, the judge and the members of the jury. The witnesses of the prosecution were an economist-a specialist on patents and transnational corporations, a professor of genetics at the University of São Paulo, and a specialist on environmental matters, and an anthropologist-a specialist on rural sustainable development. The defence witnesses were two biotechnology researchers from EMBRAPA, the national agricultural research institute, and a professor of biochemistry from the Federal University of Paraiba-a member of the national commission on biosafety.

After the presentations and cross-examination of the witnesses, the defence and prosecution made their closing arguments, after which the members of the jury went with the judge and an assistant to a closed room to proceed with the secret ballot on the 5 questions above. **The members of the jury voted unanimously against GMOs with a clear NO to all five questions** (see Campolina, 2001).

Organic Farming

Organic/biodynamic production is a farming system that contributes to healthy soils and/or people. It uses no synthetic chemicals, but rather promotes enhanced biological activity and encourages sustainability. It is a new and evolving form of agriculture.

Health, food quality, and the nature of the farming system are intimately linked. Health, whether of soil, plant, animal or man, is one and indivisible (Balfour, 1943).

Organic farming has expanded its general popularity during the past two decades. There has been a steep growth in sales of organic food, an increase in the number of farmers converting to organic production, and an increase in its media coverage. Organic food and farming may help in introducing a fundamental change in our approach to the health, animals, crops and ecosystems.

Very often organic agriculture is described in terms of what it is not. The commonest example of this is the idea that 'organic farming is farming without chemicals.' This description is untrue and misses out on several important characteristics. No doubt, the organic farming system seeks to avoid the direct and routine use of readily soluble chemicals and all biocides, including those that are naturally occurring or nature-identical; but it sometimes becomes necessary to use these materials or substances. In these situations a solution is chosen which will be least disruptive to the farming system as a whole, from the level of soil bacteria through plants and animals, to the environment in general (Woodward, 2002).

Another general misconception is that organic farming simply involves using so called 'organic' inputs inslead of 'agrochemical' ones. In fact a straight substitution of mineral fertiliser by manure can have much the same-probably adverse-effects on plant quality, disease susceptibility, and environmental pollution. This kind of approach is not organic farming. The misuse of organic materials by excess, inappropriate timing, or both, will just as effectively retard or inhibit the development and operation of the biological cycles on which the organic system seeks to build, as inorganic fertilisers or biocides.

Many farmers are exploring the idea of incorporating some of the organic principles into their farming systems. Increasing resistance to herbicides, the impoverishment of soil biota and organic matter levels, and concerns about chemical and nutrient run-off are some issues that prompt mainstream farmers to critically review all their options-including organics. Besides, changing attitudes within the farming sector towards organics and consumer buying patterns are also changing. Many rich and educated people are preferring organic foods. Some advocates of the organic farming industry think that for its expansion into a widely accepted, mainstream from of agricultural production it will have to become more large-scale, corporate and commercially focused (Lyon, 2002).

An opposing group believes that there are enough niche market opportunities for the industry to remain a loose confederation of small-scale producers primarily driven by ideological imperatives. One of the challenges facing the industry is to demonstrate the economic viability of organics on family-owned, small farms.

Worldwide, the organic industry was worth an estimated $37 billion in 2000. It has been growing at about 20-30% per year (Moore, 2002).

The biggest markets for organic products worldwide are in the USA, Europe and Japan. Consumers in these countries appear to have been particularly troubled by developments such as Mad Cow's disease, repeated food contamination scares, and resistance to irradiated and genetically modified foods.

According to Lyon, more farmers can make the conversion to organics if they are convinced that they can profit from going organic rather than continuing with the use of synthetic herbicides, fertilisers, growth promoters and food additives.

Organic agriculture also retains some features of the traditional farming practices, for example rotations, mixed farms and mechanical weeding. However, it seeks to build upon the increased understanding of mycorrhizal associations, the rhizosphere, the turnover of organic matter and other areas of soil life, and modern crop and animal husbandry practices. The organic movement is against genetic engineering, but it has no objection in using scientific knowledge and information-derived for example from genomics- because the appropriate application of science is essential for its further development (Woodward, 2002).

Papendick *et al.* (1981) defined the key practices of organic farming as they are found on the farm. According to them, it is a production system that avoids or excludes the use of synthetically compounded fertilisers, pesticides, growth regulators, and livestock feed additives. To the maximum extent feasible, organic systems depend on crop rotations, crop residues, animal manures, legumes, green manures, off-farm organic wastes, and aspects of biological pest control to maintain soil productivity and tilth, to supply plant nutrients, and to control insects, weeds, and other pests. The concept of the soil as a living system that develops the activities of beneficial organisms is central to this definition. This concept is extremely important in the context of activities on the farm as it points to the key management driver for the organic farmer, which is that everything affects everything else. On an organic farm there is no one method of weed control or of supplying nitrogen. Ley, green manures, and appropriate cultivation all contribute to both these objectives. Organic farm management focuses primarily on adjustments within the farm and farming system, in particular on rotations and appropriate manure management and cultivations, to achieve an acceptable level of output. External inputs are generally used only as supplements to this management of internal features (Lampkin, 1990). Many organic farmers have developed unique and productive, sometimes highly complex, systems of farming.

In conceptual terms the genesis of organic agriculture may arguably be traced to three quite different schools of thought, each of which originated in the first three decades of the 20th century: the biodynamic or anthroposophical school of Steiner, the organic-biological school of Muller and Rusch, and the organic school of Howard and Balfour. Also important was the work of Schuphan and Voisin, who promulgated the idea of the 'biological value' of soil, plant, and food in the early 1960s (see Balfour, 1943; Boeringa, 1980; Vogt, 2000; Woodward, 2002). Except for some significant difference between these various schools of thought, they share an essential core of agreement in respect of the following aspects:

1. The farm as a living organism, tending towards a closed system in respect of nutrient flows but responsive and adapted to its environment.
2. The 'living soil' can influence and transmit health through the food chain to plants, animals, and humans.
3. These linkages constitute a whole system within which there is a dynamic not yet clearly understood
4. Whilst the above ideas challenge orthodox scientific thinking, they need to be explored and developed; they will eventually be explained through appropriate scientific analysis (Woodward, 2002).

According to van Bueren *et al.* (1999) the available nutrients, soil structure, moisture content, air temperature, etc., and changes to that environment during a plant's growth have just as much influence on a plant's phenotype as its genes. In other words, a plant genotype is continuously interacting with its environment.

Much research done on the quality of both crops and livestock has revealed a clear trend that with appropriate crops, organic produce contains more desirable components (vitamins, dry matter, protein) and fewer undesirable substances (pesticide residues, nitrates, sodium) than conventional produce. In livestock trials, animals fed on organically grown feed usually have greater fertility and longevity than those fed on conventionally produced feed (Vogtmann 1992; Soil Association, 2000).

Organic Production

Whereas demand is expected to grow by some 25% a year supply growth is forecast at about 15% a year (Moore, 2002).

Consumers all over the world are increasingly becoming aware of the quality and safety of the food and fabrics they consume, driven by a desire to consume products, especially foods that are safe, produced from environmentally sustainable farming systems, and that are ethically and socially acceptable.

Prospective organic farmers would be well advised to have the marketing of their product negotiated before they go into production. They should at first go in for a part conversion to organic before a full conversion is undertaken.

While in many countries progress has occurred within the organic and conventional agricultural sectors, there is a fairly marked divide between organic practitioners and other farmers; the key point of contention being the issue of sustainability. Many organic farmers feel that the use of synthetic biocides, genetically modified organisms (GMOs), and chemical fertilisers will eventually result in the collapse of that system because it harms the environment and has many other adverse consequences.

On the other hand, conventional farmers see unacceptably higher risks and lower rates of return as being associated with organic farming.

Perhaps, the controversy may not be decided by agricultural producers. The final arbiter may well be the urban consumer in metropolitan centres (Moore, 2002).

Production Principles

The principles of organic production can be outlined as follows:

1. Optimal rather than maximal yields aimed for in any one species or crop.
2. Management practices rather than substance use is preferred (synthetically derived pesticides and highly soluble fertilisers are prohibited).
3. Aiming at biological cycles on-farm, with plants fed through the ecosystem of the soil, rather than via soluble mineral fertilisers.
4. Species selection, crop and livestock rotations, integration of cropping and livestock, green manuring of crops, natural mineral supplements and (non-sewage) bio-waste recycling are practised with a view to ensuring that nutrients are retained within the farm to produce healthy plants and animals.

In actual practice, however, most producers avoid using synthetic chemicals but they resort to the following farm practices:

Fertility

Use of mineral fertilisers, balancing of nutrient and micro-nutrient levels, use of (composted) manures, use of fish and other clean biological wastes, recycling of on-farm nutrients.

Pest and Disease Control

Biological agents (*Bt*, predatory insects), "soft option" oils, natural products and intercropping.

Further, the correct soil and nutrient balance is crucial to managing pest infestation.

Outcomes for the organic farm business include:

1. Enhanced, resilient crops
2. Soil protection via complex soils and good crop cover
3. Lesser disease and pest load

4. Multiple harvests: crops-diversifying risks
5. Risk management - diversification on-farm
6. Multi yields - social and physical environmental benefits (Moore, 2002).

Organic Plant Breeding

In Europe, organic agriculture greatly depends on the traditional seed industry. Organic farmers use modern productive varieties which were bred for a high-input farming system with the use of chemicals. These varieties yield better than old land races but are not adapted to specific organic conditions. They lack traits like nutrient uptake efficiency, early soil coverage against weeds, broad field tolerance against pests and diseases, etc.

The suitability of plant breeding methods can be determined as per the principles of organic farming. Organic farming is not merely the avoidance of chemical fertilisers, pesticides and GMOs: it takes the living soil as a basis and uses methods which stimulate (agro-ecological) processes, without exhausting natural resources. As it is founded on the integrity and intrinsic value of living entities like the soil, plants, animals and human beings, organic farming respects the environment, farm ecology and the complexity of nature. This attitude prevents farmers from taking actions that affect a plant's reproductive potential and impede the sustainable use of cultivars. The aim of organic plant breeding is to develop plants which enhance the potential of organic farming and biodiversity. Organic plant breeding is a holistic approach that respects natural crossing barriers and is based on fertile plants that can establish a viable relationship with the living soil.

Biodiversity: An Essential Feature

Since biodiversity is a core feature of a sustainable organic farming system, the organic sector places great value on the free exchange of the genepool. Due recognition is given to the rights of breeders but patents and techniques to make plants sterile endanger the free exchange and, consequently, the genetic diversity. Free exchange of genetic diversity can be prevented by the utilisation of cytoplasmic male sterility without restorer genes to produce hybrids. The absence of restorer genes prevents the production of seeds and hence this type of hybrids do not form. All other types of hybrids produce viable seeds which do not maintain purity after multiplication at the farm, but can still be used for developing new varieties (van Bueren and Osman, 2001).

Seed saving is not practised in the highly specialised horticultural sector in Europe. Dutch organic farmers prefer to buy their seeds, and most of them prefer hybrids. The uniformity of the plants allows for mechanical harvesting and reduces the requirement of the scarce seasonal labour.

Some of the biotechnological techniques used in modern plant breeding stay within the realm of life whereas others go beyond. If the cell is considered as the lowest organised structural entity of life, then all breeding techniques that intervene below the cellular level do not conform to the organic principles. This means that genetic modification (which interferes at DNA level) and protoplast fusion should be forbidden for the organic sector. All other cell biological techniques, including embryo rescue and in vitro-pollination, are acceptable.

Biotechnological Techniques applied in Plant Breeding (after van Bueren and Osman, 2001)

At Cell Level	
Embryo culture	Used for crossing of closely related species, such as cultivated
Ovary culture	tomatoes and wild relatives. Such crosses occur in nature but do
In-vitro pollination	not result in viable seeds because embryos abort prematurely. When these organs are separated from the plant and grown in test tubes, they develop into mature plants.
In-vitro selection	Used to select new, stress-tolerant (eg. salinity) varieties. Plants are grown in test tubes containing *e.g.*, a salt solution. Survivors are selected.
Anther culture (Microspore culture)	Pollen and anthers are grown in-vitro. These male sexual organs are not fertilised and are haploid. This set is doubled with chemicals to get (diploid) plants that are genetically identical.
Meristem culture Micro propagation	Used for rapid propagation of plants with an identical genetic make-up. Cells are multiplied in test tubes where they are regenerated
Somatic Embryogenesis	into new plants.

Contd...

Contd...

Below cell level	
Genetic modification Protoplast fusion	Genetic materials of unrelated species that do not cross in nature are inserted into cells. Protoplast fusion implies the merging of complete cells. In genetic modification only small pieces of foreign DNA are inserted into the cell.
Cytoplasmatic Male Sterility (CMS) without restorergenes	Used to produce parent lines for hybrid production which are male sterile. A plant cell is merged with a cytoplast (a plant cell of which the chromosomes are removed). The cell plasma of the cytoplast contains factors which cause male sterility. CMS does occur in nature, but is accompanied with factors that neutralise the male sterility. When CMS is transferred from an unrelated species into a crop, without the neutralising factors (restorer genes), the new male sterile plants cannot be multiplied in nature.
DNA marker assisted selection	Certain sequences of DNA can be associated with certain plant traits. These sequences (markers) can be used to select plants for traits not directly visible in the field, such as drought tolerance. This technique makes use of available DNA sequences in the plant cells, but does not change them, and is acceptable for organic agriculture. Sometimes radiation or genetically modified enzymes are used to detect these markers, which is not acceptable for organic farming. Detection can be done with substances that are permitted by organic agriculture such as fluorescence.

Cotton Production: Organic versus GM

The world's cotton production has tripled (to around 20 million tonnes) since the 1930s, largely due to the intensive use of synthetic chemicals; irrigation and higher-yielding varieties have also played their part. Large quantities of highly toxic pesticides are used on this crop. Regulatory systems are either weak or are not enforced and farmers are not properly informed of the dangers; they lack the necessary skills and equipment to protect themselves and their families.

By 2000, some 5 million hectares of GM cotton were grown, representing about 16% of the total cotton area planted worldwide. It is grown commercially in Argentina, Australia, China, South Africa and the USA. Three main types of GM cotton are herbicide tolerant (tolerant to glyphosate and bromoxynil–2.1 million ha), insect resistant (with the *Bacillus thuringiensis–Bt*-toxin genes inserted–1.5 million ha) and a third type which combines both attributes (1.7 million ha). There are grave concerns about the increasingly widespread use of this technology. Predicted reductions in herbicide use have not occurred and transfer of genes to related wild species appears inevitable (Myers, 2001). There are also concerns about the development of resistance to *Bt* . GM cotton may not have much appeal to small-scale cotton farmers in the South-mainly for economic reasons. For example, GM cotton is sold as a package that includes the herbicide which the cotton is engineered to tolerate. Farmers must buy new, more expensive seed each season.

The above concerns have prompted many small-scale farmers to seek alternatives. Growing interest on the part of Northern consumers and certain established verifiable regulatory systems for organic production have generated interest in the development of projects in many countries. Organic cotton production started in Turkey and the USA in the early 1990s. Other projects followed in South Asia, Africa and Latin America.

Certified organic cotton fibre is currently produced in widely varying production systems in 12 countries (Myers and Stolton, 1999). Global production of organic cotton was about 8000 tonnes in 1997 and it seems to have stabilized at that level since then.

Some factors that determine the success, or otherwise, of organic cotton projects include project structures, management, agronomic back-up and expertise, market conditions and access and above all the motivation of farmers which, especially for small farmers is determined by economic reasons.

Finding Local Solutions

Technically, organic cotton is a product of an organic farming system which relies on crop rotation, organic fertilisation and on non-synthetic chemical pest control methods. More specific technical details of production systems are site-specific and are determined by local conditions. Although, generally, organic systems produce less than conventional systems, this is compensated through savings on inputs, payment of premiums, and perceived improvements in health for people, their animals and their environment.

The rapid spread of GM cotton might pose a threat for organic cotton producers as the Basic Standards for Organic Agriculture prohibit the use of GM varieties. Where organic and conventional cotton are grown side by side, contamination may occur. Organic cotton farmers are accustomed to planting at a 'safe' distance between their organic fields and the conventional fields of their neighbours to avoid spray contamination. The introduction of GM cotton reinforces this requirement and probably would warrant longer distances.

Biologically Integrated Organic Farming

Agriculture today faces multiple environmental challenges, the result of a fast-growing population, the increased role of consumers in decision-making about the food system, a restrictive regulatory climate, and agriculture's contribution to non-point-source water pollution. At the same time, innovative partnerships involving growers, consumers, commodity boards, regulators and researchers have been exploring creative solutions to these challenges through biologically integrated and organic farming systems. The agricultural biotechnology industry is experiencing phenomenal growth in many industrially-advanced countries. The U.S. food industry's resistance to labeling products that contain transgenic ingredients has aroused consumer interest in organic products, which prohibit transgenics. These trends warrant the prediction that alternative farming systems would comprise an ever increasing share of all croplands in production in the coming years. Nonetheless, research investments into alternative biologically integrated and organic methods lag far behind organic product sales.

Burgeoning population growth has resulted in more contact and more conflict between farms and sprawling cities and suburbs. Production agriculture has entered a volatile political struggle over land, air, water and other resources. Motivated global consumers and environmentalists are demanding more active participation in shaping the quality and sustainability of farm and food systems. Increasing numbers of American consumers and public-interest groups are calling for disclosures on food labels regarding genetically modified organisms, countries of origin and even the labour practices employed in food production (Barham, 1997).

Agriculture continues to be a major factor in non-point-source pollution in waterways. Monitoring has revealed soil-absorbed organochlorines from historical pest-management practices and organophosphate insecticide residues-some above toxicity thresholds for resident aquatic species-in wetlands and waterways that drain into marine sanctuaries (Hunt *et al.*, 1999). Herbicides (*e.g.* simazine) and nitrates from crop and animal production threaten groundwater quality.

By bringing together diverse food-system participants, partnerships aid growers to adopt forward-looking approaches to contemporary pressures rather than waiting for change to be imposed by regulatory agencies. Model partnerships rely heavily on information-sharing to make the kinds of environmentally sound changes that consumers want (Swezey and Broome, 2001).

Farmers and ranchers are entering into partnerships with researchers and professionals to maximize the use of biologically derived sources of pest control and soil fertility to reduce their reliance on agricultural chemicals.

There is urgent need for a biological or "whole-farming-system" approach to agricultural extension and non-point-source pollution prevention. Whereas a traditional regulatory approach prescribes single-purpose "best management practices," the farming-systems approach addresses multiple environmental impacts by managing the farm as an interactive biological system.

Comprehensive management considers the linkages among the components of a farming system (tillage practices, crop rotation, nutrients, water and pest management) and the larger landscape and watershed. Emphasis on any single objective to the exclusion of others can cause unnecessary environmental trade-offs and increase costs, slow the rate of adoption and decrease the effectiveness of new technologies and management methods (NRC, 1993).

There is a general feeling that the IPM (integrated pest management) paradigm has been successful in improving the efficiency of insecticide use, but it has had less impact on weed and pathogen control. While earlier IPM strategies relied on pesticides as the primary management tool, a new ecologically based pest-management paradigm has emerged (NRC, 1996) that redesigns the farming system by using biological strategies such as naturally occurring compounds, cultural controls, biological control agents and resistant cultivars, as well as soil amendments and cover crops. The farm system is integrated into the larger landscape with hedgerows and farmscaping.

Organic agriculture promotes and enhances biodiversity, biological activity and natural cycles. It is based on minimal use of synthetic pesticides and fertilizers, and adopts management practices that restore, maintain and enhance ecological harmony. Organic farms reduce some negative impacts of conventional farming such as soil erosion and leaching of carbon and nitrogen. It may also protect wildlife. Fluesch and Sparling (1994) reported that organic apple orchards had higher numbers of nontarget insects and supported larger populations of songbirds than orchards where pesticides were used.

Cropland area increases for organic and biologically integrated farming will be fueled by consumer concerns about the environmental impacts of farming. Efforts are currently underway by grape, almond and other growers to evaluate "third-way" or "eco" labels as a means of informing consumers about products that incorporate biologically integrated farming systems (Swezey and Broome, 2001).

Transgenic Versus Organic Crops

Like organic agriculture, the agricultural biotechnology industry has experienced phenomenal growth in recent years. Transgenic crop sales totaled $236 million in 1996 worldwide and increased six-fold to between $1.2 billion and $1.5 billion in 1998, with growth projections reaching $20 billion by 2010 (James, 1997).

The food industry has resisted labeling products that contain transgenic ingredients; so future growth in the organic industry may be stimulated if transgenic ingredients are not permitted by the national organic standards. Indeed, opponents of labeling often argue that consumers who do not want to consume engineered food can simply buy organic.

A second, greener revolution may be imminent-one in which shared knowledge of biological processes that determine pest dynamics, soil health and microbial ecology will combine with the demonstrated ability of enlightened, discerning growers and agricultural researchers to innovate. Producers can expand economic and political alliances with urban consumers through direct-marketing, certified organic food and fiber, eco-labeling and educational programmes to change the rules of consumer engagement with the food system (Swezey and Broome, 2001).

Photosynthesis, Food Production and Organic Farming

Two broad advances in plant science research can contribute a great deal to enhancing public understanding of photosynthesis in the first decades of the 21st century. First, the thin green veneer of the planetary biosphere is now firmly recognized as a critical component of, and primary respondent to, global climatic change. Autotrophic functions of plants account for a significant part of global CO_2 and water vapour exchange in the atmosphere, and of nutrient fluxes in the biosphere. Second, and at the opposite pole in the scale of research, the facility of molecular genetic intervention, and information from genome projects, open up previously unimagined possibilities. However, today plant research is being adversely impacted by public anxiety about, and hostility to molecular genetic interventions, domination of plant genetic resources for food production by multinational corporations, and sustainability of biodiversity and habitat.

It is important to realize that photosynthesis research has a critical role in finding new paths to securing the global food supply and sustaining the human habitat. This realization has to sink in against the pressure of such over- statements as:

'The world today produces more food per inhabitant than ever before. Enough is available to provide sufficient quantities of grain, beans, nuts, meat, milk, eggs, fruits and vegetables to every person every day-indeed more than anyone could possibly eat.'

This type of statement appears frequently in media and is deeply offensive, even cruel in its implications and its effects. Statements such as this seem to have persuaded many people in the western world (eg. Britain) that there is no 'need' for genetic manipulation of crop plants; that 'organic' agriculture is not only desirable, but sufficient to meet all our requirements (Foyer *et al.*, 2000).

Acceptance of the associated suggestion, that it would be possible to grow all the food that the world needs, at a price that it can afford, without the use of current technology, would be not only totally illogical but absurd. The facts are self-evident but the arguments in favour of organic farming and the notion that there is 'no need' for genetically modified food plants, are based on belief, not facts. In this context, the view taken by The Food and Agriculture Organization of the United Nations is different and far more realistic.

'In the developing world, 790 million people do not have enough to eat, according to the most recent estimates (1995/97). That represents a decline of 40 million compared to 1990/92. At the World Food Summit in 1996, world leaders pledged to reduce the number of hungry people to around 400 million by 2015. At the current rate of progress, a reduction of 8 million undernourished people a year, there is no hope of meeting that goal.'

Cereals make up about one-half of the world's food. The North American wheat crop has been providing most of the world's requirements since 1950 and has tended to keep pace with demand. Even though wheat yields per unit area have doubled during the last decades its consumption has matched production. Wheat reserves have fallen as low as 10% of consumption in recent years; this low margin for error in the face of one modest drought could readily tip the scale toward deficit and famine. Yield improvements achieved by conventional plant breeding, mechanization, herbicides, pesticides and nitrogenous fertilizers have been remarkable, but a further doubling of output in the next 20 years is very unlikely. It is patently absurd to think that it might be achieved without modern agricultural practice. Given the opportunity, most people would prefer to buy carrots that have never been near lindane. But, realistically, the produce of organic farming would be beyond the reach of all except a few rich people in industrialized countries. Intensive agriculture, however deplorable and distasteful, does feed most of the world's population. Organic farming-in many ways as attractive as intensive agriculture is unattractive-does not and could not. We cannot escape from the hard fact that maintaining food production at present levels can only be achieved by continuing with intensive agricultural practices and advancing research with new technologies (Foyer *et al.*, 2000). No doubt, some of these practices have been questioned for decades because of their adverse effects on the environment in general and on human health in particular. Agricultural practices cost lives and damage the environment. New risks, real and imagined, are associated with the introduction of crops modified by new rather than traditional genetic manipulation. However, by now most or all of us have ingested some genetically modified food and pesticides. The latter are orders of magnitude more dangerous than the former. However tragic, BSE has killed fewer people than herbicides such as paraquat, and far fewer than food poisoning, or road transport of the food itself. But the crucial question that is often ignored is: how many would have starved to death, worldwide, had it not been for intensive farming? It needs to be understood clearly that intensive agriculture comes at a price, but a price that must be paid if greater evils are not to follow.

Genetically modified food plants, first derived from natural, unconscious selection by man and later by artificial selection and scientific breeding, make up the foundations of agriculture. The British Nuffield Council on Bioethics concluded that genetic engineering cannot be regarded as being more unnatural than conventional plant breeding. The US National Academy of Sciences found no evidence that unique hazards exist either in the use of recombinant DNA techniques or in the transfer of genes between unrelated organisms. With about a billion hungry people in the world and in the face of the most unlikely prospect that world hunger can be abated by present agricultural practices, we should not allow ourselves to be hamstrung by failing to explain the need to responsibly apply all of the tools at our disposal. Let us not denigrate or denouce a technology that could save the lives of millions of people and improve the lives of all of us.

Myths of Organic Farming

It is widely believed that low-yielding organic agricultural systems are more friendly to the environment and more sustainable than high-yielding farming systems. Some current aims of organic systems include maintenance of soil fertility, avoidance of pollution, crop rotation, animal-welfare concerns and wider environmental aspects. Two principles distinguish organic farming from other farming methods: Soluble mineral inputs are prohibited and synthetic herbicides and pesticides are rejected in favour of natural pesticides. But farming based on these principles is more costly because of lower yields and inefficient use of land.

Conventional agriculture is a diverse set of technologies using the best available knowledge, whose ultimate goal is the safe, efficient provision of abundant food at lowest price. As with all technologies, problems sometime arise in the practices of conventional agriculture. In organic farming there is some reduction in pesticide use. However, many current synthetic pesticides are quite unstable and cause only transient declines of most field insects even at full pesticide dosage (Trewavas, 2001). The lower levels of aphids often seen on organic farms could well reflect lower nitrogen and protein content of organic crops, and lower yield when expressed as a ratio of crop yield/aphid population, the difference becomes negligible. Some conventional mixed farming also can maintain species diversity. For example, conventional mixed farming in smaller plots (providing more field margins) or farming based on the traditional ley system (for example undersowing wheat with legumes) maintains conventional yields at low costs. The benefits for wildlife equal those provided by organic farming but at far lower cost to the consumer (see Anon, 1999).

It is also not true that organic farming practices necessarily always conserve the environment. Competitive organic farmers keep their fields clear of weeds through frequent mechanical weeding - a method that damages nesting birds, worms and invertebrates - and high use of fossil fuels, which greatly increases pollution from nitrogen oxides. A single treatment with an innocuous herbicide, coupled with no till conventional farming can avoid this damage and retain organic material in the soil surface. Similarly, although use of manure means higher, beneficial levels of earthworms in organic fields, there are numerous problems with the use of manure including possible effects on human health (Trewavas, 2001).

Organic farmers usually depend on legume nitrogen fixation, and rain water or mineral recycling in the farm, but this can lead to slow but accumulating mineral deficits, particularly of potassium and phosphate, in organic farmlands. Organic farms are required to try to balance manure and straw production with use on the farm itself. Excess organic manure or straw is often not available to meet the needs. Ultimately, many organic farms can become dependent on products that are conventionally produced with inorganic minerals.

Recent researches have shown that conventional agriculture can be much more sustainable and environmentally friendly than organic farming (see Trewavas, 2001). A conventional farm can match organic yields using only 50-70% of the farmland. Europe, having become food-surplus, many European farmers are being prompted to set aside some of their land for fast-growing willow plantations which are frequently coppiced and the wood used as fuel. This novel conventional approach is now in commercial operation throughout Europe and has resulted in much lower total fossil-fuel use and carbon dioxide production than in organic farming (Bertilsson, 1992); also because of carbon recycling it is much more sustainable. The plantation of willow trees, with its undercover of weeds, bird-nesting sites and mammal (including deer) and insect refuge, outperforms organic farms on any biological measure of environmental diversity. But this practice crucially depends on the most efficient use of land for food production (see Trewavas, 2001).

Integrated farm management combines the best of traditional farming with responsible use of modern technology, as it integrates a concern for the environment with safe, efficient methods of production. Detailed information on farm soil structure and field fertility is used to target minerals, and integrated pest management to control pesticides and avoid waste. But flexibility is crucial and should consider site-specific factors within a framework of conservation of wildlife habitat and landscape. Integrated farm management exemplifies how to enjoy the benefits of technology while minimizing the associated problems.

Alternatives to Biotechnology for Food Security

Proponents of transgenic technology argue that genetic manipulation will help us feed the world, cure diseases, and solve many other problems facing the human species. According to opponents, the projected benefits are exaggerated and the technology poses many risks that have not been adequately assessed.

But both these arguments are based on conjecture-nobody really knows for sure, whether genetic engineering will feed the world until we try it; and nobody knows of sure, whether transgenic organisms will create ecological havoc after they have been released. In the meantime, the debate is limited to our assessment of the technology's potential risks or benefits, relying on our personal or collective judgments about the technology's efficacy or on our biases about the technology's capabilities (Kirschenmann, 2001). According to Kirschenmann, it is more appropriate to consider the underlying assumptions that lead us to our conclusions about the technology's promises and problems. Should the assumptions be faulty, the conclusions may turn out to be unreliable as well.

According to Lewontin (1991), our optimism regarding the ability to solve many social, medical, and agricultural problems with transgenic technologies is based on what may be termed "biological determinism." This ideology makes the atom or individual the cause of all the properties of larger collections. The world may be studied by cutting it up into the individual bits. For biology, this world-view has resulted in a particular picture of organisms and their total life activity. Living beings are being determined by internal factors, the genes. But this ideology ignores the actual relationship that exists between organisms and their environments (Lewontin, 1991). He suggests that there are actually the following four rules of "the real relationship between organisms and their environment".

1. "Environments do not exist in the absence of organisms, but are constructed by them out of bits and pieces of the external world."
2. "The environment of organisms is constantly being remade during the life of those living beings."
3. "Fluctuations in the world matter only as organisms transform them."
4. "The very physical nature of the environment as it is relevant to organisms is determined by the organisms themselves." (Lewontin, 1991).

In the light of the above rules, organisms are not the isolated entities that we assume they are when we think about feeding the world by manipulating a few genes in a few plants or animals, or healing debilitating diseases by adjusting a few defective genes (Kirschenmann, 2001). Each individual is actually a unique outcome of both genes and the developmental environment in a constant interaction. Such interactions suggest that all problems and threats to our well-being are finally a combination of molecular specification and the unique interactions among genes, organisms, and environment. Furthermore, an organism changes throughout its life.

The idea that gene technology can, by itself, solve problems when those problems are, at least partly derived from social and environmental interactions illustrates a faith in technological fixes that is not borne out by experience. For example, it is illogical to claim that we will be able to feed the world's expanding future populations by producing more food with biotechnology when we are actually failing to feed more than 800 million malnourished people in an era of overproduction (Sen, 1981, 1984; Kirschenmann, 2001).

Molecular World as Ecosystem

Another crucial issue is whether it is at all possible to do "just one thing" at the molecular level? Ecologists have shown repeatedly that it is not possible to do "just one thing" in the ecosystems in which we live because species within ecosystems are interdependent. One of the best examples of this is the report of Baskin (1997). Attempting to boost the numbers of salmon that swim upstream from Montana's Flathead Lake to spawn in Glacier National Park's McDonald Creek, state fisheries officials stocked the upstream portions of the watershed with exotic opossum shrimp, thinking that by so doing they world provide extra food for the salmon. Extra salmon, if was supposed would, in turn, provide more food for eagles, gulls, goldeneyes, coyotes, minks, otters, and several other species that feed on the salmon and their eggs. What actually happened, however, was exactly the opposite. The plan had overlooked an important point about natural history of both shrimp and fish. The salmon feed on zooplankton near the surface during the day at a time when the shrimp are living near the bottom, quite out of reach of the fish. At night the shrimp move upwards to feed on zooplankton themselves–the same zooplankton, unfortunately, that is the major food for the salmon. Consequently instead of supplying a new food resource for the salmon, the fisheries officials had unwittingly introduced a competitor-so the zooplankton including *Daphnia*-a favoured food of both the salmon and the shrimp quickly declined, in turn leading to the collapse of salmon population in the lake. About a hundred km upstream in McDonald Creek, the disappearance of the spawning salmon eliminated a food resource that had once fortified eagles for their winter migration and fattened bears for hibernation (Baskin, 1997).

Within just one decade, the population of 100,000 salmon was reduced to 50. The lesson that emerges from this story is that when our judgment can be so faulty, how can we think of tampering with a genome? It is quite likely that the same ecosystem dynamics that are at work on the organism level may also be at work on the molecular level. Ignoring the ecological dimensions of any genome modification work, as is being done often so with genetic engineering, is quite likely to bring many unpleasant surprises.

The finding that ecosystems dynamics are at work at the molecular level warrants a more cautious approach than most molecular biologists have done thus far. It underscores the need to pay greater attention to basic ecological principles in the process of our modifications. We should no longer just assume that the contemplated modifications are "safe" simply because we have convinced ourselves that:

1. Genetic engineering is no different from ordinary sexual reproduction,
2. Nature will always keep all populations in balance,
3. Transgenic organisms will always be ecologically competent, or
4. Because the host has been domesticated, it is so genetically debilitated that the transgenic organism will not pose an ecological problem (Kirschenmann, 2001).

Regal (1994) proposed a suitable set of ecological principles that can help in assessing the risk of releasing transgenic organisms. These principles are based on work done on patterns and mechanisms of adaptation to natural environments in some plants and animals.

According to Giampietro (1994), the decision to release a transgenic organism into the environment should be based on critical analysis at three different levels *viz.*, the individual, the social, and the biospherical. For instance, at the individual level, we should judge whether a transgenic organism would be beneficial to the company that develops it, to the individual who will use it, or to the organism that has been altered. At this level it is fairly easy to quantify risks and benefits. Most industries wish to make decisions at this level.

Judgements at the social level are more difficult because we have to assess if the release of the transgenic organism will contribute to the general welfare of society or to the economic welfare of the community in which it is released, and whether it poses unacceptable health risks to human beings.

At the biospheric level the issues are extremely complicated and too difficult to analyze through conventional risk/benefit analysis. The tremendous complexity of ecological systems makes it virtually impossible to quantify outcomes (see Kirschenmann, 2001). According to Giampietro (1994), if we wish to promote sustainability, we have to give greater attention to the biospheric level.

We should be clear about which problems we intend to solve with transgenic organisms. If our aim is only to make more food immediately available to help feed a growing population, we may support the development of genetically engineered organisms that promise to improve yield (the individual level). On the other hand, if we are concerned about the social and political dimensions or compulsions that prevent people from having access to food despite adequate production (the social level), or if our concern relates to the magnitude of the ecological impact that rising populations of overconsuming humans may exert on the planet, causing a degradation of the environment and loss of the ecosystem services on which food production depends (the biospheric level), then we should address the problem of hunger from a different perspective.

Current applications of biotechnology in agriculture are chiefly tailored to tackle the problems of monoculture farming (*i.e.*, production system based on just one or two species of crops or animals within a bioregion). These applications help to strengthen monocultures and the industrial food system they serve. But monocultures are inherently unstable and generate grave pest problems because they are highly specialized and un-natural-nature is diverse and complex. All organisms in nature have learned to adapt to biodiversity. Nature always can overcome the specialization and simplification of monocultures. A recent study on the benefits of biodiversity (CAST, 1999) concluded that "the development and increased use of high-diversity cropping systems, which currently are greatly underutilized, could substantially contribute to agricultural productivity, sustainability, and stability". When considered in this light there is no warrant to infer that molecular biology will be any more successful at solving monoculture's inherent weaknesses than *e.g.*, toxic chemicals have been.

Ethical Issues

The various ethical issues relating to the production and use of transgenic organisms are highly complex. For the past few centuries, human culture has insisted on separating facts from values, the latter having been relegated to the realm of personal opinion and private faith (Kirschenmann, 2001). Neither ethics nor values have anything to do with science and facts. There are no disciplined tools that allow us to make ethical decisions as a society. The modern technologies, however, are rapidly pushing us into a world in which we can no longer relegate ethics to the sphere of private and personal choice. Such current technologies as robotics, genetic engineering, and nanotechnology not only are self-replicating, but can also radically change the physical world and potentially inflict substantial damage in the physical world (Joy, 2000). Although they have the potential to prolong our average life span and improve the quality of our lives, they lead to an accumulation of great power and, concomitantly, great danger. In the 20th century, the technologies underlying the weapons of mass destruction were very powerful. But building nuclear weapons required access to rare or effectively unavailable raw material and highly protected information. In contrast, the 21st century technologies-genetics, nanotechnology, and robotics-are so powerful that they can lead to entirely new classes of accidents and abuses. Most dangerously, for the first time in history, these accidents and abuses are now within easy reach of individuals or small groups. They will not require large facilities or rare raw materials. Knowledge alone will facilitate their use. This has raised the frightening possibility not just of weapons of mass destruction but of knowledge-enabled mass destruction, with the this destructiveness hugely amplified by the power of self-replication. According to Joy (2000), we may well be on the cusp of the further perfection of extreme evil!

The possibility looms large that within two decades, the physician would be sufficiently well equipped and trained in karyotype analysis and computers to be able to inform a pregnant woman that her foetus is carrying a genetic marker for severe manic depression. This may prompt the lady-and others like her-to abort (see Shenk, 1997). This exemplifies just one of many ethical decisions we will be forced to make, including what kind of children we will decide to bring forth into the world. And what happens if a new culture emerges that prompts millions of people to choose identical offspring-another monoculture-with all of its accompanying risks and pitfalls? The pertinent question here is whether to allow these powerful technologies to be available to anyone who wants them, or should they be controlled as to who uses them and for what purpose? If the technologies are made freely available they will inevitably fall into the hands of certain unscrupulous people who will use them for destructive purposes. Likewise, according to Shenk, free markets and consumer choice can become even more dominant forces in society than they already are, and the prospect of individuals or elite groups of individuals buying genetic advantages for themselves is real. The long-held notion that we *all* have to be considered equal at some fundamental level in order to sustain a peaceful, just, and functional society, may evaporate (Shenk, 1997).

Many farmers have toiled hard to develop and supply markets for crops that do not contain genetically modified organisms (GMO). These farmers face another, more immediate ethical problem. As transgenic crops spread throughout the landscape, it is becoming very difficult for farmers to produce GMO-free crops. Martens (2001, in press, cited by Kirschenmann, 2001) explored the difficulty farmers are having with the production of non-GMO crops. She discovered that virtually all of the year 2000 non-GMO corn crop produced in the Midwest USA that was tested revealed GMO contamination at an average level of 0.25 percent.

The controversies surrounding genetically modified organisms (GMOs) and genetically modified (GM) foods have arisen from differing views concerning the products of recombinant DNA (rDNA) technology. This new technology has enabled scientists to move genes across species' boundaries, to create traits in plants, animals, and microorganisms that could otherwise not be accomplished using traditional cross breeding techniques. For instance, genes from cold-water fish can be inserted into tomato plants to make them more tolerant to colder weather. Work on transgenic technology has prompted some people to raise questions about the nature and consequences of GMOs. The questions concern scientific and legal-political issues. Also, some controversies have arisen regarding the *ethics* of GMOs and GM foods: whether GMOs and GM foods are morally and ethically acceptable. Should they be ethically acceptable, then people can produce, use, or consume them. If they are not acceptable, they should not be produced or, those people who find them unacceptable should be able to avoid them.

Judgements about ethical acceptability are made on the basis of answering certain questions. Some people for philosophical or religious (not scientific) reasons reject transgenic technology. But we must admit that scientifically also differences exist among the products of biotechnology. This prompts the first question regarding acceptability "What GMO are we talking about?" Different products have different ethical dimensions. For example, an early GM product *viz.*, bovine growth hormone (recombinant bovine somatotropin, or rBST), was designed to increase the efficiency of milk production by getting cows to produce more milk without increasing their feed intake. Comstock (1989) drew attention to its possible negative effects on cows, potential impact on human health, and economic effects on small-scale dairy operations. In contrast, Roundup-ReadyÒ crops, such as soybeans and cotton, were designed to permit a farmer to spray a herbicide on his or her field, killing weeds but not affecting the Roundup-ReadyÒ crops at all (Burkhardt, 2001; 2001a). Analyses were made of the potential cost savings resulting from farmers not having to till weeds or use several herbicides to kill the different kinds of weeds invading their field. Lappe and Bailey (1998) pointed out the potential human health risks and economic effects on small farms. *Bt* corn was designed to make a substance in the plant that is toxic to insect pests. The product was intended to reduce the need for spraying insecticides; however, it has been claimed that the pollen from *Bt* crops kills monarch butterfly larvae that consume it. Yet another, more recent example is that of the so-called "golden rice", a transgenic product with greatly enhanced beta carotene (pro-vitamin A) content, intended to provide a more nutritious food staple for people in Third World rice-consuming countries where vitamin A deficiency is widespread and causes blindness in children. The golden rice is several years away from the market but it is being debated in terms of both its major health benefits as well as its potentially prohibitive cost to poor people (Burkhardt, 2001).

A second set of concerns bearing on ethical acceptability is the context in which the analysis or argument is set. One party focuses on a set of issues in one context that are quite different from the issues and context that concern another party. For instance, whereas much of the scientific community has focused on the role of the new biotechnology in contributing to food quantity, quality, and affordability, others have focused on contexts such as health, environmental safety, social justice or fairness, or different implications of GM technology for the developed versus the developing world. Doubtless, each of these general areas is important in the ethical appraisal of GMOs and GM foods. By focusing chiefly on one area, however, parties involved in the debates often ignore other relevant issues or considerations that appear in a different context (Burkhardt, 2001). Blackburn (2001) has critically discussed the major differences among, and subtle naunces within, various types of ethical orientations.

Alternatives to Biotechnology?

Proponents of agricultural biotechnology often argue that although some risks may be involved in using this technology, we have no alternative but to move ahead (see, *e.g.*, Klaus Leisinger, 2000 unpublished, quoted by Kirschenmann, 2001). Leisinger points out that the global population will grow another 50 percent by the year 2050 (to nine billion people). Most of that population growth will take place in the developing world, much of it in urban centers. According to him people living in cities cannot feed themselves through subsistence food production in the same manner that people living in rural areas do. This will have a chain effect. Burgeoning urban populations of developing countries will necessitate production of higher yields. Because the eating patterns of urban people are very different from those of rural people, we will also have to produce different food. Urban dwellers usually eat more high-value foods, more animal proteins, and more vegetables. This will divert cereals from food to feed and increase the need to produce even more grain because of the loss of protein involved in the conversion of plant food to meat (Kirschenmann, 2001).

According to Leisinger, besides achieving this higher productivity only with biotechnology, biotechnology will also have positive ecological effects. His calculations show that if average annual per hectare productivity increases just 1 percent, there will be a global need to bring more than 300 million hectares of new land into agriculture by 2050 to meet the expected demand. A productivity increase of 1.5 percent could double output without using any additional cropland. Failure to achieve that productivity through biotechnology will mean that fragile lands and wilderness areas will have to be pressed into service for agricultural production, with much attendant ecological devastation. Leisinger is silent on the land that will go out of production due to urban sprawl if his scenario materializes, or the potential for increased production through successful urban farming ventures such as the urban gardens in Cuba, where 50,000 tons of food are now being produced annually in the city of Havana–without the aid of genetic engineering (Kirschenmann, 2001).

There exists considerable potential for increasing food availability by decreasing waste, both in the developed countries and the developing ones. Also, it is possible to boost yields merely by improving soil quality in some cases (see NRC, 1993). Also, Leisinger does not tell us how the poor people crammed into urban centers will be able to buy the food produced with biotechnology. According to him, as the economies of developing nations grow, people will eat higher on the food chain. But he does not point out the well-established fact that as economies grow, the gap between rich and poor increases (see Korten, 1995).

Leisinger does refer to the additional problems associated with maintaining current levels of productivity, such as declining water resources, declining soil quality, unpredictable climate changes, and poor governance-issues that biotechnology proponents often overlook. But he does not point out that many of these problems were actually caused by the industrial farming methods that he wishes to perpetuate. He also does not realize that food security is usually most strongly affected by two consequences of modern industrial agriculture; *viz.* (1) the pest infestations that occur because of the lack of biodiversity and genetic variability that is integral to modern industrial farming practices, and (2) the failure to initiate land reforms that could give possession of land to local farmers who can produce food for local populations (Kirschenmann, 2001).

Leisinger acknowledges that genetic engineering should be judged in the context of a wider technological pluralism and that biotechnology may be used only if it proves superior to other technologies with regard to cost-effectiveness but this latter factor has to include the potential ecological and social costs which most proponents of agricultural biotechnology have failed to include.

Above all Leisinger's analysis does not give sufficient attention to alternatives for achieving the goals of providing adequate food and fiber within a robust economy, a healthy ecology, and vibrant communities (see Kirschenmann, 2001).

Prior to the publication of Rachel Carson's *Silent Spring* in 1962, many agricultural scientists used to proclaim that pesticides did not contaminate the environment-they now admit that they do. When asked about the presence of pesticides in food and food quality, they assured the public that if a pesticide was applied in compliance with the label, agricultural products would be free of pesticides; today most agronomists admit they're not (Jim Davidson, quoted in Pesek, 1990). The list of similar assurances given by scientists from time to time, that later turned out to be untenable, is quite long: they assured us that there was no connection between organophosphates and pesticide poisoning; and between the release of chlorofluorocarbons (CFCs) and the hole in the ozone; they mistakenly assured us that nuclear energy was safe and would be "too cheap to meter" and that thalidomide was a safe drug. All these instances prove that it is not possible to make any accurate predictions, based on isolated data collected in laboratories about how a technology will perform in the world of interconnected and interdependent relationships of living systems. The world of social and ecological relationships always comes across surprises-surprises that will be greatly magnified when we introduce technologies into ecosystems with which they did not evolve.

According to Schneider (1976), "The bigger the technological solution, the greater the chance of extensive, unforeseen side effects and, thus, the greater the number of lives ultimately at risk". Long before him, Leopold (1949) had proclaimed, "The greater the rapidity of human-induced changes, the more likely they are to destabilize the complex systems of nature." According to Kirschenmann, we should follow the advice of ecologists who have carefully observed the workings of nature rather than the advice of Leisinger, who may have observed only the potential promises of a largely untested technology. Daily *et al.* (2000) warned that "the level of uncertainty in our understanding of ecological processes suggests that it would be prudent to avoid courses of action that involve possibly dramatic and irreversible consequences and, instead, to wait for more reliable information.

It is now emerging that the present application of biotechnology in agriculture conforms to the same paradigm that failed us in chemical technology. Lewis *et al.* (1997), who have intensively studied pest-management problems in agriculture, argue that the chief trouble with industrial pest management is that we are operating out of a paradigm that may be termed "therapeutic intervention". That approach addresses pest problems by applying a "direct external counterforce" against the problem-the problem of a pest within a complicated, interconnected system is attacked by intervening in that system with an external force geared simply to eradicate the pest. Though that approach has killed some target pests, it has failed to solve the problem of crop losses due to pests. Some studies, in fact, show that crop losses have actually increased with the continued intensification of pesticide applications (Lewis *et al.*, 1997).

This kind of therapeutic intervention approach is now being questioned, not only in agriculture but also in medicine, social systems, and business management; it is now widely appreciated that using a counterforce from outside the system to solve a problem that is intrinsic to the system exacerbates, rather than solves, the problem (Kirschenmann, 2001). Trying to solve a pest problem by applying a pesticide kills not only some of the target pest but also the predators of the target pest that previously kept other pests in check. Besides, it creates resistant varieties of the target pest, making the original pest even more difficult to manage.

Because so far the application of biotechnology has largely followed this same interventionist paradigm, it is likely to experience the same problems. Instead of using the technology to better understand how systems work and perhaps using it as one tool within a whole-systems approach, we are using the technology to intervene in the system to "fix" the problem. Genetically inserting *Bacillus thuringiensis* (*Bt*) into the corn plant to control the corn borer is one good example; actually, the corn borer will develop resistance to *Bt* sooner or later. There are some indications that genes encoding resistance to *Bt* in the European corn borer are dominant rather than recessive as hitherto believed; if so, the high dose/refuge strategy that farmers have been told to use to postpone resistance may have little effect (Huang *et al.*, 1999). The high dose/refuge strategy involves inserting high doses of *Bt* into the transgenic plants to obtain maximum kill and simultaneously requiring that farmers plant at least 20 percent of their crop to conventional, non-transgenic varieties on which no pesticides at all are used to serve as a breeding ground for insects unaffected by *Bt* .

Fortunately, some simpler alternatives exist to the "qsuick-fix" applications of biotechnology. One was mentioned by Ho (1999); a Japanese farmer has developed a method of producing rice, called the Aigamo method, that increases rice yields by up to 50 percent. About 200 ducklings are released into each hectare of rice paddy. The ducks, it seems, eat insects and snails that attack rice plants; eat weed seeds and seedlings; and oxygenate the water, thereby stimulating the growth of the roots of rice plants. The mechanical agitation of their paddling makes rice plants sturdier. Using this method, the farmer's two-hectare farm annually produces "seven tonnes of rice, 300 ducks, 4,000 ducklings and enough vegetables to supply 100 people" (Ho, 1999). The Aigamo method is now being adopted in several developing countries and can potentially make Japan-which currently imports 80 percent of its food-food self-sufficient again. Halweil (2001) gives another example of an alternative to transgenic crops. He reports that some African farmers have managed to control the *Striga* weed by planting leguminous trees prior to planting corn. He feels this may be a more useful technology than herbicide-resistant corn because the corn and the herbicide are too expensive for African farmers.

Another interesting example of approaches to food security (other than through biotechnology) is the development of perennial polycultures from wild grasses that could reduce soil erosion, use water more efficiently, and reduce planting and tillage costs (Land Institute, 2000). John Jovens who designed the so-called double digging method (a method of cultivation that loosens the soil at both the topsoil and subsoil levels) has obtained phenomenal yield increases in vegetable production (see Madden and Chaplowe, 1997). Manning (2000), on the basis of his extensive studies in developing countries, believes that we will never succeed in our attempts to feed the world if we do not take the complexity and diversity of local cultures and local ecologies into consideration. According to Manning, genetic engineering may be a limited tool that can be used effectively in these whole-systems approaches to food production in an expanding human population, but it cannot be the solution.

References

Ahmed, F.E. Detection of genetically modified organisms in foods. *TiBTECH* 20: 215-223 (2002).

Altieri, M.A., Rossett, P. Ten reasons why biotechnology will not ensure food security, protect the environment and reduce poverty in the developing world. *AgBioForum* 2(3&4): 155-162 (1999).

Anon. House of Lords Select Committee on European Communities. 16th Report: Organic Farming and the European Union. HMSO, London (1999).

Anonymous. Commission regulation (EC) No. 49/2000, amending council regulation (EC) No. 1139/98 concerning the compulsory indication on the labelling of certain foodstuffs produced from genetically modified organisms of particulars other than those provided for in directive 79/112/EEC. *Official J. Eur. Communities: Legislation* 6: 13-14 (2000).

Anonymous. Commission regulation (EC) No. 50/2000, on the labelling of foodstuffs and food ingredients containing additives and flavourings that have been genetically modified or have been produced from genetically modified organisms. *Official J. Eur. Communities: Legislation* 6: 15-17 (2000a).

Anonymous. Council regulation (EC) No. 285/97, concerning novel foods and novel food ingredients. *Official J. Eur. Communities: Legislation* 43: 1-5 (1997).

Balfour, E.B. *The Living Soil.* Faber and Faber, London (1943).

Barham, E. What's in a name? " Eco-labeling in the global food system. www.pmac.net/bbarham.htm. (1997).

Baskin, Y. *The World of Nature.* Island Press, Washington, D.C. (1997).

Beachy, R.N. Facing fear of biotechnology. *Science* 285: 335 (1999).

Bertilsson, G. Environmental consequences of different farming systems using good agricultural practices. *Proceedings of the Fertiliser Society* No. 332, London (1992).

Blackburn, S. *Being Good: An Introduction to Ethics.* Oxford University Press, Oxford (2001).

Boeringa, R. (Ed.). *Alternative Methods of Agriculture*. Elsevier, Amsterdam (1980).

Buckingham, D., Phillips, P. Hot potato, hot potato: Regulating products of biotechnology by the international community. *JWT* 35: 1-31 (2002).

Burkhardt, J. Agricultural biotechnology and the future benefits argument. *Agricult. Environ. Ethics* 14: 2-10 (2001).

Burkhardt, J. The genetically modified organism and genetically modified foods debates: Why ethics matters. *Transactions* (Wisconsin Academy) 89: 63-82 (2001).

Campolina, A. Brazilian small-scale farmers and poor consumers reject GMOs. *LEISA* Magazine 17(4): 29 (Dec., 2001).

CAST (Council for Agricultural Science and Technology). *Benefits of Biodiversity*. Council for Agricultural Science and Technology Task Force Report no. 133. Council for Agricultural Science and Technology, Ames, Iowa (Feb. 1999).

Comstock, G. The case against BGH. *Agriculture and Human Values* 5: 1-5 (1989).

Conway, G. *The Doubly Green Revolution: Food for All in the 21st Century*. Cornell University Press, Ithaca, NY (1998).

Daily, G.C., Soderqvist, T., Aniyar, S., Arrow, K., Dasgupta, P., Erlich, P. R., Folke, C., Jansson, A.M., Jansson, B.O., Kautsky, N. Levin, S., Lubchenco, J., Maler, K., Simpson, D., Starrett, D., Tilman, D., Walker, B. The value of nature and the nature of value. *Science* 395-396 (July 21, 2000).

Fluetsch, K.M., Sparling, D.W. Avian nesting success and diversity in conventionally and organically managed apple orchards. *Env. Tox. & Chem.* 13: 1651-1659 (1994).

Foyer, C., Osmond, B., Walker, D. Introduction. *Phil. Trans. R. Soc. Lond.* B 355: 1333-1335 (2000).

GeneScan Europe. *GMO Chip: Test Kit for the Detection of GMOs in Food Products*. Cat No. 5321300105, Bremen, Germany (2001).

Giampietro, M. Sustainability and technological development in agriculture: A critical appraisal of genetic engineering. *BioScience* 44: 677-690 (1994).

Gupta, A. Governing trade in genetically modified organisms. The Cartagena Progocol on Biosafety. *Environment* 42: 22-23 (2000).

Halweil, B. Biotech, African corn and the vampire weed. *World Watch* Magazine. 14 (5): 26-31 (Sept.Oct. 2001).

Haslberger, A.G. Monitoring and labeling for genetically modified products. *Science* 287: 431-432 (2000).

Ho, M.W. One bird-Ten thousand treasures. *The Ecologist* p. 339 (Oct. 1999).

Huang, F., Buschman, L.L., Higgins, R.A., McGaughey, W.H. Inheritance of resistance to *Bacillus thuringiensis* toxin (Dipel ES) in the European corn borer. *Science* 284: 965-967 (1999).

Hunt, J.W., Anderson, B.S., Phillips, B.M. Patterns of aquatic toxicity in an agriculturally dominated coastal watershed in California. *Ag. Econ. Envir.* 75: 75-91 (1999).

James, C. *Global Status of Transgenic Crops in 1997.* International Service for the Acquisition of Agri-biotech Applications (ISAAA) Brief #5. www.isaaa.org (1997).

Joy, B. Why the future doesn't need us. *Wired* (Magazine) 238-262 (April, 2000).

Kalaitzandonakes, N. (ed.) *The Economics of Transgenic Crops.* Kluwer Academic Publishers, Dordrecht (2003).

Kirschenmann, F. Questioning biotechnology's claims and imagining alternatives. *Transactions* (Wisconsin Academy) 89: 35-61 (2001).

Knoppers, B., Mathios, A. *Biotechnology and the Consumer.* Kluwer Academic Publishers, Dordrecht (1998).

Korten, D. *When Corporations Rule the World.* Kumarian, West Hartford, Conn. (1995).

Lampkin, N. Organic farming, pp. 2-6, Farming Press Books, Ipswich (1990).

Land Institute. *Land Report* (quarterly newsletter). No. 67, Summer. Land Institute, Salina, Kansas (2000).

Lappé, M., Bailey, B. *Against the Grain.* Common Courage Press, Monroe, Maine (1998).

Leopold, A. *A Sand County Almanac.* Oxford University Press, London (1949).

Lewis, W.J., Lenteren, J.C., Phatak, S.C., Tumlinson, J.H. A total system approach to sustainable pest management. *PNAS* (USA) 94: 12243-12248 (1997).

Lewontin, R.C. *Biology as Ideology.* House of Anansi, Concord, Ontario (1991).

Lyon, N. Consolidating organic gains. *Austral. Farm J.* 11 (12) page 1: (Feb. 2002).

Madden, J.P., Chaplowe, S.G. (eds.). *For All Generations.* Glendale, Calif., OM. (1997).

Magn, K. *et al. Methods for Detection of GMO Grain in Commerce.* American Crop Protection Association, Washington, D.C. (2000).

Manning, R. *Food's Frontier: The Next Green Revolution.* New York: North Point, New York (2000).

Medvitz, A.G., Sokolow, A.D. Population growth theatens agriculture, open space. *Cal. Ag.* 49 (6): 11-27 (1995).

Moore, S. Organic farming - for all the right reasons. *Austral. Farm J.* 11(12): 6-8 (Feb. 2002).

Myers, D. Organic cotton production. *LEISA* Magazine 17(4): 21-22 (Dec., 2001).

Myers, D., Stolton, S. (eds.). *Organic Cotton: From Field to Final Product.* Intermediate Technology Publications, London (1999).

NRC (National Research Council). *Soil and Water Quality.* National Academy Press, Washington, D.C. (1993).

NRC (National Research Council). *Soil and Water Quality: An Agenda for Agriculture.* National Academy Press, Washington, D.C. (1993).

NRC. (National Research Council). *Ecologically Based Pest Management: New Solutions for a New Century.* National Academy Press, Washington, D.C. (1996).

ODI. (Overseas Development Institute). Briefing Paper: The Debate on Genetically Modified Organisms: Relevance for the South, ODI, London (1999).

Papendick, R.I. *et al.* Report and recommendations on organic farming. U.S. Department of Agriculture, Washington, D.C. (1981).

Pesek, J. Research findings on alternative methods: Based on the NRC Report on alternative agriculture. Unpublished manuscript, presented at Northwest Farm Managers Association, (February 26, 1990).

Phillips, P., Khachatourians, G.G. *The Biotechnology Revolution in Global Agriculture: Invention, Innovation and Investment in the Canola Sector.* pp. 360, CABI (2001) (cited in Phillips, 2002).

Phillips, P.W.B. Biotechnology in the global agri-food system. *TiBTECH.* 20: 376-381 (2002).

Pimbert, M., Wakeford, T., Satheesh, P.V. Citizens' juries on GMOs and farming futures in India. *LEISA* Magazine 17(4): 27-30 (Dec., 2001).

Regal, P.J. Scientific principles for ecologically based risk assessment of transgenic organisms. *Molecular Ecology* 3: 5-13 (1994).

Rossett, P. Genetically engineered crops. *LEISA* Magazine 17(4): 6-8 (2001).

Schneider, S. *The Genesis Strategy.* Plenum, New York (1976).

Schreiber, G.A. Challenges for methods to detect genetically modified DNA in foods. *Food Control* 10: 351-352 (1999).

Sen, A. *Poverty and Famines: An Essay on Entitlement and Deprivation.* Clarendon Press, Oxford (1981).

Sen, A. *Resources, Values and Development.* Harvard University Press, Cambridge, Mass. (1984).

Serageldin, I. Biotechnology and food security. *Science* 285: 387-389 (1999).

Shenk, D. Biocapitalism: What price the genetic revolution? *Harper's Magazine,* 37-45 (Dec., 1997).

Soil Association. Organic farming, food quality and human health. A review of the evidence. Soil Association, Bristol (2001).

Swezey, S.L., Broome, J.C. Growth predicted in biologically integrated and organic farming in California. *Global Pesticide Campaigner* 11: 6-9 (2001).

Trewavas, A. Urban myths of organic farming. *Nature* 410: 409-410 (2001).

van Bueren, E.L., Osman, A. Stimulating GMO-free breeding for organic agriculture: a view from Europe. *LEISA* Magazine 17(4): 12-14 (Dec., 2001).

van Bueren, E.T.L. *et al.* Sustainable organic plant breeding. pp. 16-18. Louis Bolk Instituut, Driebergen (1999).

Vogt, G. Origins, development, and future challenges of organic farming. In *Proc. 13th IFOAM Scientific Conf.* pp. 708-711. Vdf Hochschulverlag AG an der ETH, Zürich (2000).

Vogtmann, H. New approaches to the determination of food quality. In *Food Quality, Concepts and Methodology*. pp. 44-47, Elm Farm Research Centre, Newbury (1992).

Wolf, S., Zilberman, D. (eds.). *Knowledge Generation and Technical Change: Institutional Innovation in Agriculture.* pp. 177-192, Kluwer-Plenum Academic Publishers, New York (2001).

Woodwand, L. The scientific basis of organic farming. *Interdiscip. Sci. Rev.* 27: 114-119 (2002).

Chapter 7

Terminator Technology and Its Antidote

Introduction

Some genetically modified organisms (GMOs) are now a commercial reality in agriculture. In 1998 over 18 million acres in the United States were planted in Roundup Ready soybeans, which were first introduced in 1996 (Horstmeier, 1998). These soybeans were engineered by Monsanto Corporation to contain a bacterial gene that confers tolerance to the herbicide glyphosate (Roundup), also made by Monsanto. Only two years after the introduction of Roundup Ready soybeans, over 30% of the corn and soybeans planted in the US and nearly 50% of the canola planted in Canada were genetically engineered to be either herbicide or pesticide resistant (Crouch, 1998).

It takes huge investments and several years of research to develop the biotech crops that deliver superior value to growers. Future investment in biotech research depends on companies' ability to make profits. If growers save and replant patented seed, there is less incentive for companies to invest in future technology, such as the development of seeds with traits that produce higher-yielding, higher-value and drought-tolerant crops. The growers who save and replant patented seed jeopardize the future availability of innovative biotechnology for all growers. This prompted the development of a procedure that builds up plant variety protection into the plants themselves.

In March 1998, Monsanto, Delta and Pine Land Company, in collaboration with the United States Department of Agriculture, was awarded U.S. Patent Number 5,723,765; entitled *Control of Plant Gene Expression*. Although the patent is broad and covers many applications, one important application is a scheme to engineer crops to kill their own seeds in the second generation, thus making it impossible for farmers to save and replant seeds. This invention has been termed the "Terminator Technology" by Rural Advancement Foundation International (RAFI).

Cotton seed has previously been genetically engineered with a unique trait, herbicide tolerance. Cotton is not usually sold as a hybrid seed, and is thus a likely candidate for Terminator protection. By contrast, corn is usually planted as a hybrid, and thus already contains some variety protection, because the first generation of a hybrid is genetically fairly uniform, and has been bred to have desired characteristics that are not present in either parent alone. When these hybrids produce seeds, however, the second generation is quite variable because of the reshuffling of genes. Industrial (mechanized) agriculture requires uniformity. Therefore, industrial farmers who grow corn usually buy new seed every year (see Rissler and Mellon, 1996).

Four major crops that are usually not grown from hybrid seeds include wheat, rice, soybeans, and cotton. Farmers often save the seeds from these crops, and do not go back to the seed company for several years to purchase a new variety.

The profits of a seed company will increase greatly if growers who now grow non-hybrid crops would have to buy new seed every year. This was the chief incentive for developing the Tereminator Technology. Another reason for developing Terminator was related to the way in which Terminator's effect differs from hybridization. With Terminator the second generation is killed. With hybridization, the second generation is variable, but alive; and any genes present in the hybrid will be present in the second generation, although in unpredictable combinations. Therefore, a plant breeder who wishes to use the genetic material from the hybrid in his own breeding program can retrieve it from these plants. With Terminator, the special genes, such as the herbicide tolerance of cotton would not be easily available for use by competitors.

In the cotton example, the goal is to develop a variety that will grow normally until the crop is almost mature. Then, and only then, a toxin will be produced in the (seed) embryos, specifically killing the entire next generation of seeds.

There are three key components here:

1. A gene for a toxin that will kill the seed late in development, but which will not kill any other part of the plant.
2. A method for allowing a plant breeder to grow several generations of cotton plants, already genetically engineered to contain the seed-specific toxin gene, without any seeds dying. This is required to produce enough seeds to sell for farmers to plant.
3. A method for activating the engineered seed-specific toxin gene after the farmer plants the seeds, so that the farmer's second generation will be killed (Crouch, 1998).

To accomplish these three, a series of genes, which are all transferred permanently to the plant are engineered so that they are passed on via the normal reproduction of the plant.

The process of moving genes between species is called transformation, and the result is a transgenic organism. Lately, transgenic organisms are being called genetically modified organisms, or GMOs.

The key to Terminator is the ability to make a lot of a toxin that will kill cells, and to confine that toxin to seeds.

In the case of our cotton example, the promoter from a gene normally activated late in seed development is taken and fused to the coding sequence for a protein that will kill an embryo going through the last stages of development.

In the Terminator patent, the authors used a promoter from a cotton LEA (late embryogenesis abundant) gene which is one of the last to be activated. Its protein is not made until the seed is full sized, has accumulated most of its storage oil and protein and is drying down in preparation for the dormant interregnum between leaving the parent plant and germinating in the soil. If the engineered gene has the same pattern of expression, LEA-promoter-directed proteins should be made in high quantities, only in seeds, and late in development. The cotton seeds need to go through most of their growth before the toxin acts, because the cotton fibre is an outgrowth of the seed coat and is made as the cotton develops. Further, after the cotton fibres are removed (for human use), the seed is crushed for oil and protein, both of which are eaten by people and livestock.

As for the toxin, a ribosome inhibitor protein (RIP) from the plant *Saponaria officinalis* is used. This protein stops the synthesis of all proteins. Since cells need proteins for almost everything, they die quickly when they are unable to make them. The RIP is claimed to be non-toxic to organisms other than plants.

The manipulations of DNA required to engineer a seed-specific promoter/toxin coding sequence gene are done in test-tubes and through bacteria. The modified gene is then transferred into a cotton plant, using one of several possible methods.

The Terminator patent provides an ingenious method for preventing the toxin gene from activating until long after the farmers plant their crops. This feat is accomplished by inserting a small segment of DNA in between the seed-specific promoter and the toxin coding sequence that blocks it from being expressed to make protein.

At either end of the blocking DNA are added special DNA segments recognizable by a recombinase enzyme. Whenever the recombinase encounters these DNA segments the DNA is cut precisely at the outside of each piece, and the cut ends of the DNA fuse together, so that the blocking DNA is removed. When this happens, the seed-specific promoter is right next to the toxin coding sequence, and can function in making the toxin. However, this does not happen immediately. Toxin is not produced until the end of the next round of seed development, because it is at that time that the LEA promoter is active (Crouch, 1998).

After the action of recombinase enzyme, the plant grows normally from germination until most of seed development. It is only then that, on cue, the seeds die.

When all this is accomplished, one important problem still remains (for the company). How to grow several generations of the genetically engineered variety so that its seed can be multiplied for selling to farmers?

This is done by blocking recombinase action until just before the farmers plant their seeds. A recombinase coding sequence can be put next to a promoter that is always active - in all cells, at all times - but remains repressed. The promoter can be actived again (*i.e.* derepressed) by a chemical treatment. Therefore, the seed merchants can treat the seeds right before planting, thus allowing the recombinase to be made then, but not before (Crouch, 1998).

One of the repressible promoter systems is controlled by the antibiotic tetracycline. A gene that makes a repressor protein all of the time may be put into the cotton plant, along with a recombinase gene that has a promoter engineered to be inactivated by the repressor protein. Under most conditions, then, the repressor would interact with the recombinase gene; no recombinase would be made; the toxin gene is blocked; and no toxin is produced even during seed development when the LEA promoter normally would be active (Crouch, 1998).

To activate the toxin gene seeds are treated with tetracycline, just before being sold to farmers. The tetracycline interacts with the repressor protein, keeping it from interfering with production of recombinase. Recombinase is made, cutting out the blocking DNA from the toxin gene. The toxin gene can then express itself and make toxin but does not actually do so until the end of seed development. The next generation thus would be killed.

To achieve the Terminator effect in cotton, then, three engineered components must all be transferred into a cotton plant's DNA:

1. a toxin gene controlled by a seed-specific promoter, but blocked by a piece of DNA in between the promoter and the coding sequence;
2. a repressor protein coding sequence with a promoter that is active all of the time; and
3. a recombinase coding sequence, controlled by a promoter that would be active at all times, except that it is also regulated by repressor protein, which can be overridden with tetracycline (Crouch, 1998).

The actual transfer of genes into the plant is not a very precise operation. The genetically-engineered DNA may be injected into the nucleus of a cotton cell with a tiny needle, or plants cells can be soaked in the DNA and electrically shocked, or the DNA can be attached to small metal particles and shot into the cells with a gun, or viruses and bacteria can be engineered to infect cells with the

DNA. Some of the genetically-engineered DNA must find its way to the nucleus, and become incorporated into the plant chromosomes. The number of copies of the inserted genes and their locations on the plant chromosomes are unpredictable, and how well the new genes will function depends in the balance (Crouch, 1998). Much effort is needed to find cells that have incorporated DNA in significant amounts and in workable locations. Whole plants have to be regenerated from the cells or tissues that were transformed with the foreign DNA, and each plant has to be tested for the presence and function of the new genes.

After plants with well-functioning new genes are identified, they are then crossed in suitable combinations that result in a line of cotton where both sets of chromosomes (in all of the offspring) have all the components necessary for Terminator to function. These plants are crossed together to make a large quantity of seed for sale.

Terminator Technology enables the seed producer to determine when to set Terminator in motion. Until the recombinase is made, the cotton plants grow normally. After recombinase is made, the second generation of seeds is killed, protecting the patented variety.

Preventing the Spread of Transgenes into the Environment

According to Radin (1999), this technology should be viewed as a tool to introduce beneficial biotechnology into crops because the control of seed germination is an important step in the long-standing evolution of crop improvement.

The Technology Protection System (TPS) is the only genetic system currently capable of preventing the spread of transgenes into the environment. Not only does activated TPS prevent seeds from germinating, but pollen from a TPS plant also leads to non-viable seeds when it fertilizes a non-TPS plant. Consequently naturally-occurring hybrids between crops and wild plants are not perpetuated. This means that the TPS prototype being developed might be useful primarily with self-pollinated crops, to avoid problems of sterility in nearby fields of the same crop. In fact, the patented system is intended for use in cotton, a self-pollinated crop with little field-to-field spread of pollen. But for applications in cross-pollinated crops, TPS will obviously need to be modified to fit their specific needs. It represents a new way to protect valuable agricultural biotechnology and promote its applications to agriculture. Much of the investment in agricultural biotechnology has occurred in the private sector, which must get some profit on its investment if R&D is to continue. The TPS simplifies protection of technology and removes it from the legal ambit. It may allow companies to increase their research with the expectation of a fair return on the investment, just as development of hybrids opened the door for companies to invest in maize several decades ago (Radin, 1999).

The world can greatly benefit in multiple ways. The first products of agricultural biotechnology research include insect-resistant and herbicide-resistant crops. In 1997, transgenic insect-resistant maize saved American farmers an estimated US$ 190 million and substantially reduced insecticide use in the USA. According to critics of TPS, the technology will victimize poor farmers in developing countries. They speculate that non-TPS varieties will disappear, forcing farmers to buy seeds from the large seed companies each season, instead of saving seeds for replanting. However, Radin feels that the channels for development and distribution of non-TPS germplasm will not disappear because agricultural research is multifaceted, with many organizations pursuing different goals. In the case of TPS, the technology will be transferred to the private sector, which can create specific products. Although USDA-ARS is co-owner of the patent it does not intend to develop and release varieties that contain TPS. Internationally, the *Consultative Group on International Agricultural Research* (CGIAR) has stated that it will not introduce TPS into its breeding material. Therefore, TPS may be applied exclusively by the private sector, which will do so only when there is a good market for the product. Non-TPS germplasm will continue to be made available through public and private channels.

According to Radin, TPS should be viewed as a tool that will promote the continuing trend toward technology by facilitating the wider introduction of beneficial biotechnology into crops. In this sense, TPS is part of a continuing evolution of modern improved crops, rather than a revolution in technology. The public discussion of TPS should focus not on whether to shun technological advances in crops, but how to manage them to the advantage of humanity.

Types of GURT

A substantial number of patent applications relate to various GURT (genetic use restriction technology) concepts and elements. Two types of technologies can be distinguished: (i) T-GURT technologies can restrict the use of a specific trait by regulating its expression. One or more genes conferring a single trait are switched on or off at will through chemical inducers. The seed itself remains viable. (ii) V-GURT technologies can restrict the use of the entire variety through interference with reproduction. Three types of V-GURT strategies are:

1. The seeds are fertile but, through the application of a chemical, a dormant lethal gene can be activated. This gene inhibits full seed development. Consequently, seeds are fit for consumption but infertile.
2. A lethal gene expressed in the seed results in sterility. Breeders can apply a chemical compound that activates another gene to safeguard the fertility of the seed and the reproduction of the variety, but they will stop doing so before selling the seed;
3. This concerns vegetatively reproducing crops, *e.g.* root and tuber crops and ornamentals. It attempts prevention of growth during storage, which may be in the interest of breeders, growers and/or consumers. Normally, a gene that blocks growth is expressed, but growth can be restored by activation of a second gene. Regulation of hormone metabolism and function forms the basis of this strategy.

Although current patent applications apply to plants, GURTs can be built into any organism, including farm animals, fish and forest trees.

Other methods that share some degree of use restriction with V-GURTs are hybrid technology, triploidisation and the introduction of male or female sterility. Differences between these latter technologies and V-GURTS are that the germplasm remains available to farmers and competing breeders for further breeding. In other words, if V-GURTs can be made sufficiently effective, they largely rule out the Breeder's Exemption and the Farmer's Privilege on the use of protected varieties as prescribed in Plant Breeder's Rights.

Motives, Applications and Impacts

The application of GURTs can be inspired by different, not mutually exclusive motives which fall in the following categories.

Property Protection Interests

Breeding companies wish to safeguard their investments in improved varieties, whether produced by classical breeding or genetic engineering. GURTs may offer a better-biological-insurance against the free use of genetic innovations than patents, plant breeder's rights or licences. Hybrid seed, produced from two genetically different paternal and maternal lines, offers a certain protection of germplasm against reproduction by farmers, because the seeds normally result in different and heterogeneous plants in the next generation. For cross-fertilizing crops, this prevents easy regeneration of elite varieties. For many self-fertilizing crops, hybrids are not feasible or effective. It is these crops that are the primary targets for the use of V-GURTs in order to obtain biological property protection, and include rice, wheat, soybean and cotton; and horticultural crops as well as ornamentals that are vegetatively multiplied. T-GURTs to protect high-value added traits in newly released commercial varieties may be applied to virtually all crops.

Environmental Interests

GURTs are used for the environmental containment of entire transgenic varieties (V-GURT) or of specific transgenes contained in transgenic varieties (T-GURT). In both cases, the focus tends to be on species that establish themselves in ecological niches, and/or for which wild relatives exist in involved crop production systems.

T-GURTs are used to contain traits with suspected health risks for the farmer or consumer, or traits that may somehow pose a threat to biodiversity or other aspects of the environment. An example is a pathogen resistance gene that may outcross to weedy relatives and thus increase the fitness of those weedy relatives.

Production Interests

It can be in the producer or farmer's interest to restrict the expression of a trait, for example, to a specific phase in the development of the plants or animals or during biotic or abiotic stress. Stress response takes a lot of resources and may diminish the quality of the processed product; it is only desirable when really needed. Alternatively, V-GURTs may be used to control successful reproduction of farm animals in order to safeguard the integrity of adapted crossbreeds, resulting from the mating between local and high-yielding commercial breeds.

It is important to distinguish between the impacts of GURTs on farmers and those on breeders. It is more likely that access to technology circumventing GURT protection will be more available to breeders than to farmers. Farmers will have fewer options to choose from.

GURT and Genomics

The increasing availability of genetic information from genomics research is likely to result in several effective GURT prototypes within 5-10 years. Still, several bottlenecks in current concepts will prevent the application of GURTs in some cases. A major one is the requirement for watertight regulation of the genes. For instance, if a lethal gene is not only expressed after induction but also continuously at a low level, this can decrease seed viability.

Certain technical factors will determine for what purposes, on which organisms/fish, and on what time scale will GURTs be developed and applied. Costs to solve technical bottlenecks as well as costs stemming from regulatory requirements will be juxtaposed against expected return when decisions are made on investments in GURT development. Application of GURT to crop breeding is likely to precede its applications in the breeding of trees, fish and farm animals. Options for introduction of GURTs in tree species are technically similar to those in crops. Applications in aquatic species may also materialize soon, although economic and environmental issues, such as the threat to survival of local indigenous fish populations with which the transgenic fish can interbreed, may slow down or restrict development. The application of GURTs in animals faces much higher technological and ethical barriers. Currently, breeding efforts focus on crops and animal breeds for which hybrid technology, triploidisation, male or female sterility and/or the legal provisions of intellectual property rights (IPRs) are available. A major influence of GURT availability might be a shift towards those crops and animals which command the largest market shares or highest profit margins.

Furthermore, it appears that T-GURTs will be more widely acceptable and technologically less demanding than V-GURTs because control may be less absolute.

Impact on Agrobiodiversity

Stakeholders in agrobiodiversity encompass the public and private sector, as well as farmers and supportive non-governmental organizations (NGOs). Each of these stakeholders has different roles and interests. Thus the private sector largely maintains its own *ex situ* collections of germplasm and other commodities or has ready access to such stocks, while autonomous farming systems largely maintain their own knowledge system. Stakeholders in the private sector are not much concerned with maintaining agrobiodiversity in autonomous farming systems, or to ensure the survival of wild relatives in natural ecosystems as a much wider source of genetic resources. Although the private sector may regard genetic diversity in such farming systems as highly relevant in the long run, it is often considered as being mainly a public, long-term responsibility.

With increasing GURT application, access for autonomous farming systems to novel industrial innovations under GURT control is likely to become limited or even absent. The extent to which this phenomenon will affect these farming systems will depend on their current use of such new traits and varieties. The issue at stake here is one of genetic overlap-which varieties or breeds form major constituents of both industrial and autonomous farming systems? To what extent does industrial and autonomous production under similar agroecological conditions result in breeding investments in varieties and breeds used by both sectors? Which of the traits developed by the private sector are relevant to small-scale producers? (Visser *et al.*, 2001).

Environmental Effects

Adverse effects on yields of outcrossing from V-GURT containing varieties to neighbouring crop fields may occur. The effects will also occur in those cases in which V-GURT strategies would be consciously employed to prevent unwanted gene flow from GMOs into the environment, cultivated or wild, as an approach to avoid undesirable effects of a deliberate release of GMOs. Such bad effects are most likely to occur in outcrossing crops and trees, especially the latter because of the ability of pollen to disperse over long distances. However, adverse impact may be low if tree seeds do not form an economically important product.

Table 7.1: Some Potential Targets of GURT Applications (after Visser *et al.*, 2001)

Target	*Trait Examples*	*Remarks*
Wheat, rice, maize	nutrient quality; taste; yield; disease resistance; drought avoidance; cold tolerance	staple crops
Soybean	nutrient quality; feed quality	
Cotton	agronomic; colour	
Oilcrops	fatty acid composition	sunflower, olive, oil palm; canola: gene flow containment (V)
Horticultural crops	quality	V-GURTs for non hybrids
Plantation crops	agronomic	coffee, banana
Cattle	meat quality; feed conversion efficiency	specialty products (pharmaceuticals)
Trees	environmental concerns health benefits; lignin content	*Eucalyptus, Populus, Pinus, Acacia*
Fish, other aquatic animals	environmental concerns yield low-temperature tolerance disease resistance	salmonids, carp, tilapia, crustaceans, molluscs

It is easier to avoid potential negative environmental effects in animal husbandry because of the high level of domestication and the current farm practices to control reproduction. In contrast, the transfer of GURT constructs into wild fish populations with potential negative effects on population survival might be substantial because of the low level of domestication and the high probability of escapes of fish varieties (Visser *et al.*, 2001).

For T-GURTS, the impact of outcrossed constructs will be limited in most cases. Most GURT-protected traits may be under the positive control of a chemical inducer. If such a construct outcrosses to other crops to which the chemical is not applied, the construct usually remains unnoticed. But these constructs may still survive, and there may be exceptions to this rule: (i) the trait could be induced by a range of related compounds or triggered by naturally occurring events, such as steroids, pest and disease infestations; (ii) the receiving organisms may already contain a GURT construct controlled by related inducers so that the inducer would activate both constructs. Such undesirable effects might be avoided by choosing highly specific inducers.

Yet another more problematic scenario is the outcrossing of GURT constructs that show negative control of the involved trait. This means that a trait is expressed unless it is blocked by the application of an inducer. In the case of outcrossing, the chemical compound would very probably not be applied to the organism that freshly integrated the construct and the trait would be continuously suppressed, even when not desired. Such outcrossing could affect not only domesticates but also wild relatives.

In conclusion, to safeguard long-term on-farm conservation and development of plant genetic resources, increased investments in public plant breeding, especially participatory plant breeding may be required to narrow the increasing gap in absorption of innovations by autunomous farming systems. This is very important for all staple crops and major vegetables and fruits grown in industrial and autonomous farming systems. Similar assumptions are valid for applications of GURTs in the other agricultural sectors; particularly in aquaculture, some species such as tilapia and crustaceans are grown in both small-scale and large-scale systems. The growing dependence of the small-scale sector on seed developed and provided by industry may have adverse effects on the biodiversity of species maintained in aquaculture. Applications of V-GURTs in crops, whether to protect investments or to protect the environment should be assessed for their negative yield effects (Visser *et al.*, 2001).

Patenting Issues

The U.S. Department of Agriculture (USDA) has licensed the Terminator technology to its seed industry partner, Delta & Pine Land (D&PL) (see Anon., 2001). As a result of joint research, the USDA and D&PL are co-owners of three patents on the controversial technology that genetically modifies plants to produce sterile seeds, preventing farmers from re-using harvested seed. A licensing agreement establishes the terms and conditions under which a party can use a patented technology. Although many of biotechnology corporations hold patents on Terminator technology, D&PL is the only company that has publicly declared its intention to commercialize Terminator seeds.

With a view to pacifying its critics, the USDA has also imposed the following restrictions on how the Terminator technology could be deployed by Delta & Pine Land.

The licensed Terminator technology will not be used in any heirloom varieties of garden flowers and vegetables and it will not be used in any variety of plant available in the marketplace before January 1, 2003.

All royalties accruing to USDA from the use of Terminator will be earmarked to technology transfer efforts for USDA's Agricultural Research Service innovations that will be made widely available to the public.

Ironically, the agency promotes Terminator as a "green" technology that will prevent gene flow from transgenic plants. The notion that Terminator is a biosafety "bandage" for genetically engineered crops has been rejected by several experts in agricultural biotechnology (see Anon, 2001). But even if it were, biosafety at the expense of food security may be unacceptable. In 2000, the UN Food and Agriculture Organization's Panel of Eminent Experts on Ethics in Food and Agriculture concluded that Terminator seeds are unethical.

Economic Aspects

The potential of developing seeds that produce non-viable offspring led to the terms terminator technology; or genetic use restriction technology (GURT). Two types of GURT mechanisms are known: V-GURTs, which produce sterile seeds; and T-GURTs that only manifest their added traits if treated with a specific chemical inducer.

The GURT may provide a biological means to strengthen intellectual property protection on novel crop varieties or animal breeds. By controlling access to a necessary inducer compound, breeding companies could restrict farmers or other breeders from reproducing their innovation. Second, GURTs might be used to contain transgenes in genetically modified varieties, thereby addressing biosafety concerns. Third, GURTs could facilitate precision agriculture in which certain traits such as stress response or vegetative development are turned on or off by the farmer, precisely as and when needed. Reinforcing the protection on innovations raises important issues in terms of economic effects and policy responses (Eaton, 2001; Eaton, *et al.*, 2002).

The potential economic consequences of GURTs can be summarized in terms of benefits, costs and the risks involved for various groups (see Tables 7.1,7.2).

Farmer Seed Systems

In many countries two distinct kinds of seed delivery systems are met with: (i) the formal (regulated) seed supply system and (ii) the farmer's own seed supply system. Globally, most of seed is actually produced by farmers themselves as they tend to be very particular about the seed that they use. Many women are engaged in seed selection and storage. The gene pool that is used in farmers' seed systems is dynamic. Genetically diverse land races or farmers' varieties evolve with changing conditions, requiring a regular influx of genes, and farmers attach great importance to new materials as a source of influx.

In general modern varieties do not perform well in marginal conditions but their characteristics do enrich the genetic base of farmers' varieties in more marginal production systems. Therefore, alternative approaches to plant breeding and seed supply have emerged in the 1990s in response to the limitations of the Green Revolution in highly marginalized farming systems. These include breeding for specific adaptation, participatory variety selection and participatory plant breeding, combining scientific and farmers' knowledge and materials (Eaton *et al.*, 2002). As free access to a wide variety of plant genetic resources is crucial for the success of this approach, by reducing or blocking this genetic diffusion, GURTs could have several adverse impacts on those who depend on farmer seed systems.

GURTs provide greater scope for breeders to obtain the benefits of their innovations than the current intellectual property right (IPR) systems.

Table 7.2: Potential Economic Benefits, Costs and Risks of Genetic Use Restriction Technology (GURT) (after Eaton *et al.*, 2002)

	Benefits	*Costs*	*Risks*
Farmers	Increased productivity from improved inputs due to increased research and development (R&D) investment	Increased input costs from seed purchase (incl. transaction costs)	Misuse of monopoly powers by breeders. Reduced seed security and access to genetic improvements (marginalized farmers)
Breeders (especially private sector)	Increased appropriation of research benefits from products	Increased cost for access to gene pools of other new breeders	
Governments	Reduced investment requirements in breeding. Fewer enforcement costs for plant variety protection	Complementary R&D investment requirements. Other regulatory measures required	
Society	Increased agricultural productivity		Reduced genetic diversity in fields

Breeders can earn profits in two ways: legally, through IPR legislation and licence fees; and biologically through hybrids. IPR protection systems for agricultural crops are functioning in industrialized countries in the form of patents or/and plant variety protection (PVP) which are less restrictive. In recent years, however, the scope of protection has increased in several countries with more and more restrictions being placed on the use of farm-saved seed (Eaton, 2001).

Hybridization has been a good way of earning profits from growing several important crops such as maize, sorghum, rice and vegetables. Private sector investment in these crops has greatly increased in some countries. In developing countries, the private sector investment for hybrid crops has been higher than in crops that cannot be hybridized easily. This means that GURTs might provide the incentive for the private sector to increase investments in seed breeding. It appears likely that GURTs will first be developed for such major crops as wheat and cotton where hybridization has not been very successful.

Indeed, GURTs provide a good rationale for a strengthening of the trend towards vertical integration in the seed breeding and agrochemical sectors. T-GURTs for instance need specific chemical inducers to activate the desired traits. The need for co-ordinating product development between the seed and the inducers strengthens the benefits from vertical integration. But while vertically coordinated structures are necessary to foster the development of GURTs, this may also deter competitors from entering either the seed and/or the inducer market. According to Eaton *et al.*, whether this proves beneficial or detrimental to farmers and consumers will depend on the balance between efficiency savings on the one hand, and the extent of cartelisation and restricted entry for new competitors, on the other.

Besides vertical integration, there has also been a horizontal concentration in the sense of fewer suppliers of the same product in breeding and agricultural input industries because of increasing economies of scale. In R&D work also there is an increasing focus in the seed industry on only a few crops rather than many crops.

Governments have very few options to restrict or prohibit the use of GURTs. Biosafety legislation is not quite valid for prohibiting the introduction of GURTs, because most GURTs in themselves do not pose any specific threat to food or environmental safety. Biosafety laws can be used to ban GURTs from the market only where countries can specify socioeconomic reasons for restricting access to GMOs (see Srinivasan and Thirtle, 2000).

In conclusion, while GURTs may be attractive for increased private sector innovation in the agricultural breeding sector, on balance, the development implications of GURTs generate much concern, particularly from the perspective of the more vulnerable and marginalized farmers. Like many other technological innovations, wealthy farmers and richer farming countries stand to gain most of the benefits (Eaton *et al.*, 2002).

Apomixis

Apomixis is a natural, asexual type of reproduction in which plant embryos grow from ova without being fertilized by pollen. It offers a means of cloning plants through seed. The progeny are genetically identical to the mother plant. Unlike normal sexual hybrids or open-pollinated varieties, apomictic seed is genetically uniform from generation to generation.

Apomixis occurs naturally in many plant species and wild relatives of some crops. The challenge lies in inserting the trait for apomixis into such sexually propagated crops as rice, wheat, millet and sorghum. Plant breeders have succeeded in transferring the genes that confer apomixis from a wild grass species, *Tripsacum dactyloides*, to maize. Maize is the first sexual species successfully transformed into an apomictic form (see Becker, 1998).

Many multinational seed and agrochemical corporations are interested in apomixis research, particularly because of its potential to reduce the cost of hybrid breeding programs. Once a superior variety is created by the combination of inbred lines, the apomictic hybrid plant and its genetically uniform offspring can produce asexual seeds more conveniently than inbred lines. Using apomixis to produce hybrid seeds, companies can greatly cut costs associated with maintaining inbred lines, including land and labour-intensive practices such as detasseling to prevent cross-pollination.

Four different scientific approaches to the study of apomixis are:

1. Generation of hybrids between crops and apomictic wild varieties. This strategy has been applied on maize, pearl millet, beet, apple, wheat grasses and wild rye. The lines obtained are agronomically unsuitable, only partially apomictic and even infertile.
2. Identifying and mapping the genes regulating apomixis, with a view to transferring them to crops through genetic engineering. Gene mapping work has shown that apomixis is closely related to a supergene: a group of genes that are always transmitted together as a package.
3. Mutagenesis in model plants through the use of mutagens. Applied to wild mustard and rice.
4. Crossing plant varieties at different stages of ovum development. Attempts are underway to obtain apomictic plants by crossing two plant varieties of the same species but with different periods for egg maturation. The idea is that the offspring of these matings might produce confusing signals about when to develop the egg, causing

the plant to skip egg formation and produce an apomictic embryo.

Biology is probably just as important as are genes for apomixis to occur. According to Savidan (2000) apomictic plants seem to be generated by the ectopic expression of the genes regulating embryo development, which all plants with flowers share. In other words, the genes that regulate embryo development are turned on earlier than in sexual reproduction, and they are turned on in cells that would otherwise not be turned on at all. But this ectopic expression only results in apomixis when there is also a favourable genetic background. Savidan hopes that the tools of functional genomics will elucidate the genetic switches triggering the ectopic expression of regulatory genes. These mechanisms could be introduced in crops through genetic engineering. The favourable genomic background is likely to be much harder to pinpoint. The supergene may play a role. Yet the fact that apomixis only occurs in polyploid plants suggests that polyploidy might be a prerequisite for apomixis expression. Although some crops such as potatoes are polyploid, many others, including maize, rice and pearl millet, are not.

If apomixis is linked to polyploidy, then developing apomictic maize or pearl millet would only be possible after radically transforming their genomes-this is why scientists forecast that developing apomictic maize may take up to two decades.

As apomixis is the asexual production of seeds, apomictic seeds are clones of the mother plant. Although quite rare in crop plants, apomixis is a most cherished dream of plant breeders who believe that everyone would gain from apomixis-breeders would be able to produce new varieties of seeds more quickly and more cheaply; seed companies stand to benefit from the increased breeding capabilities and their ability to produce new, cheaper varieties faster than their competitors. Even more important benefit is that farmers would be able to save hybrid seed for the following crop, thereby saving themselves money while keeping their yields high. Unfortunately, despite earlier claims that its transfer into crops was imminent, apomixis is proving elusive; apomictic crops may still be some two decades away from reaching the market.

Apomixis is widespread in nature but infrequent: it occurs in around 10% of the 400 families of flowering plants but only in 1% of the 40,000 species belonging to those families (Van Dijk and Van Damme, 2001). It is most frequent in Poaceae, Compositae, Rosaceae and Asterceae. Only a few crops are apomictic: citrus, mango, some tropical forages and a few others. Apomixis can arise in two ways. Apomictic seeds can arise from a plant's sexual cells which fail to divide by meiosis. Alternatively, they can be generated from non-sexual (somatic) cells. Occasionally, both sexual and asexual seeds develop from the same flower. Apomictic plants produce cloned seed, enabling them to reproduce asexually. But their pollen is often viable, so that apomixis can also be transmitted through the more common mechanism of sexual reproduction (Vieille Calzada *et al.*, 1996; Savidan, 2000).

The most striking benefit of introducing apomixis into crops is to allow selecting an individual plant and propagating it as clones through its seeds. A second benefit can be to expand the range of wild relatives that could be integrated into breeding programmes. This is because asexual seeds can contain two sets of chromosomes of different sizes and still be viable, while equivalent sexual seeds may not develop (Anonymous, 2001).

Like genetic engineering, apomixis can potentially demolish some of the species barriers that have acted as speed breakers in the evolution of crops. The capacity of apomixis to create and stabilise new genetic combinations on the one hand, and to break the species barriers on the other, could lead to on asexual revolution, which could even dwarf the Green Revolution.

To begin with, apomixis would substantially decrease the costs of hybrid production. Farmers cannot save the seeds from hybrids because they do not breed true. Companies need land and labour to maintain the parental inbred lines that are crossed every year. With apomixis, hybrid seed could be developed from hybrid seed leading to savings in time and costs.

The implications of the apomixis-led acceleration of plant breeding could be dramatic. Lower costs and much shortened time frames could change the focus of plant breeding. Breeders would be able to genetically adapt plants to specific micro-environments, rather than the current practice of adapting the overall cultivation environment to the crop plants' requirements. Such precision farming would further benefit from research on plant genomics, which aims to identify the genes confering complex plant traits.

Apomixis and plant genomics would also combine to quickly deliver uniform crops that are better tailored for end uses, be it as food, fibre, pharmaceuticals, plastic or any raw material (Anonymous, 2001).

If made available in the public domain, apomixis is likely to result in a general decrease of seed prices, and in a steep increase in the varieties available to farmers. But if one or a few companies control apomixis, the effects would be radically different. Apomixis-led savings in variety development could allow the companies in control of the technology to push competitors out or impose abusive licensing conditions.

Introduction of Apomixis into Crop Plants

Introducing apomixis into sexually propagated crops can involve two approaches.

Introgression

Involves wide hybridization between a crop species and an apomictic relative, followed by a backcrossing programme using the sexually propagated species as the recurrent parent.

Synthesis

Apomixis may be achieved by crossing plants with mutations in key processes which, in combination, may create apomixis. Examples of such mutations include those that produce an egg with an unreduced number of chromosomes (meiotic mutants), and those that develop an embryo without fertilization (parthenogenetic mutants).

As apomixis is the asexual formation of seed, apomictic varieties can be easy to maintain but difficult to breed (Bicknell and Bicknell, 1999).

There is some concern that mass production of low-cost clonal varieties might promote genetic uniformity and crop monoculture. The introduction of genetically uniform cultivars could unintentionally reduce genetic diversity in agriculture if it is widely introduced in areas where farmers grow traditional crop varieties. Proponents of apomixis respond to these misgivings by stating that apomixis will also permit the rapid development of new, resistant varieties on a more regular basis. The simplicity and low cost of apomictic breeding would encourage the introduction of a wider range of varieties that could be uniquely suited to a particular micro-environment, and hence encourage genetic diversity in farm communities (Anonymous, 2000).

Merits of Apomixis

There are numerous potential benefits of clonal plant reproduction by apomixis. Farmers can profit from apomictic hybrid varieties that proliferate their superior characteristics; they can obviate the need for regular purchase of new hybrid seed. Commercial plant breeders can employ apomixis to economize variety development.

For agricultural purposes, both sexual and asexual crop reproduction suffer from certain limitations. It would be valuable if the merits of clonal uniformity could be combined with the cost effectiveness and utility of seed propagation. In some plant species this happens naturally through apomixis which involves embryo formation without the fertilization of an egg cell by a pollen cell. In this process, the viable seeds are produced asexually, so are genetically identical to each other and to the mother plant, forming clonal populations. In some plants that reproduce by 'autonomous apomixis', the asexual formation of seed takes place without a requirement for pollination. Other plants continue to require either cross- or self-pollination to stimulate seed formation and/or to ensure the development of the nutritive embryo sac ('pseudogamous apomixis') (Bicknell and Bicknell, 1999). In facultative apomictics both apomixis and sexual reproduction occur in the same plant. Natural apomicts include plants such as *Taraxacum* spp. and *Potentilla argentea*. Among crop plants very few are apomictic. The majority of apomictic crops are either tropical fruit trees, such as citrus and mango, or forage species, such as *Poa pratensis* and *Brachiaria decumbens*.

Six chief benefits of apomixis have been noted:

Rapid Development of New Hybrid Varieties

It allows 'fixing' the desirable genetic make-up of any individual plant immediately without the creation of inbred lines, thereby significantly reducing the costs of a hybrid breeding programme.

Economic Hybrid Seed Production

Apomixis allows drastic cuts in the cost of hybrid seed

production, because the apomictic hybrid plant and its identical offspring produce seeds asexually at a higher rate than inbred lines.

Propagation of Hybrid Seed

Whereas the F1 generation of hybrid crops does not render the same genotype or performance as the parent plants, apomictic varieties do not change their genetic make-up and hence 'breed true'. This means that, instead of purchasing new hybrid seed every year, farmers can save and sow seed of apomictic hybrid varieties without losing hybrid vigour.

Resistance Against Pathogens

Vegetatively propagated plants (such as sweet potato) transfer viruses and viroids from one asexual generation to the next. The pathogen load of the clone increases with each successive generation as plants are subjected to new strains and species of pathogen. This 'horizontal transfer' of pathogens leads to progressive degeneration of a clone. By contrast, very few pathogens are transferred through seed.

Handling of Propagation Material

Being perishable, many asexual propagules, such as potato tubers, are awkward to store. In contrast, seed is typically both small and robust, and is easier to transport and sow.

Increased Reproduction Efficiency

Some crop losses are caused by difficulties/limitations of fertilization or pollination caused by incompatible varieties, inadequate pollinator activity, or some biotic/abiotic stress. These limitations can be overcome by 'autonomous' apomictic species without a requirement for pollination.

Distinction Between Terminator Technology and Apomixis Technology

In contrast to the terminator technology which is designed to prevent farmers from saving second generation seed, apomixis technology has the potential to dramatically expand and decentralize plant breeding opportunities, especially for resource-poor farmers. Apomictic hybrid seed could potentially provide tremendous benefits to poor farmers because desirable traits could be maintained indefinitely with no loss of hybrid vigour, and farmers would be able to save their hybrid seed for replanting year after year. Apomixis technology offers fast, flexible and low-cost plant breeding strategies responsive to locally-targeted crop breeding needs.

Apomixis and Farmers

Farmers can benefit from apomixis in two ways: saving hybrid seeds and stabilising their best plants.

Apomixis would allow farmers to save the seeds from hybrid plants but still conserve the superior yields. Farmers would no longer need to access or purchase new hybrid seed for every planting season in order to ensure a marketable surplus. Apomixis would serve farmers living in remote areas where neither the seed industry nor governments can guarantee an assured and timely supply of hybrid seed. But apomictic seeds would still be hybrids, which means they would demand dependence on fertilisers and pesticides. Like other hybrids, apomictic hybrids would still be designed to perform their best under certain environmental conditions, which small farmers are unlikely to achieve and maintain. Apomixis might increase farmers' access to hybrids, but not their control of them (Anonymous, 2001).

Perhaps more important in the long run, apomixis would allow farmers to fix the genetic characteristics of any of their individual crop plants, by crossing them with an apomictic line. It would allow farmers to become faster breeders. It would give farmers greater control of their local agro-environment. It would theoretically guarantee yield and uniformity (and therefore, marketability) of their own selected varieties.

However, some experts are not sure about the merits of apomixis-according to them existing approaches to participatory plant breeding are probably more helpful. Angela Cordeiro who worked on participatory breeding with Brazilian family farmers feels that skilled farmers breed for variability rather than yield, to afford security in the face of unpredictable environmental conditions. The stability and fixation entailed by apomixis is alien to small farmers' traditional strategies. Besides, Cordeiro's work has shown that yields in traditional, open-pollinated maize varieties are limited by bottlenecks in soil management and seed storage, rather than by genetic potential.

Commercial farmers are expected to benefit from the acceleration in formal plant breeding that apomixis would create: cheaper seed, more varieties adapted to their particular growing conditions and more potential end-product markets to choose from. This would depend on whether apomixis is accessible to all seed companies or only to a few. In the latter case, companies might at first lower seed prices to wipe competitors out, and then raise them.

Apomixis as an Antidote to Terminator

One major threat to farmers benefiting from apomixis is the potential use of Traitor/Terminator Technologies, also known as Genetic Use Restriction Technologies (GURTs). Because in principle apomixis would allow farmers to save hybrid seed, it has been touted as the antidote to Terminator, which renders seeds sterile. However, seed companies will only bank on apomixis if they can prevent farmers and competitors from obtaining clones from apomictic varieties. It seems likely that companies will use GURTs in conjunction with their apomictic varieties, just as they plan to do with other seeds. So apomixis would not be an antidote to the Terminator-it would rather complement it (Anonymous, 2001).

Apomixis and the Environment

The introduction of apomixis into crops can alter the genetic diversity of both crops and their wild, non-apomictic relatives. This can lead to unforeseen results because apomixis is so poorly understood. Predictions on apomixis' impact on biodiversity are highly speculative and even contradictory. The first image apomixis brings to mind is of uniform fields of cloned plants. More uniform than current monocultures, these crops may be more fragile, and more susceptible to pest and disease pressures. This will demand an accelerated genetic treadmill to ward off pest and diseases.

Several issues surround the possible dangers of the apomixis mechanism spreading to wild populations and the impact that could have on genetic diversity and plant evolution. In nature apomixis seems to be limited to polyploid species, which are relatively few in number. But if an apomixis supergene were successfully transferred into diploid plants, the supergene could spread to wild diploid species. The competitive advantage this might afford these plants could lead to the genetic erosion of many of the wild, non-apomictic relatives.

Apomixis may remain a most cherished dream of plant breeders for many years to come. Notwithstanding the hurdles, key actors are motivated because the rewards could be immense. Companies are likely to continue investing in apomixis research because of the technology's potential to transform agriculture completely. Apomixis could put an end to variability within plant varieties and hence make them more predictable. This would open the way to the use of plants as bioreactors to extract uniform, high-quality substances for any kind of industrial use. Agriculture could then compete with industry in producing organic polymers, for example. Servicing of many new markets promises spectacular profit gains to the seed industry (Anonymous, 2001).

Small farmers working with traditional varieties might have more to fear from apomixis than to gain from it. The ability of apomixis to genetically freeze particular plants would thwart attempts to maximise variability, which is an important farmer strategy. Moreover, should polyploidy turn out to be a prerequisite for apomixis and the seed industry turns exclusively to polyploid seeds, farmer's diploid varieties will become further marginalised. Indeed, apomixis might be marginalising farmers' varieties already. By promising to solve farmers' loss of hybrid vigour, it indirectly promotes hybrids at the expense of other technological options which are more appropriate to farmers' objectives of minimising risk and promoting diversity in their crops (Anonymous, 2001).

References

Anonymous. Apomixis: the plant breeder's dream. *Seedling* (Newsletter of GRAIN Foundation Barcelona) 18(3): 9-17 (Sept., 2001).

Anonymous. Seeding Solutions. Vol. 1. Internat. Develop. Res. Centre, Ottawa/Dag Hammarskjold Foundation, Uppsala (2000).

Becker, H. Revolutionizing hybrid corn production. *Agricul. Res.* (USDA) (Dec. 1998).

Bicknell, R.A., Bicknell, K.B. Who will benefit from apomixis? *Biotech. Develop. Monitor* 37: 17-20 (1999).

Eaton, D. *TRIPS and Plant Varietal Protection: Economic Analysis and Policy Choices.* Agricultural Economic Research Institute (LEI) Report 7.02.01, The Hague (2001).

Savidan, Y. L'apomixie, ou le clonage par les graines. *Biofutur* No. 198 (March, 2000).

Srinivasan, C.S., Thirtle, C. *Impact of Terminator Technologies in Developing Countries: A Framework for Economic Analysis.* Paper presented at the 4th International Conference on the "Economics of Agricultural Biotechnology" organised by the International Consortium on Agricultural Biotechnology Research (ICABR), Ravello, Italy, 24-28 August (2000).

Van Dijk, P., Van Damme, J. Apomixis technology and the paradox of sex. *Trends in Plant Science* 5 (2): 81-84 (2001).

Vieille Calzada *et al.* Apomixis-the asexual revolution. *Science* 274: 1322 (1996).

Eaton, D., van Tongeren, F., Louwaars, N., Visser, B., van der Meer, I. Economic and policy aspects of 'terminator' technology. *Biotech. Develop. Monitor* 49: 19-22 (March, 2002).

Anon. U.S. Department of Agriculture says yes to Terminator. *Global Pest. Camp.* 11(2): 18-19 (2001).

Visser, B., van der Meer, I., Louwaars, N., Beekwilder, J., Eato, D. The impact of 'terminator' technology. *Biotech. Develop. Monitor* 48: 9-12 (2001).

Radin, J.W. The technology protection system: Revolutionary or evolutionary? *Biotech. Develop. Monitor* 37: 24 (1999).

Horstmeier, G.D. Lessons from year one: experience changes how farmers will grow Roundup Ready beans in '98. *Farm Journal* p. 16 (January 1998).

Rissler, J., Mellon, M. *The Ecological Risks of Engineered Crops.* The MIT Press, Cambridge (1996).

Crouch, M.L. How the Terminator terminates: an explanation for the non-scientist of a remarkable patent for killing second generation seeds of crop plants. Occasional paper of the Edmonds Institute, Edmonds, Washington (1998).

Glossary

Accession or entry: A population or line in a breeding programme or germplasm collection; also an individual sample in a germplasm bank.

Agenda 21: A multi-faced intergovernmental Plan of Action to address sustainable development issues in the 21st Century adopted at the United Nations Conference on Environment and Development ('Earth Summit') in Rio de Janeiro, Brazil, in June 1992.

Antibody: An immunoglobulin protein produced by B lymphocytes of the immune system that binds to a species antigen molecule.

Antigen: Any foreign substance, such as a virus, bacterium, or protein, that elicits an immune response by stimulating the production of antibodies.

Antisense RNA: A complementary RNA sequence that binds to a naturally occurring (sense) mRNA molecule, thus blocking its translation.

Antisense RNA: A complementary RNA sequence that binds to a naturally occurring (sense) mRNA molecule, thus blocking its translation.

Apomixis: An asexual breeding system involving seed production without meiosis and without fertilization. Obligate apomicts necessarily reproduce by apomixis. Facultative apomicts can reproduce either by apomixis or sexually.

Backcrossing: A procedure for introducing characteristics from one donor variety or species into a second recipient variety or species. It consists of crossing hybrids with the recipient variety, selecting the offspring, expressing the targeted character, and crossing them with the recipient species again and again until a new line is obtained that resembles the donor species or variety in terms of the selected character.

BioPiracy: The expropriation of any biological material for commercial purposes, in the absence of effective intergovernmental regulation and the fully informed consent of those who have nurtured and developed the material.

Biotechnology: The scientific manipulation of living organisms, especially at the molecular genetic level, to produce useful products on a large scale.

Breeder's seed: Seed stock produced by the breeder or by the agency licensed to produce or maintain a variety.

Callus: A mass of undifferentiated plant cells manipulated in plant tissue culture.

Central dogma: Francis Crick's idea that genetic information flows from DNA to RNA to protein.

Cloning: The mitotic division of a progenitor cell to give rise to a population of identical daughter cells or clones.

Codon: A group of three nucleotides that specifies addition of one of the 20 amino acids during translation of an mRNA into a polypeptide.

Compatible interaction: Interaction between susceptible host and virulent pathogen.

Conference of the Parties (COP) to the Convention on Biological Diversity (CBD): All the national governments which have ratified the Biodiversity Convention. The COP meets periodically to debate and carry out the Convention's mandate.

Consensus sequence: A DNA sequence that is highly conserved in different species.

Cooperative Innovation System(s): (Also known as the Informal or Community Innovation System). The collective process of scientific research traditionally carried out by indigenous peoples and other farming communities.

Cosmid: A plasmid vector containing a COS site that enables it to be packaged into infective bacteriophage particles. Used for cloning DNA sequences of 35,000-45,000 bp.

Crop evolution: The adaptation of a crop over generations of association with man, to forms more advantageous to man and brought about by generally unconscious selection, provision of nutrients and protection from pests and diseases. It may occur to the extent that the domesticated form loses the ability to survive in nature.

Cross-hybridization: The hydrogen bonding of a single-stranded DNA sequence that is partially but not entirely complementary to a single-stranded substrate. Often, this involves hybridizing a DNA probe for a specific DNA sequence to the homologous sequences of different species.

Crossing-over: The exchange of DNA sequences between chromatids of homologous chromosomes during meiosis.

Cross-pollination: Transfer of pollent o the stigma of a different plant or clone.

Crown gall disease: A tumorous growth of certain dicot plants caused by *Agrobacterium tumefaciens.*

Cultivar: A variety of a plant produced by selective breeding.

Defense signalling pathways: The plant senses entry of the pathogen. This signal activates defense mechanisms through different signalling pathways mediated by such molecules as salicylate, jasmonate and ethylene. These different pathways are interconnected.

Directional cloning: DNA insert and vector molecule are digested with two different restriction enzymes to create noncomplementary sticky ends at either end of each restriction fragments. This allows the insert to be ligated to the vector in a specific orientation and prevents the vector from recircularizing.

Disease: A deleterious alteration in the dynamic interaction between an individual and the environment, caused by a biotic or an abiotic factor.

Diversity: The existence of alternate forms (genetic or otherwise). (also: variability)

Double cross hybrids: Hybrid resulting from a cross between two other (single cross) hybrids.

Ecosystem: All the biotic (living organisms) and abiotic (non-living components) and the total environment within which the organisms naturally occur.

Ectopic expression: Expression of a gene out of its expected time or place.

Electroporation: A method for transforming DNA, especially useful for plant cells, in which high-voltage pulses of electricity are used to open pores in the cell membrane through which foreign DNA can pass.

Elicitors: Molecules (generally from pathogen cell wall), that trigger defense reactions in the host plant.

Elite germplasm: Germplasm that has been manipulated for use in a breeding programme.

Embryology: The study of the morphological and biochemical development of the fertilized egg into an adult organism.

Enhancer: A regulatory DNA sequence that greatly increases transcription of a gene. An enhancer retains its influence regardless of orientation and can be located several thousand base pairs upstream or downstream from the gene it influences.

Enhancer: A regulatory DNA sequence that greatly increases transcription of a gene. An enhancer retains its influence regardless of orientation and can be located several thousand base pairs upstream or downstream from the gene it influences.

Evolution: The long-term process through which a population of organisms accumulates genetic changes that enable its individuals to adapt to environmental conditions and to better exploit food resources.

Exon. The portion of a gene found in mature mRNA. Exons of a gene are linked together by mRNA splicing.

Farmers' Rights: Rights arising from the past, present and future contributions of farmers in conserving, improving and making available plant genetic resources, particularly those in the centres of origin/diversity.

Farmers' Rights: The recognition of farmers as past, present and future *in situ* agricultural innovators who collectively conserve and develop agricultural genetic resources. Farmers are recognised as innovators entitled to intellectual integrity and to compensation whenever their innovations are commercialised. They have the right to Germplasm, Information, Funds, Technologies and Farming/Marketing Systems (GIFTS). Compensation is anticipated via a global Gene Fund, paid into by the North for genetic conservation and improvement in the South. Agenda 21 and the Biodiversity Convention have also adopted the principle of Farmers' Rights.

First Farmers: Indigenous peoples around the world who were the first to domesticate plant and livestock species and who continue to conserve and enhance the majority of the world's agricultural biodiversity.

Flanking region: The DNA sequences extending on either side of a specific locus or gene.

Food: In the EU, any substance or product, whether processed, partially processed, or unprocessed, intended to be ingested by humans (chewing gum and water included). In the US, an article used for food or drink for humans or other animals, chewing gum, and articles used for components of any such article.

Food additive: In the EU, A substance that does not occur normally in consumed foods (flavorings, minerals, and vitamins excluded). In the US, a food component not generally accepted as safe under the conditions of its intended use (pesticides, drugs, and color additives excluded).

Food supplement: In the EU, a concentrated source of nutrients (vitamins, minerals) marketed in the form of capsules, tablets, pills, sachets of powder, ampules of liquids, and drop-dispensing bottles. Its purpose is to supplement the intake of those nutrients in the normal diet. In the US, a vitamin, mineral, herb or other botanical, amino acid, and/or dietary substance for use by humans to supplement the diet by increasing the total dietary intake; may take the shape of a tablet, capsule, powder, softgel, gelcap, or liquid form, or if not intended for ingestion in such a form, is not represented as conventional food and is not represented for use as a sole item of a meal or diet.

Formal seed system: Seed production, control and distribution activities carried out by the public and commercial sector. This may include breeding.

Functional food: Resembles a conventional food in appearance and is consumed as part of a usual diet. Has physiological benefits and/or reduces the risk of chronic disease beyond basic nutritional functions.

Functional Genomics: The science of how the genes in organisms interact to express complex traits.

Gene (or seed) bank: A form of *ex situ* conservation for plant germplasm. Gene banks may be humidity- and temperature-controlled facilities where seeds (or other reproductive material) are stored for future use in research and breeding. Many gene banks are nothing more than deep freezers or refrigerators. Banks should only be seen as a back-up for the maintenance of crop genetic diversity *in situ* or *on farm*.

Gene: A locus on a chromosome that encodes a specific protein or several related proteins.

Gene: The basic unit of active transmission of the genetic information that determines patterns of inheritance.

Gene amplification: Production of multiple copies of a gene within a single genome.

Gene amplification: The presence of multiple copies of a gene. Amplification is one mechanism through which proto-oncogenes are activated in malignant cells.

Gene expression: The process of producing a protein from its DNA- and mRNA-coding sequences.

Gene for gene relationship: For every resistance gene in the host there is a corresponding avirulence gene in the pathogen.

Gene targeting (gene trapping): A method to select for transfected cells in which a DNA sequence has integrated at a homologous site on the chromosomes.

Genebank: Storage facility where germplasm is stored in the form of seeds, pollen or *in-vitro* culture, or in the case of a field genebank, as plants growing in the field.

Genetic code: The three-letter code that translates nucleic acid sequence into protein sequence.

Genetic diversity: Total amount of genetic variation present in a population or species.

Genetic engineering: Manipulation of an organism's genome by introducing or eliminating specific genes. A broad definition of genetic engineering also includes selective breeding and other means of artificial selection.

Genetic erosion: Gradual loss of genetic diversity between and within populations of the same species over time; or reduction of the genetic base of a species due to human intervention, environmental change, etc.

Genetic resource: Germplasm in plants, animals or other organisms, containing useful characters of actual or potential value.

Genome: The collective term for all genes carried by a single representative of each of all the chromosome pairs.

Genomic library: A library composed of fragments of genomic DNA.

Genotype: 1. The genetic constitution of an organism. 2. A group of organisms with similar genetic constitutions.

Germplasm: The total genetic variability, represented by germ cells or seeds, available to a particular population of organisms.

Green revolution: Advances in genetics, agrochemicals, and machinery that resulted in a dramatic increase in crop productivity during the third quarter of the 20th century.

GxE interaction: Genotype by Environment interaction. Phenomenon that two (or more) varieties will react differently to a change of environment.

Herbicide: Any substance that is toxic to plants: often used to kill specific unwanted plants.

Heteroduplex: A double-stranded DNA molecule or DNA-RNA hybrid, where each strand is of a different origin.

Human Genome Project: A project to determine the entire nucleotide sequence of the human chromosomes.

Hybrid: The progeny of genetically dissimilar parents.

Hybrid vigour (heterosis): The exhibition by a hybrid of a more vigorous growth, greater yield, or increased disease resistance than either parent.

Hypersensitive response (HR): Small brown necrotic lesions produced by the host plant during incompatible interaction resulting from localized cell death. Associated with resistance.

***In situ* conservation:** Conservation on site. It is the conservation of ecosystems and natural habitats, and the maintenance, recovery and development of visable populations of species in their natural surroundings. In the case of domesticated livestock or cultivated crop species, it is their conservation on-farm.

Inbred line: Genetically (nearly) homozygous population, derived through several cycles of selfing; also used for hybrid seed production.

Incompatible interaction: Interaction between resistant host and avirulent pathogen.

Informal seed system, local seed system or farmers' seed systems: Seed production and exchange activities by farmers and grassroots organizations.

Interspecific hybrid: It results from crossing two related, but different, plant species. One example is crossing maize with its wild relative, *Tripsacum* sp.

Intraspecific hybrid: It results from crossing two different varieties of the same species.

Intron: A nucleotide sequence that intervenes between exons, which is excised from pre-mRNA during RNA processing.

Keystone Dialogue: The Keystone International Dialogue on Plant Genetic Resources (1988-91) conducted by the Keystone Center of Keystone, Colorado, USA.

Labelling: Attaching labels to seed lots with information on variety identity, purity and seed quality.

Landrace: Farmer-developed cultivars of crop plants which are adapted to local environmental conditions.

Life industry: With the development of biotechnology and the expansion of patent-like regimes over living material, the agrochemical, seed, pharmaceutical, animal health care, human genomics, and food industries are merging into a new Life Industry dependent upon a common set of technologies and monopoly laws.

Lineage: A chart that traces the flow of genetic information from generation to generation.

Locus: The term generally refers to the position of a gene on a chromosome.

Mass selection: Selection of individual plants from a population. Mass selection may be positive or negative selection. Seeds from mass selection form the next generation.

Megabase cloning: The cloning of very large DNA fragments of 100,000 or more nucleotides.

Microinjection: Introducing a solution of DNA, protein, or other soluble material into a cell using a fine microcapillary pipet.

Modern varieties (MVs): Varieties developed by breeders in the formal system.

Molecular biology: The study of the biochemical and molecular interactions within living cells.

Molecular cloning: The biological amplification of a specific DNA sequence through mitotic division of a host cell into which it has been transformed or transfected.

Molecular genetics: The study of the flow and regulation of genetic information between DNA, RNA, and protein molecules.

Monoculture: The growing of a single plant species in one area, usually the same type of crop grown year after year.

mRNA splicing (processing): The excision of introns and subsequent joining of exons of pre-mRNA to form a mature mRNA transcript.

Natural selection: The differential survival and reproduction of organisms with genetic characteristics that enable them to better utilize environmental resources.

Nitrogen fixation: The conversion of atmospheric nitrogen (N2) to biologically usable nitrates.

Nutraceutical: Any product isolated from foods that is generally sold in medicinal forms not usually associated with food and shown to be physiologically beneficial or that protects against chronic disease.

Off-type plant: Plant that is morphologically different from the variety or the types that constitute a given population.

Open pollination: Pollination by wind, insects, or other natural mechanisms.

Open reading frame: A long DNA sequence that is uninterrupted by a stop codon and encodes part or all of a protein.

Oxidative burst: Rapid production of active oxygen species *e.g.*, superoxide anion (O-2), hydroxide radical (OH), or H_2O_2. An early defense mechanism triggered by infection.

Participatory crop improvement (PCI): includes PPB and PVS.

Participatory plant breeding (PPB): Plant breeding (crossing and subsequent selection in the heterogeneous progeny) involving farmers.

Participatory variety selection (PVS): Selection between stable lines or populations involving farmers.

Patent: A legal instrument granting an inventor the exclusive right, for a limited time, to exploit the invention in exchange for disclosure about it.

Pathogen: An organism capable of causing a disease in another organism (a host).

Pathogenesis-related (PR) proteins: A class of proteins induced by many biotic and abiotic stresses. First discovered in tobacco after TMV infection, hence called PR proteins. Have a role in defense.

Pest: Any form of plant or animal life or any agent pathogen to plants or plant products.

Pesticide: A substance that kills harmful organisms (for example, an insecticide or fungicide).

Pharmaceutical: Single, well-defined, medicinal, therapeutic compound that is administered for curing, mitigating, or diagnosing disease.

Plant breeding: The application of genetic principles and practices to the development of individuals, cultivars or varieties, more suited to the needs of man.

Plant Breeder's Rights: A legal instrument granting the developer of a plant variety the exclusive right to market it for a limited time. Varieties protected by such legislation may be used by others in the development of new varieties.

Plant genetic resources: Genetic material of plants, including modern cultivars, landraces and wild relatives of crop plants, of value as a resource for present and future generations of people.

Polygenes: The genes responsible for the genetic components of variation in a quantitative character, each individual gene exerting only a slight effect on the phenotype.

Polyploid: A multiple of the haploid chromosome number that results from chromosome replication without nuclear division.

Population: In genetics: A group of individuals which share a common genepool and have the potential to interbreed.

Primer: A short DNA or RNA fragment annealed to single-stranded DNA, from which DNA polymerase extends a new DNA strand to produce a duplex molecule.

Progeny: Offspring.

Pyramiding: Backcrossing of several genes into a certain background.

Qualitative trait or discontinuous variation: Variation between individuals of a population in which differences are marked and do not grade into each other, normally brought about by the effects of different alleles of a few major genes. The traits exist only in a limited number of specific forms or are absent.

Quality-declared seed: A seed system in which a proposed 10 per cent of the seed produced and distributed is checked by an autonomous seed control agency and the rest by the seed producing organisation.

Quantitative trait or continuous variation: Variation between individuals of a population in which differences are slight and grade into each other. Quantitative traits are usually those that are determined by a large number of genes and/or considerable environmental influences, *e.g.* weight of yield or plant height.

Recalcitrant seed: Seed that cannot be dried and so cannot be stored at low temperatures without damage.

Recognition sequence (site): A nucleotide sequence-composed typically of 4, 6, or 8 nucleotides, that is recognized by a restriction endonuclease. Type II enzymes cut (and their corresponding modification enzymes methylate) within or very near the recognition sequence.

Recombinant DNA: The process of cutting and recombining DNA fragments as a means to isolate genes or to alter their structure and function.

Regulator proteins: Molecules which bind to specific base sequences (operator sites) and prevent (inhibitors) or facilitate (activators) the activity of mRNA polymerase. Inducers cause the former to detach from, and the latter to bind to, DNA: negative and positive control respectively.

Repeated sequence: A nucleotide sequence whose multiple copies occur throughout the genome. Usually, structural genes are not repeated.

Resistance gene clusters: *Arabidopsis* genome sequencing revealed that resistance genes are not spread throughout the genome but are clustered at specific loci.

Reverse genetics: Using linkage analysis and polymorphic markers to isolate a disease gene in the absence of a known metabolic defect, then using the DNA sequence of the cloned gene to predict the amino acid sequence of its encoded protein.

RNA polymerase: Transcribes RNA from a DNA template.

Seed certification: Assurance of varietal identity and purity through generation control (*i.e.* control of origin and class in the multiplication from breeder seed to basic seed), inspection and labelling.

Seed quality control: Control of physiological, sanitary and genetic seed quality characteristics.

Seed regulation: The total set of rules and protocols related to variety development and release, seed production, quality control and delivery.

Selfing: The process whereby members of a particular generation resulting from a specific cross are allowed to breed among themselves, but not with individuals from other crosses.

Signal transduction: The biochemical events that conduct the signal of a hormone or growth factor from the cell exterior, through the cell membrane, and into the cytoplasm. This involves a number of molecules, including receptors, G proteins, and second messengers.

Silking: Female flowering in maize, *i.e.* the moment that silks emerge from the husk.

Single cross hybrid: Hybrid resulting from a cross between two inbred lines.

Site-directed mutagenesis: The process of introducing specific base-pair mutations into a gene.

Source community: Community from which a local variety or a seed lot originated.

Species: A group of actually or potentially interbreeding natural populations which normally are reproductively isolated from other such groups and/or show common characteristics.

Split gene: A configuration common to eukaryotic genes where exons (sequences in mRNA) are interrupted by introns (sequences removed from precursor mRNA during mRNA splicing).

Stratified mass selection: Mass selection in which the population is split into subpopulations that are grown under different environmental conditions (*i.e.* in different fields or in different parts of a field). Plants for next generation seed are selected from the different subpopulations.

Subcloning: The process of transferring a cloned DNA fragment from one vector to another.

Supergene: A group of genes that are always transmitted together as a package.

Systemic acquired resistance (SAR): Following infection of a plant by a pathogen and its recovery, the plant may develop resistance to future infections by the same or other pathogens. Analogous to immunity in animals.

Three-way cross: Production of a hybrid over two generations. A single cross hybrid is produced in the first season and crossed with an inbred line in the following season. This additional production step requires resources in terms of time, land and money, but it guarantees more reliable yields than inbred lines.

Ti (tumor-inducing) plasmid: A giant plasmid of *Agrobacterium tumefaciens* that is responsible for tumor formation in infected plants. It is used as vector to introduce foreign DNA into plant cells.

Transcription: Copying of the nucleotide sequence of DNA into a complementary sequence in mRNA.

Transcriptional control: Control of enzyme production through regulation of mRNA synthesis.

Transduction: The transfer of DNA from one cell to another by means of a virus.

Transfection: The uptake and expression of a foreign DNA sequence by cultured eukaryotic cells.

Transformation: The conversion of a bacterial cell to a new genotype by direct uptake of DNA.

Transgenic: An organism in which a foreign DNA gene (a transgene) is stably incorporated into its genome early in embryonic development. The transgene is present in both somatic and germ cells, is expressed in one or more tissues, and is inherited by offspring in a Mendelian fashion.

Translation: Decoding of mRNA into a polypeptide sequence at the ribosomes by aminoacyl tRNAs.

Transposon (transposable, or movable genetic element): A relatively small DNA segment that can move from one chromosomal position to another.

Transposon: A mobile genetic element consisting of a structural gene flanked by DNA sequences that facilitate its transposition.

Truthfully labelled seed: Seed with label of the producer with information on the seed quality.

Variability: The state of being variable, *i.e.* being able to change characteristics, form or nature.

Variety: A subdivision of a species below subspecies.

Vector: An autonomously replicating DNA molecule into which foreign DNA fragments are inserted and then propagated in a host cell.

Wild relative: Uncultivated relative of a crop species.

Bibliography

Aerni, P. Public attitudes towards agricultural biotechnology in developing countries: A comparison between Mexico and the Philippines. Center for International Development at Harvard University, Cambridge, Mass. (2001).

Ahmed, F.E. Detection of genetically modified organisms in foods. *TiBTECH* 20: 215-223 (2002).

Altieri, M.A. *Agroecology: The Scientific Basis of Alternative Agriculture.* Westview Press, Boulder (1987).

Altieri, M., Rossett, P., Thrupp, L. The potential of agroecology to combat hunger in the developing world. Institute for Food and Development Policy. *Food First Policy Brief* No. 2 (1998).

Altieri, M., Rossett, P. Ten reasons why biotechnology will not ensure food security, protect the environment and reduce poverty in the developing world. *AgBioForum* 2(3&4): 155-162 (1999).

Altieri, M.A. *Genetic Engineering in Agriculture: The Myths, Environmental Risks and Alternatives.* Special Report No. 1, Food First, Oakland, CA (2001).

Ammann, K., Jacot, Y., Simonsen, V., Kjellsson, G. (eds.). *Methods of Risk Assessment of Transgenic Plants.* Birkäuser Verlag, Basel (1999).

Anon. House of Lords Select Committee on European Communities. 16th Report: Organic Farming and the European Union. HMSO, London (1999).

Anon. U.S. Department of Agriculture says yes to Terminator. *Global Pest. Camp.* 11(2): 18-19 (2001).

Anon. Genetic engineering: not the only option. *LEISA Magazine* 17(4): 4-5 (Dec., 2001).

Anonymous. Council regulation (EC) No. 285/97, concerning novel foods and novel food ingredients. *Official J. Eur. Communities: Legislation* 43: 1-5 (1997).

Anonymous. Seeding Solutions. Vol. 1. Internat. Develop. Res. Centre, Ottawa/Dag Hammarskjold Foundation, Uppsala (2000).

Anonymous. Commission regulation (EC) No. 49/2000, amending council regulation (EC) No. 1139/98 concerning the compulsory indication on the labelling of certain foodstuffs produced from genetically modified organisms of particulars other than those provided for in directive 79/112/EEC. *Official J. Eur. Communities: Legislation* 6: 13-14 (2000).

Anonymous. Commission regulation (EC) No. 50/2000, on the labelling of foodstuffs and food ingredients containing additives and flavourings that have been genetically modified or have been produced from genetically modified organisms. *Official J. Eur. Communities: Legislation* 6: 15-17 (2000a).

Anonymous. Apomixis: the plant breeder's dream. *Seedling* (Newsletter of GRAIN Foundation Barcelona) 18(3): 9-17 (Sept., 2001).

Anonymous. Missing the big picture. *Nature* 421: 675 (2003).

Atkinson, H.J., Urwin, P.E., Hansen, E., and McPherson, M.J. Designs for engineered resistance to root-parasitic nematode. *Trends Biotechnol.* 13: 369-374 (1995).

Auberson-Huang, L. The dialogue between precaution and risk. *Nature Biotech.* 20: 1076-1078 (2002).

Balfour, E.B. *The Living Soil.* Faber and Faber, London (1943).

Barham, E. What's in a name? " Eco-labeling in the global food system. www.pmac.net/bbarham.htm. (1997).

Baskin, Y. *The World of Nature.* Island Press, Washington, D.C. (1997).

Baulcombe, D.C., Saunders, G.R., Bevan, M.W., Mayo, M.A., Harrison, B.D. *Nature* 321: 446-449 (1986).

Beachy, R.N. Facing fear of biotechnology. *Science* 285: 335 (1999).

Beachy, R. plus 17 others. Divergent perspectives on GM food. *Nature Biotech.* 20: 1195-1196 (2002).

Beck, U. *Risk Society*. Sage, London (1992).

Becker, H. Revolutionizing hybrid corn production. *Agricul. Res.* (USDA) (Dec. 1998).

Bent, A.F. *Plant Cell* 8: 1757-1771 (1996).

Bertilsson, G. Environmental consequences of different farming systems using good agricultural practices. *Proc. Fertiliser Society* No. 332, London (1992).

Betz, F.S., Hammond, B.G., Fuchs, R.L. Safety and advantages of *Bacillus thuringiensis*-protected plants to control insect pests. *Regulatory Toxicol. Pharmacol.* 32 (2): 156-173 (2000).

Bhat, A.I., Jain, R.K., Varma, A. Naresh Chandra and Lal, S.K. *Indian J. Phytopathol.* 54: 112-116 (2001).

Bicknell, R.A., Bicknell, K.B. Who will benefit from apomixis? *Biotech. Develop. Monitor* 37: 17-20 (1999).

Blackburn, S. *Being Good: An Introduction to Ethics*. Oxford University Press, Oxford (2001).

Boeringa, R. (Ed.). *Alternative Methods of Agriculture*. Elsevier, Amsterdam (1980).

Bonetta, L. GM crops under new US scrutiny. *Current Biol.* 11: 201 (2001).

Borlaug, N.E. Ending world hunger: the promise of biotechnology and the threat of antiscience zealotry. *Plant Physiol.* 124: 487-490 (2000).

Broekaert, W.F., Terras, F.R.G., Cammune, B.P.A., Osborn, R.W. *Plant Physiol.* 108: 1353-1358 (1995).

Brown, I.R., Renner, M., Flavin C. *Vital Signs 1998: The Environmental Trends That Are Shaping our Future.* W.W. Norton, New York (1998).

Buckingham, D., Phillips, P. Hot potato, hot potato: Regulating products of biotechnology by the international community. *JWT* 35: 1-31 (2002).

Burkhardt, J. Agricultural biotechnology and the future benefits argument. *Agricult. Environ. Ethics* 14: 2-10 (2001).

Burkhardt, J. The genetically modified organism and genetically modified foods debates: Why ethics matters. *Transactions Wisconsin Acad.* 89: 63-82 (2001a).

Buttel, F.H., Goodman, R.M. Introduction to the scientific, political, and ethical dialogue on genetically modified organisms. *Trans. Wisconsin Acad.* 89: 1-13 (2001).

Campolina, A. Brazilian small-scale farmers and poor consumers reject GMOs. *LEISA* Magazine 17(4): 29 (Dec., 2001).

Cao H, Li, X., Dong, X. Generation of broad-spectrum disease resistance by overexpression of an essential regulatory gene in systemic acquired resistance. *PNAS* (USA) 95: 6531-6536 (1998).

Carmona, M.J., Molina, A.,Fernandwz, J.A., Lopez-Fando, J.J., Garcia-Olmedo, F. *Plant J.*, 3: 457-462 (1993).

Cassman, K.G. Ecological intensification of cereal production systems: Yield potential, soil quality, and precision agriculture. *PNAS* (USA) 96: 5952 (1999).

CAST (Council for Agricultural Science and Technology). *Benefits of Biodiversity.* Council for Agricultural Science and Technology Task Force Report no. 133. Council for Agricultural Science and Technology, Ames, Iowa (Feb. 1999).

CCC. (Canola Council of Canada). An agronomic and economic assessment of transgenic canola. Canola Council of Canada, Winnipeg (Jan., 2001).

Charles, D. *Lords of the Harvest*. Perseus Publishing, Cambridge, Mass. (2001).

Chopra, V.L. (Ed.). *Breeding Field Crops.* Oxford & IBH, New Delhi (2001).

Choudhary, B. Legislating for a genetic heritage. *Biotech. Develop. Monitor* Issue No. 48: 19-21 (Dec., 2001).

Choudhary, B., Laroia, G. Technological developments and cotton production in India and China. *Current Science* 80: 925-932 (2001).

Christoforous, T. *Environ. Law J.* 8: 622-648 (2000).

Clark, N., Yoganand, B., Hall A.J. New science, capacity development and institutional change: the case of the Andhra Pradesh-

Netherlands Biotechnology Programme (APNLBP)'. *Internat. J. Technol. Management and Sustainable Develop.* 1 (3): 196-212 (2002).

Cohen, M.B., Gould, F., Bentur, J.S. *Int. Rice Res. Notes* 25: 4-10 (2000).

Commoner, B. Failure of the Watson Crick theory as a chemical explanation of inheritance. *Nature* 220: 334-340 (1968).

Comstock, G. The case against BGH. *Agriculture and Human Values* 5: 1-5 (1989).

Conway, G. *The Doubly Green Revolution: Food for All in the 21st Century.* Cornell University Press, Ithaca, NY (1998).

Cornelissen, B.J.C., Does, M.P., Melchers, L.S. In *Rhizoctonia Species: Taxonomy, Molecular Biology, Ecology, Pathology and Control* (eds Sneh, B., Jabaji-Hare, S., Neate, S., Deist, G.), pp. 529-536, Kluwer, Dordrecht (1996).

Crawley, M.J., Brown, S.L., Hails, R.S., Kohn, D.D., Rees, M. Transgenic crops in natural habitats. *Nature* 409: 682-683 (2001).

Crick, F.H.C. On protein synthesis. In *Sympos. Soc. Exp. Biol. XII.* p. 153 Academic Press, New York (1958).

Crick, F.H.C. The Central Dogma of molecular biology, *Nature* 227: 561-563 (1970).

Crouch, M.L. How the Terminator terminates: an explanation for the non-scientist of a remarkable patent for killing second generation seeds of crop plants. Occasional paper of the Edmonds Institute, Edmonds, Washington (1998).

Cryan, P. GE pollution in Mexico: Native corn contaminated. *Global Pesticide Campaigner* 11(3): 16 (Dec. 2001).

Daily, G.C., Soderqvist, T., Aniyar, S., Arrow, K., Dasgupta, P., Erlich, P. R., Folke, C., Jansson, A.M., Jansson, B.O., Kautsky, N. Levin, S., Lubchenco, J., Maler, K., Simpson, D., Starrett, D., Tilman, D., Walker, B. The value of nature and the nature of value. *Science* 395-396 (July 21, 2000).

Damodaran, A. Regulating transgenic plants in India-Biosafety, plant variety protection and beyond. *Economic and Political Weekly.* 34(13): A-34 - A-42 (1999).

Dasgupta, I., Malathi, V.G., Mukherjee, S.K. Genetic engineering for virus resistance. *Current Science* 84: 341-354 (2003).

Dawkins, R. *The Selfish Gene.* Oxford University Press, Oxford (1976).

Dawkins, R., Krebs, J.R. Arms races between and within species. *Proc. Royal Soc. London* Series B 205: 489-511 (1979).

Delaney, T.P. *Science* 266: 1247-1250 (1994).

Denison, R.F., Kiers, E.T., West, S.A. Darwinian agriculture: When can humans find solutions beyond the reach of natural selection? *Quart. Rev. Biology* 78: 145-168 (2003).

Dhar, B. Regulations, negotiations and campaigns: introducing biotechnology into India. *Biotech. Develop. Monitor* 47: 19-21 (2001).

Ding, X. *et al. Transgenic Res.* 7: 77-84 (1998).

Dinham, B. GM cotton-farming by formula? *Biotech. Develop. Monitor* No. 44/45: 7-9 (March, 2001).

Dirie, A.M. Regulatory oversight in developing nations. *Nature Biotech.* 20: 547-548 (2002).

Does, M.P., Cornelissen, B.J.C., in *Crop Productivity and Sustainability: Shaping the Future* (eds. Chopra, V.L., Singh, R.B., Verma, A.) pp. 233-244. Oxford and IBH, New Delhi (1998).

Dominguez, C.A. Genetic conflicts of interest in plants. *Trends in Ecology & Evolution* 10: 412-416 (1995).

Dove, A. Survey raises concerns about *Bt* resistance management. *Nature Biotech.* 19: 293-294 (2001).

DRR. High yielding rice varieties of India - 2000, Bulletin 2001-1, Directorate of Rice Research, Hyderabad, pp. 1-102 (2001).

Duvick, D.N. Plant breeding, an evolutionary concept. *Crop Science* 36: 359-548 (1996).

Dyson, T. *PNAS* (USA) World food trends and prospects to 2025. 96: 5929 (1999).

Eaton, D. *TRIPS and Plant Varietal Protection: Economic Analysis and Policy Choices.* Agricultural Economic Research Institute (LEI) Report 7.02.01, The Hague (2001).

Eaton, D., van Tongeren, F., Louwaars, N., Visser, B., van der Meer, I. Economic and policy aspects of 'terminator' technology. *Biotech. Develop. Monitor* 49: 19-22 (March, 2002).

Ellis, J., Dodds, P., Pyror, T. *Trends in Plant Sci.* 5: 373-378 (2000).

Ellis, R.J., Hemmingsen, S.M. Molecular chaperones. *Trends in Bioch. Sci. (TIBS)* 14(8): 339-342 (1989).

Elomaa, P., Helariutta, Y., Griesbach, R.J., Kotilainen, M., Seppanen, R., Teeri, T.H. Transgenic inactivation in *Petunia hybrida* is influenced by the properties of the foreign gene. *Mol. Gen. Genet.* 248: 645-649 (1995).

EPA (US Environmental Protection Agency), FIFRA Scientific Advisory Panel, Subpanel on *Bacillus thuringiensis (Bt)* Plant-Pesticides and Resistance Management, EPA, Washington, D.C. (1998).

EPA. *Insect Resistance Management in Bt Crops.* EPA, USDA, Washington, D.C. (1999).

Estruch, J.J., Warren, G.W., Mullins, M.A., Nye, G.J., Craig, J.A., Koziel, M.G. *PNAS* (USA) 93: 5389-5394 (1996).

European Commission (EC). *Proposals COM 2001-182 final and COM 2001-425 final.* (EC, Brussels, Belgium (25 July, 2001).

Evans, L.T. *Crop Evolution, Adaptation and Yield.* Cambridge Univ. Press, Cambridge (1993).

Evans, L.T. *Feeding the Ten Billion. Plant and Population Growth.* Cambridge Univ. Press, Cambridge, (1998).

Ewel, J.J. Natural systems as models for the design of sustainable systems of land use. pp. 1-21 in *Agriculture As a Mimic of Natural Ecosystems,* Lefroy, E.C. *et al.* (eds.) Kluwer, Dordrecht (1999).

Firbank, L.G., Forcella, F. Genetically modified crops and farmland biodiversity. *Science* 289: 1481-1482 (2000).

Flor, A.H. *Phytopathology* 45: 680-685 (1955).

Fluetsch, K.M., Sparling, D.W. Avian nesting success and diversity in conventionally and organically managed apple orchards. *Env. Tox. & Chem.* 13: 1651-1659 (1994).

Foissac, X., Loc, N.T.,Christou, P., Gatehouse, A.M.R., Gatehouse, J.A. *J. Insect Physiol.* 46: 573-583 (2000).

Foyer, C., Osmond, B., Walker, D. Introduction. *Phil. Trans. R. Soc. Lond.* B 355: 1333-1335 (2000).

Fransman, M. Designing Dolly: interactions between economics, technology and science and the evolution of hybrid institutions. *Research Policy* 30 (2): pp. 263-273 (2001).

Fritig, B., Legrand, M. (Ed.) *Mechanisms of Plant Defense Responses.* Kluwer, Dordrecht (1993).

Frutos, R., Rang, C., Royer, M. *Crit. Rev. Biotechnol.* 19: 227-276 (1999).

Fukai, S. Intercropping-bases of productivity. *Field Crops Research* 34: 239-245 (1993).

Gatehouse, A.M.R., Boulter, D. *J. Sci. Food Agric.* 34: 345-350 (1983).

GeneScan Europe. *GMO Chip: Test Kit for the Detection of GMOs in Food Products.* Cat No. 5321300105, Bremen, Germany (2001).

Ghosh, P.K. Impact of industrial policy and trade related itnellectual property rights on biotech industries in India. *J. Sci. Indus. Res.* 54: 217-230 (1995).

Giampietro, M. Sustainability and technological development in agriculture: A critical appraisal of genetic engineering. *BioScience* 44: 677-690 (1994).

Ginns, D. Timely lesson in genetically modified canola case. *Austral. Farm J.* 11(12): 48-49 (2002).

Giri, A.P., Harsulkar, A.M., Deshpande, V.V., Sainani, M.N., Gupta, V.S., Ranjekar, P.K. *Plant Physiol.* 116: 393-401 (1998).

Goklany, I.M. From precautionary principle to risk-risk analysis. *Nature Biotech.* 20: 1075 (2002).

Goklany, I.M. *The Precautionary Principle: A Critical Appraisal of Environmental Risk Assessment.* Cato Institute, Washington, D.C. (2001).

Gordon, K.H.J., Johnson, K.N., Hanzlik, T.N. *Virology* 208: 84-98 (1995).

GRAIN. Blinded by the gene (Editorial). *Seedling* (GRAIN, Barcelona) pp. 1-5 (July, 2003).

Green, M.B., LeBaron, H.M., Moberg, W.K. (eds.) *Managing Resistance to Agrochemicals.* pp. 3-16 ACS, Symposium Series, Washington (1990).

Grover, A., Gowthaman, R. Strategies for development of fungus-resistant transgenic plants. *Current Science* 84: 330-340 (2002).

Grover, A., Pental, D. Breeding objectives and requirements for producing transgenics for major field crops of India. *Current Science* 84: 310-320 (2003).

Gupta, A. Governing trade in genetically modified organisms. The Cartagena Progocol on Biosafety. *Environment* 42: 22-23 (2000).

Gutierrez-Campos, R., Torres-Acosta, J.A., Saucedo-Arias, L.J., Gomez-Lim, M.A. *Nature Biotech.* 17: 1223-1226 (1999).

Hain, R.H.J. *et al., Nature* 361: 153-156 (1993).

Hallauer, A.R. (Ed.). *Specialty Corns.* CRC Press, Boca Raton, Fl pp. 80-121 (1994).

Halweil, B. Biotech, African corn and the vampire weed. *World Watch* Magazine. 14 (5): 26-31 (Sept.Oct. 2001).

Hamilton, A.J., Baulcombe, D.C. *Science* 286: 950-952 (1999).

Haribabu, E. Cognitive empathy in inter-disciplinary research: the contrasting attitudes of plant breeders and molecular biologists towards rice. *J. Biosci.* 25: 323-330 (2000).

Harlan, J.R. *The Living Fields. Our Agricultural Heritage.* Cambridge Univ. Press, Cambridge (1995).

Harrison, B.D.,Robinson, D.J. *Annu. Rev. Phytopathol.* 37: 369-398 (1999).

Haslberger, A.G. Monitoring and labeling for genetically modified products. *Science* 287: 431-432 (2000).

Hayama, R., Yokoi, S., Tamaki, S., Yano, M., Shimamoto, K. Adaptation of photoperiodic control pathways produces short day flowering in rice. *Nature* 422: 719-722 (2003).

Hedden, P. Constructing dwarf rice. *Nature Biotech.* 21: 873-874 (2003).

Hedden, P., Phillips, A.L. Gibberellin metabolism: new insights revealed by the genes. *Trends Plant Sci.* 5: 523-530 (2000).

Hedden, P., Phillips, A.L. Manipulation of hormone biosynthetic genes in transgenic plants. *Curr. Opin. Biotechnol.* 11: 130-137 (2000a).

Hedley, C., Richards, R.L., Khokhar, S. (eds.) *Agri Food Quality: An Interdisciplinary Approach.* pp. 23-26. Special Publication No. 179, Royal Society of Chemistry, Cambridge (1996).

Hickey, E. U.S. EPA approves *Bt* corn despite lack of testing. *Global Pesticide Campaigner* 11(3): 17 (Dec. 2001).

Hilder, V.A. *et al. Transgenic Res.* 4: 18-25 (1995).

Ho, M.W. One bird-Ten thousand treasures. *The Ecologist* p. 339 (Oct. 1999).

Ho, M-W. Perils amid promises of genetically modified food. *Financing Agriculture* 33(3): 5-9 (2001).

Holmberg, N., Lilius, G., Bailey, J.E., Bulow, L. Transgenic tobacco expressing Vitreoscilla hemoglobin exhibits enhanced growth and altered metabolite production. *Nature Biotech.* 15: 244-247 (1997).

Holton, T.A., Brugliera, F., Tanaka, Y. Cloning and expression of flavonol synthase from *Petunia hybrida. Plant J.* 4: 1003-1010 (1993).

Horstmeier, G.D. Lessons from year one: experience changes how farmers will grow Roundup Ready beans in '98. *Farm Journal* p. 16 (January 1998).

Horvath, H., Rostoks, N., Brueggeman, R., Steffenson, B., von Wettstein, D., Kleinhofs, A. Genetically engineered stem rust resistance in barley using the *Rpg1* gene. *PNAS* (USA) 100: 364-369 (2003).

Houlb, E.B. *Nature Genet.* 2: 516-527 (2001).

Huang, F., Buschman, L.L., Higgins, R.A., McGaughey, W.H. Inheritance of resistance to *Bacillus thuringiensis* toxin (Dipel ES) in the European corn borer. *Science* 284: 965-967 (1999).

Huang, J., Rozelle, S., Pray, C., Qang, Q. *Science* 295: 674 (2002).

Hull, R. *In Methods in Molecular Biology: Plant Virology Protocols.* Foster, G.D., Taylor, S.C. (eds.). pp. 547-555, Humana Press, New Jersey (1998).

Hull, R. *Matthew's Plant Virology.* 4th ed. Academic Press, New York (2002).

Hunt, J.W., Anderson, B.S., Phillips, B.M. Patterns of aquatic toxicity in an agriculturally dominated coastal watershed in California. *Ag. Econ. Envir.* 75: 75-91 (1999).

IHGSC (International Human Genome Sequencing Consortium). Initial sequencing and analysis of the human genome. *Nature* 409: 860-921 (2001).

IRRI (Internat. Rice Res. Inst.). *Fragile Lives in Fragile Environments.* Int. Rice Res. Inst. Manila pp. 369-394 (1995).

Ishimoto, M., Sato, T., Chrispeels, M.J., Kitamura, K. *Entomol. Expt. Appl.* 79: 309-315 (1996).

Ismael, Y., Bennett, R., Morse, S. Farm level impact of *Bt* cotton in South Africa. *Biotech. Develop. Monitor* Issue No. 48: 15-19 (Dec., 2001).

Jackson, D.L., Jackson, L.L. (eds.) *The Farm as Natural Habitat: Reconnecting Food Systems with Ecosystems.* Island Press, Washington, D.C. (2002).

Jackson, *I.E.*, Koch, G.W. The ecophysiology of crops and their wild relatives, pp. 3-37 in *Ecology in Agriculture,* Jackson, *I.E.* (Ed.) Academic Press, San Diego (1997).

Jackson, W., Piper, J. The necessary marriage between ecology and agriculture. *Ecology* 70: 1591-1593 (1989).

James, C. *Global Status of Transgenic Crops in 1997.* International Service for the Acquisition of Agri-biotech Applications (ISAAA) Brief #5. www.isaaa.org (1997).

James, C.A. *Global Review of Commercialized Transgenic Crops: 1998.* No. 9-98. International Service for the Acquisition of AgBiotech Applications, (ISAAA) Ithaca, N.Y. (1998).

James, C. *Global Review of Commercialized Transgenic Crops.* ISAAA Briefs No. 21 (2000).

James, C. *Global Review of Commercialized Transgenic Crops: 2001 (Feature: Bt Cotton).* ISAAA Briefs No. 26, Ithaca, N.Y. (2002).

James, C. Global review of commercialized transgenic crops. *Current Science* 84: 303-309 (2003).

Jayaraman, K.S. *Nature Biotech.* 19: 1090 (2001).

Jayaraman, K.S. Poor crop management plagues *Bt* cotton experiment in India. *Nature Biotech.* 20: 1069 (2002).

Jenner, H.L. *Trends Biotechnol.* 21: 190-192 (2003).

Jones, M.M. Monsanto: Rewriting the script. *Biotech. Develop. Monitor* 48: 13-14 (2001).

Jonsen, A.R. *The Birth of Bioethics.* Oxford University Press, Oxford (1998).

Jorgensen R.A. Cosuppression, flower color patterns and metastable gene expression states. *Science* 268: 686-691 (1995).

Joy, B. Why the future doesn't need us. *Wired* (Magazine) 238-262 (April, 2000).

Kalaitzandonakes, N. Biotechnology and identity-preserved supply chains: A look at the future of crop production and marketing. *Choices* (Fourth Quarter), 15-18 (1998).

Kalaitzandonakes, N. (ed.) *The Economics of Transgenic Crops.* Kluwer Academic Publishers, Dordrecht (2003).

Keller, H. *et al., Plant Cell* 11: 223-236 (1999).

Khachatourians, G.G., McHughen, A., Scorza, R., Nip, W., Hui, Y. *Transgenic Plants and Crops.* Marcel Dekker, New York (2002).

Khush, G.S. Green revolution: preparing for the 21st century. *Genome* 42: 646-655 (1999).

Kirschenmann, F. Questioning biotechnology's claims and imaging alternatives. *Trans. Wisconsin Acad.* 89: 35-61 (2001).

Kleter, G.A., van der Krieken, W.M., Kok, E.J., Bosch, D., Jordi, W., Gilissen, L.J.W.J. Regulation and exploitation of genetically modified crops. *Nature Biotech.* 19: 1105-1110 (2001).

Knoppers, B., Mathios, A. *Biotechnology and the Consumer.* Kluwer Academic Publishers, Dordrecht (1998).

Kong, Q.X., Richter, L., Yang, Y.F., Arntzen, C.J., Mason, H.S., Thanavala, Y. Oral immunization with hepatitis B surface antigen expressed in transgenic plants. *PNAS* (USA) 98: 11539-11544 (2001).

Korten, D. *When Corporations Rule the World.* Kumarian, West Hartford, Conn. (1995).

Kumar, H.D. *A Textbook on Biotechnology.* 2ed. Affil. East West Press, New Delhi (1998).

Lampkin, N. *Organic Farming*, pp. 2-6, Farming Press Books, Ipswich (1990).

Land Institute. *Land Report* (quarterly newsletter). No. 67, Summer. Land Institute, Salina, Kansas (2000).

Lappé, M., Bailey, B. *Against the Grain.* Common Courage Press, Monroe, Maine (1998).

Leah, R., Tommerup, H., Svendsen, I., Mundy, J. *J. Biol. Chem.* 266: 1467-1573 (1991).

Lefroy, E.C., Hobbs, R.J., O'Connor, M.H., Pate, J.S. (eds.). *Agriculture as a Mimic of Natural Ecosystems.* Kluwer, Dordrecht (1999).

Leopold, A. *A Sand County Almanac.* Oxford University Press, London (1949).

Lev-Yadun, S., Abbo, S., Doebley, J. Wheat, rye, and barley on the cob? *Nature Biotech.* 20: 337-338 (2002).

Lewis, W.J., Lenteren, J.C., Phatak, S.C., Tumlinson, J.H. A total system approach to sustainable pest management. *PNAS* (USA) 94: 12243-12248 (1997).

Lewontin, R.C. *Biology as Ideology.* House of Anansi, Concord, Ontario (1991).

Lobell, D.B., Asner, G.P. Climate and management contributions to recent trends in U.S. agricultural yields. *Science* 299: 1032 (2003).

Lurquin, P.F. *High Tech Harvest: Understanding Genetically Modified Food Plants.* Westview (*Perseus) Press, Boulder, CO (2002).

Lutman, P.J.W. (Ed.). *Gene Flow and Agriculture. Relevance for Transgenic Crops.* British Crop Protection Council, Nottingham (1999).

Lyon, N. Consolidating organic gains. *Austral. Farm J.* 11 (12) page 1: (Feb. 2002).

Madden, J.P., Chaplowe, S.G. (eds.). *For All Generations.* Glendale, Calif., OM. (1997).

Magn, K. *et al. Methods for Detection of GMO Grain in Commerce.* American Crop Protection Association, Washington, D.C. (2000).

Mallik, S. in *Sustaining Crop and Animal Productivity - The Challenge of the Decade* (ed. Deb, D.L.), Associated Publishing Co., New Delhi, pp. 37-46 (1995).

Mallik, S., Mandal, B.K., Sen, S.N., Sarkarung, S. Shuttle-breeding: An effective tool for rice varietal improvement in rainfed lowland ecosystem in eastern India. *Curr. Sci.* 83: 1097-1102 (2002).

Manicka Velu, A., Durai, R.S.R., Malar, R.P.G. Golden rice. *LEISA India* 3: 4 (Dec., 2001).

Mann, C.G. Genetic engineers aim to soup up crop photosynthesis. *Science* 283: 314-316 (1999).

Manning, R. *Food's Frontier: The Next Green Revolution.* New York: North Point, New York (2000).

McBride, K.E., Svab, Z., Schaaf, D.J., Hogan, P.S., Stalker, D.M., Maliga, P. *Bio/Technology* 13: 362-365 (1995).

McGarvey, P.B., Kaper, J.M. In *Transgenic Plants* (ed. Kung, S.D., Wu, R.) pp. 277-296, Academic Press, New York (1993).

McHughen, A. *Pandora's Picnic Basket: The Potential and Hazards of Genetically Modified Foods.* Oxford University Press, Oxford (2000).

McNeil, S., Nuccio, M.L., Hanson, A.D. Betaines and related osmoprotectants. Targets for metabolic engineering of stress resistance. *Plant Physiol.* 120: 945-949 (1999).

Medvitz, A.G., Sokolow, A.D. Population growth theatens agriculture, open space. *Cal. Ag.* 49 (6): 11-27 (1995).

Mehra, S., Pareek, A., Bandypadhyay, P., Sharma, P., Burma, P.K., Pental, D. Development of transgenics in Indian oilseed mustard *(Brassica juncea)* resistant to herbicide phosphinothricin. *Current Science* 78: 1358 (2000).

Meldolesi, A. Italy performs GMO trial about-face. *Nature Biotech.* 19: 293 (2001).

Mellon, M., Rissler, J. (eds.). *Now or Never: Serious New Plans to Save a Natural Pest Control.* Union of Concerned Scientists, Cambridge, MA (1998).

Mertz, E.T. (Ed.). *Quality Protein Maize.* Am. Assoc. Cereal Chemists, St. Paul, MN (1990).

Miller, H., Conko, G. *Nature Biotech.* 18: 697 (2000).

Moellenbeck, D.J. +28 others. Insecticidal proteins from *Bacillus thuringiensis* protect corn from corn rootworms. *Nature Biotech.* 19: 668-672 (2001).

Mohanty, H.K., Mallik, S., Grover, A. *Curr. Sci.* 78: 132-137 (2000).

Moore, S. Organic farming - for all the right reasons. *Austral. Farm J.* 11(12): 6-8 (Feb. 2002).

Moss, J.P. (ed.). *Biotechnology for Crop Improvement in Asia* (ed.). ICRISAT, Hyderabad (1992).

Mourgues, F., Brisset, M-N., Chevreau, E. Strategies to improve plant resistance to bacterial diseases through genetic engineering. *Trends Biotechnol.* 16: 203-210 (1998).

Muir, S.R. *et al. Nature Biotech.* 19: 470-474 (2001).

Mwangi, P.N., Ely, A. Assessing risks and benefits: *Bt* maize in Kenya. *Biotech. Develop. Monitor* 48: 6-9 (2001).

Myers, D. Organic cotton production. *LEISA* Magazine 17(4): 21-22 (Dec., 2001).

Myers, D., Stolton, S. (eds.). *Organic Cotton: From Field to Final Product.* Intermediate Technology Publications, London (1999).

Nakajima, H., Muranaka, T., Ishige, F., Akutsu, K., Oeda, K. Fungal and bacterial resistance in transgenic plants expressing human lysozyme. *Plant Cell Rep.* 16: 674-679 (1997).

NRC (National Research Council). *Soil and Water Quality: An Agenda for Agriculture.* National Academy Press, Washington, D.C. (1993).

NRC. *Ecologically Based Pest Management: New Solutions for a New Century.* National Academy Press, Washington, D.C. (1996).

NRC. *The Future Role of Pesticides in US Agriculture.* National Academy Press, Washington, D.C. (2000).

ODI. (Overseas Development Institute). Briefing Paper: The Debate on Genetically Modified Organisms: Relevance for the South, ODI, London (1999).

OECD. *Safety Evaluation of Foods Derived by Modern Biotechnology: Concepts and Principles.* OECD, Paris (1993).

Oerke, E.-C., Dehne, H.-W., Schönbeck, F., Weber, A. *Crop Production and Crop Protection: Estimated Losses in Major Food and Cash Crops.* Elsevier, Amsterdam (1994).

Olson, R.A., Frey, O.J. *Nutritional Quality of Cereal Grains: Genetic and Agronomic Improvement.* pp. 183-236 Am. Soc. Agron., Madison, WI (1987).

Osorio, L.G. Plants protecting other plants. *LEISA Magazine* 17(4): 23-24 (2001).

Ostlie, K. Crafting crop resistance to corn rootworms. *Nature Biotech.* 19: 624-625 (2001).

Oud, J.S.N., Schneiders, H., Kool, A.J., van Grinsven, M.Q.J.M. Breeding of transgenic orange *Petunia hybrida* varieties. *Euphytica* 84: 175-181 (1995).

Paarlberg, R.L. *The Politics of Precaution: Genetically Modified Crops in Developing Countries.* Johns Hopkins Univ. Press, Baltimore (2001).

Papendick, R.I. *et al.* Report and recommendations on organic farming. U.S. Department of Agriculture, Washington, D.C. (1981).

Park, C.M., Berry, J.O., Bruenn, J.A. *Plant Mol. Biol.* 30: 359-366 (1996).

Pattanayak, D., Ananda Kumar, P. Plant biotechnology: Current advances and future perspectives. *Proc. Indian Natn. Sci. Acad. (PINSA)* B66: 265-310 (2000).

Paula, L. Ethics: the key to public acceptance of biotechnology? *Biotech. Develop. Monitor* 47: 22-23 (2001).

Pearce, F. Reaping the rewards. *New Scientist*, p. 12 (Feb. 2, 2002).

Peng, J. *et al.* 'Green revolution' genes encode mutant gibberellin response modulators. *Nature* 400: 256-261 (1999).

Perlak, F.J., Fuchs, R.L., Dean, D.A., McPherson, S., Fischhoff, D.A. *PNAS* (USA) 88: 3324-3328 (1991).

Pesek, J. Research findings on alternative methods: Based on the NRC Report on alternative agriculture. Unpublished manuscript, presented at Northwest Farm Managers Association, (February 26, 1990).

Phillips, P., Khachatourians, G.G. *The Biotechnology Revolution in Global Agriculture: Invention, Innovation and Investment in the Canola Sector.* pp. 360, CABI (2001) (cited in Phillips, 2002).

Phillips, P.W.B., Khachatourians, G.G. *The Biotechnology Revolution in Global Agriculture: Invention, Innovation and Investment in the Canola Sector.* CABI, Wallingford (2001).

Phillips, P.W.B. Biotechnology in the global agri-food system. *TiBTECH*. 20: 376-381 (2002).

Phillipson, M. Agricultural law: containing the GM revolution. *Biotech. Develop. Monitor* Issue No. 48: 3-5 (2001).

Pimbert, M., Wakeford, T., Satheesh, P.V. Citizens' juries on GMOs and farming futures in India. *LEISA* Magazine 17(4): 27-30 (Dec., 2001).

Pinstrup-Andersen, P., Schiøler, E. *Seeds of Contention: World Hunger and the Global Controversy over GM Crops*. Johns Hopkins Univ. Press, Baltimore (2001).

Prasad, G.S.V., Prasadarao, U., Rani, N.S., Rao, L.V.S., Pasalu, I.C., Muralidharan, K. Indian rice varieties released in countries around the world. *Current Science* 80: 1508-1511 (2001).

Prasanna, B.M., Vasal, S.K., Kassahun, B., Singh, N.N. Quality protein maize. *Current Science* 81: 1308-1319 (2001).

Price, E., Hamburger, J. Genetically engineered crops expand in China. *Global Pesticide Campaigner* 11(3): 7 (Dec. 2001).

Qaim, M., Krattiger, A.F., von Braun J. (Eds.). *Agricultural Biotechnology in Developing Countries: Towards Optimizing the Benefits for the Poor*. Kluwer, Boston, MA (2000).

Qaim, M., Zilberman, D. Yield effects of genetically modified crops in developing countries. *Science* 299: 900-902 (2003).

Radin, J.W. The technology protection system: Revolutionary or evolutionary? *Biotech. Develop. Monitor* 37: 24 (1999).

Raffensperger, C., Tickner, J. (eds.). p. 8. *Protecting Public Health & the Environment. Implementing the Precautionary Principle* (Island Press, Washington, D.C., 1999).

Raina, R.S. Innovation capability for agrobiotechnology: Policy issues. *Biotech. Develop. Monitor* 50: 29-31 (2003).

Ram, A.S., Indira, M. Agricultural biotechnology - the science and fiction. *LEISA* India 3: 1-2 (Dec., 2001).

Ranjekar, P.K., Patankar, A., Gupta, V., Bhatnagar, R., Bentur, J., Kumar, P.A. Genetic engineering of crop plants for insect resistance. *Current Science* 84: 321-329 (2003).

Regal, P.J. Scientific principles for ecologically based risk assessment of transgenic organisms. *Molecular Ecology* 3: 5-13 (1994).

Reid, W.V., Laird, S.A., Meyer, C.A., Gamez, R., Sittenfeld, A., Janzen, D.H., Gollin, M.A., Juma, C. *Biodiversity Prospecting*. World Resources Inst., Washington, D.C. (1993).

Reilly, J.M., Fuglie, K.O. Future yield growth in field crops: What evidence exists. *Soil and Tillage Research* 47: 275-290 (1998).

Risch, S.J., Andow, D., Altieri, M.A. Agroecosystem diversity and pest control: data, tentative conclusions, and new research directions. *Environ. Entomol.* 12: 625-629 (1983).

Rissler, J., Mellon, M. *The Ecological Risks of Engineered Crops*. The MIT Press, Cambridge (1996).

Rommens, C.M.T., Salmeron, J.M., Oldroyd, G.E., Staskawicz, B.J. *Plant Cell* 7: 1537-1544 (1995).

Rosenzweig, C., Parry, M.L. *Nature* 367: 133 (1994).

Rossett, P. Genetically engineered crops. *LEISA* Magazine 17(4): 6-8 (Dec., 2001).

Roush, R.T. In *Insecticide Resistance: From Mechanisms to Management* (eds. Denham, I., Pickett, J.A., Devenshire, A.), pp. 101-110 CAB Internat. Publishing, Wallingford (1999).

Ruivenkamp, G. Monitoring biotechnological developments: Looking back for finding new perspectives. *Biotech. Develop. Monitor* 50: 3-5 (2003).

Ruttan, V.W. Biotechnology and agriculture: A skeptical perspective. *Transactions* (Wisconsin Academy) 89: 83-92 (2001).

Ruttan, V.W. *Technology, Growth and Development: An Induced Innovation Perspective*. Oxford University Press, New York (2001a).

Sakamoto, T., Morinaka, Y., Ishiyama, K., Kobayashi, M., Itoh, H., Kayano, T., Iwahori, S., Matsuoka, M., Tanaka, H. Genetic manipulation of gibberellin metabolism in transgenic rice. *Nature Biotech.* 21: 909-913 (2003).

Sasaki, A. *et al.* A mutant of gibberellin-synthesis gene in rice. *Nature* 416: 701-702 (2002).

Savan, K.W., Baudinette, S.C., Graham, M.W., Michael, M.Z., Nugent, G.D., Lu, C-Y., Chandler, S.F., Cornish, E.C. Antisense ACC oxidase dalays carnation petal senescence. *HortSci.* 30: 970-972 (1996).

Savidan, Y. L'apomixie, ou le clonage par les graines. *Biofutur* No. 198 (March, 2000).

Saxena, D., Flores, S., Stotzky, G. Transgenic plants: Insecticidal toxin in root exudates from *Bt* corn. *Nature* 402: 480 (1999).

Saxena, D., Stotzky, G. *Bt* toxin uptake from soil by plants. *Nature Biotechnol.* 19: 199 (2001).

Schenkelaars, P. Agronomic and environmental effects of growing glyphosate soybean, Centrum voor Landbouw en Milieu and Schenkelaars Biotechnology Consultancy (June 2001).

Schmucker, D. *et al. Drosophila* Dscam is an axon guidance receptor exhibiting extraordinary molecular diversity. *Cell* 101: 671-684 (2000).

Schneider, S. *The Genesis Strategy.* Plenum, New York (1976).

Schreiber, G.A. Challenges for methods to detect genetically modified DNA in foods. *Food Control* 10: 351-352 (1999).

Schubert, D. A different perspective on GM food. *Nature Biotechnol.* 20: 969 (2002).

Schuler, T.H., Poppy, G.M., Denholm. I. *TiBTECH.* 16: 168-175 (1998).

Sen, A. *Poverty and Famines: An Essay on Entitlement and Deprivation.* Clarendon Press, Oxford (1981).

Sen, A. *Resources, Values and Development.* Harvard University Press, Cambridge, Mass. (1984).

Serageldin, I. Biotechnology and food security. *Science* 285: 387-389 (1999).

Sharma, D. The introduction of transgenic cotton in India. *Biotech. Develop. Monitor* 44/45: 10-13 (March, 2001).

Sharma, H.C., Ortiz, R. Transgenics, pest management, and the environment. *Current Science* 79: 421-437 (2000).

Sharma, M., Charak, K.S., Ramanaiah, T.V. Agricultural biotechnology research in India: Status and policies. *Current Science* 84: 297-302 (2003).

Shenk, D. Biocapitalism: What price the genetic revolution? *Harper's Magazine*, 37-45 (Dec., 1997).

Shiva, V. Biotechnology: The failed miracle. *Focus WTO* 1(6): 5-11 (2001).

Simmonds, N.W. "Shades of Green." Review of *The Doubly Green Revolution: Food For All in the Twenty-First Century*, by Conway, G. *Nature* 391: 139 (1998).

Singh, U. *Quality Protein Maize Products for Human Nutrition*. Directorate of Maize Research, New Delhi (2001).

Sivamani, E., Huet, H., Shen, P., Ong, C.A., De Kochko, A., Fauquet, C., Beachy, R.N. *Mol. Breeding* 5: 177-185 (2002).

Slusarenko *et al.* (eds.). *Mechanism of Resistance to Plant Diseases*. p. 428, Kluwer, Dordrecht (2000).

Smyth, S., Khachatourians, G.G., Phillips, P.W.B. Liabilities and economics of transgenic crops. *Nature Biotech.* 20: 537-541 (2002).

Snow, A.A. Transgenic crops-why gene flow matters. *Nature Biotech.* 20: 542 (2002).

Soil Association. Organic farming, food quality and human health. A review of the evidence. Soil Association, Bristol (2001).

Soule, J.D., Piper, J.K. *Farming in Nature's Image. An Ecological Approach to Agriculture*. Island Press, Corelo, Calif. (1992).

Spillane, C. Could agricultural biotechnology contribute to poverty alleviation? *AgBiotechNet* 2: ABN042 (2000).

Srinivasan, C.S., Thirtle, C. *Impact of Terminator Technologies in Developing Countries: A Framework for Economic Analysis*. Paper presented at the 4th International Conference on the "Economics of Agricultural Biotechnology" organised by the International Consortium on Agricultural Biotechnology Research (ICABR), Ravello, Italy, 24-28 August (2000).

Stirling, A. *et al. On Science and Precaution in the Management of Technological Risk: A Synthesis Report of Case Studies*. Vol. 1. Inst. Prospective Technol. Studies, Sevilla (1999).

Stokstad, E. Study shows richer harvests owe much to climate. *Science* 299: 997 (2003).

Stotz, H.U., Contoa, J.J.A., Powell, A.L.T., Bennett, A.B., Labavitch, J.M. *Plant Mol. Biol.* 25: 607-617 (1994).

Strittmatter, G., Goethals, K., Van Montagu, M. Strategies to engineer plants resistant to bacterial and fungal diseases. In *Subcellular Biochemistry* 29: 191-213, (eds.) Biswas, B.B., Das, H.K. Plenum Press, New York (1998).

Sundquist, W.B., Menz, K.M., Neumeyer, C.F. *A Technology Assessment of Commercial Corn Production in the United States.* Bulletin 546. University of Minnesota Agricultural Experiment Station, St. Paul (1982).

Suzuki, K., Zue, H., Tanaka, Y., Fukui, Y., Mizutani, M., Kusumi T. Molecular breeding of flower color of *Torenia fourieri. Plant Cell Physiol.* 38: 538-540 (1997).

Svitashev, K., Somers, D.A. *Genome* 44: 691-697 (2002).

Swaminathan, M.S. Biotechnology and agricultural betterment in the developing countries; in *Biotechnology in Agriculture.* Natesh, S., Chopra, V.L., Ramachandran, S. (eds.) Oxford and IBH (New Delhi) pp. 3-11 (1987).

Swezey, S.L., Broome, J.C. Growth predicted in biologically integrated and organic farming in California. *Global Pesticide Campaigner* 11: 6-9 (2001).

Takken, F.L.W., Joosten, M.H.A.J. *Eur. J. Plant Pathol.* 106: 699-713 (2000).

Tanaka, Y., Tsuda, S., Takaaki, K. Application of recombinant DNA to floriculture; in *Applied Plant Biotechnology* pp. 181-235 (eds.) Chopra, V.L., Malik, V.S., Bhat, S.R. Oxford & IBH, New Delhi (1998).

Taverne, D. Food for thought and action. *Nature* 409: 559-560 (2001).

Taylor, M.R. Rethinking US leadership in food biotechnology. *Nature Biotech.* 21: 852-854 (2003).

Tester, M. Some GM facts. *Science* 298: 1341 (2002).

Tousch, D., Jacquemond, M., Tepfer, M. *J. Gen. Virol.* 75: 1009-1014 (1994).

Trewavas, A. Urban myths of organic farming. *Nature* 410: 409-410 (2001).

Trudel, J., Potvin, C., Asselin, A. *Plant Sci.*, 55-62 (1995).

UNESCO. *Vital Water Graphics. Water Use and Management.* United Nations Educational Scientific & Cultural Organization, (UNESCO) Paris (2002).

Uphoff, N. The system of rice intensification: agro-ecological opportunities for small farmers? *LEISA Magazine* 17(4): 15-16 (Dec. 2001).

US Food and Drug Administration (FDA). *FDA announces proposal and draft guidance for food developed through biotechnology.* (FDA, Center for Food Safety and Applied Nutrition, Washington, DC (2001).

van Bueren, E.L., Osman, A. Stimulating GMO-free breeding for organic agriculture: a view from Europe. *LEISA Magazine* 17(4): 12-14 (Dec., 2001).

van Bueren, E.L. *et al.* Sustainable organic plant breeding. pp. 16-18. Louis Bolk Instituut, Driebergen (1999).

van Dijk, P., Van Damme, J. Apomixis technology and the paradox of sex. *Trends in Plant Science* 5 (2): 81-84 (2001).

van Dommelen, A. *Hazard Identification of Agricultural Biotechnology: Finding Relevant Questions.* International Books, Utrecht (1999).

van Dommelen, A. A transgene-centred approach to the transgenic herbicide-tolerant crops. *Biotech. Develop. Monitor* 38: 6-7 (1999a).

van Loon, L.C., Van Kammen, A. *Virology.* 40: 199-211 (1970).

Varma, A., Jain, R.K., Bhat, A.I. *Indian J. Biotechnol.* 1: 73-86 (2002).

Vasil, I.K. (ed.) *Plant Biotechnology 2002 and Beyond.* Kluwer, Dordrecht (2003).

Vasil, I.K. The science and politics of plant biotechnology-a personal perspective. *Nature Biotech.* 21: 849-851 (2003).

Venter, C. *et al.* The sequence of the human genome. *Science* 291: 1304-1351 (2001).

Verma, H.N. *Anti-viral Proteins in Higher Plants.* (eds. Chessin *et al.*) pp. 1-37, CRC Press, Boca Raton (1995).

Vieille Calzada *et al.* Apomixis-the asexual revolution. *Science* 274: 1322 (1996).

Visser, B. Biotechnology. *LEISA* Magazine 17(4): 9-11 (2001).

Visser, B., van der Meer, I., Louwaars, N., Beekwilder, J., Eato, D. The impact of 'terminator' technology. *Biotech. Develop. Monitor* 48: 9-12 (2001).

Vogt, G. Origins, development, and future challenges of organic farming. In *Proc. 13th IFOAM Scientific Conf.* pp. 708-711. Vdf Hochschulverlag AG an der ETH, Zürich (2000).

Vogtmann, H. New approaches to the determination of food quality. In *Food Quality, Concepts and Methodology*. pp. 44-47, Elm Farm Research Centre, Newbury (1992).

von der Weid, J.M., Tardin, J.M. Genetically modified soybeans. *LEISA* Magazine 17(4): 19-20 (Dec., 2001).

Vos, P., Simons, G., Jesse, T., Wijbrandi, J., Heinen, L., Hogers, R., Fritzers, A., Groenendijk, J., Diergaarde, P., Reijans, M., Fierens-Onstenk, J., de Both, M., Peleman, J., Liharska, T., Hontelez, J., Zabeau, M. The tomato *Mi-1* gene confers resistance to both root-knot nematodes and potato aphids. *Nature Biotech.* 16: 1365-1369 (1998).

Watkinson, A.R., Freckleton, R.P., Robinson, R.A., Sutherland, W.J. Predictions of biodiversity response to genetically modified herbicide-tolerant crops. *Science* 289: 1554-1556 (2000).

Williams, N. Europe opens the door to GM crops. *Current Biol.* 11: 199-200 (2001).

Williams, N. Seeking balance in the GM crop debate. *Current Biol.* 13: R163-R164 (2003).

Windels, P. *et al.* Characterisation of the roundup ready soybean insert. *Eur. Food Res. Technol.* 213: 107-112 (2001).

Wolf, S., Zilberman, D. (eds.). *Knowledge Generation and Technical Change: Institutional Innovation in Agriculture.* pp. 177-192, Kluwer-Plenum Academic Publishers, New York (2001).

Woodwand, L. The scientific basis of organic farming. *Interdiscip. Sci. Rev.* 27: 114-119 (2002).

Wossink, G.A.A., van Kooten, G.C., Peters, G.H. (Eds.). *Economics of Agro-Chemicals: An International Overview of Use Patterns, Technical and Institutional Determinants, Policies and Perspectives.* Ashgate, Aldershot, UK (1998).

Ye, X., Al-Babili, S., Kloti, A., Zhang, J., Lucca, P., Beyer, P., Potrykus, I. Engineering the provitamin-A (b-carotene) biosynthetic pathway into (carotene-free) rice endosperm. *Science* 287: 303-305 (2000).

Yoshida, K., Kondo, T., Okazaki, Y., Katou K. Cause of blue petal colour. *Nature* 373: 291 (1995).

Zhang, B.H. *Cotton Biotechnology and its Application*. Chinese Agricultural Press, Beijing (1997).

Zhang, B.H., Liu, F., Yao, C.B., Wang, K.B. Recent progress in cotton biotechnology and genetic engineering in China. *Current Science* 79: 37-44 (2000).

Index